The
Travels and Adventures
of
Serendipity

THE
TRAVELS AND ADVENTURES
OF

Serendipity

ROBERT K. MERTON • ELINOR BARBER

A STUDY IN
SOCIOLOGICAL
SEMANTICS
AND THE
SOCIOLOGY
OF SCIENCE

PRINCETON UNIVERSITY PRESS
PRINCETON AND OXFORD

Second printing, and first paperback printing, 2006
Paperback ISBN-13: 978-0-691-12630-2
Paperback ISBN-10: 0-691-12630-5

The Library of Congress has cataloged the cloth edition of this book as follows

Merton, Robert King, 1910–
The travels and adventures of serendipity : a study in sociological semantics and the
sociology of science / Robert K. Merton, Elinor Barber.
p.   cm.
Includes bibliographical references and index.
ISBN 0-691-11754-3
1. Serendipity in science.   2. Science—social aspects.   I. Barber, Elinor G.   II. Title.

Q172.5.S47M47 2004
501—dc21   2003056327

British Library Cataloging-in-Publication Data is available

This book has been composed in Galliard with Centaur & Copperplate-Twenty Display

Printed on acid-free paper. ∞

pup.princeton.edu

Printed in the United States of America

3   5   7   9   10   8   6   4   2

for Harriet and Bernard

Contents

Preface

ROBERT K. MERTON

*T*his book traces the word *serendipity* from its coinage in 1754 by
the English man of letters, Horace Walpole, almost to our own immediate times, precisely two and a half centuries later. That it was still an
esoteric, not to say mysterious, word known only to a few bibliophiles,
antiquarians, and a handful of scientists in the 1950s when the core of
the book was virtually completed, is shown by their common practice of
defining the word when they first used it.

I make no effort to detail the checkered career of this now extended
book for that is amply described in the Introduction and the Afterword.
Suffice it to say that, at my probably mistaken request, my ever-generous
collaborator Elinor Barber agreed in 1958 to set our manuscript aside
"for a while." When my editors at Il Mulino, Laura Xella and Giovanna
Movia, approached us in the 1990s, Elinor and I agreed to have it appear
in print for the first time in Italian. After all, the "silly fairy tale" that led
Walpole to his coinage first appeared in sixteenth-century Venice under
the title *Peregrinaggio di tre giovani, figliuoli del re di Serendippo, tra-
datto dalla lingua persiana in lingua italiana de M. Christoforo Armeno*
and was not translated into English, via a French intermediary, until
1722; this under the title *Travels and Adventures of Three Princes of Serendip.* Lamentably, Elinor has not lived to see our own *Travels and Adventures* in print.

Since our text of the 1950s was never published before the Italian
edition appeared in 2002, it has become, without our intent, a sort of
time capsule indicating how the past and present of the word *serendipity*
seemed to us as observers back then. This means, of course, that we
could not modify that text if it were to serve the time-capsule function.
Indeed, the only change, apart from the re-division of chapters, is in the

subtitle of the book. That change reflects our retrospective awareness that the original text did not, as alleged by the original subtitle, actually make use of the research procedures of what was then the already quite developed technical specialty known as "*historical* semantics" but was, rather, an early exercise in what can be better described as a barely emerging "*sociological* semantics" that examines the ways in which the word *serendipity* acquired new meanings as it diffused through different social collectivities.

Since the Introduction and Afterword provide extensive context for the 1958 text, no more by way of context here. But I do want to acknowledge with thanks those who helped bring this recalcitrant book into print. To begin with, of course, I thank Laura Xella and Giovanna Movia for their remarkable patience and for having seen to it that the footnotes in the Afterword, many of them quite discursive, appear at the foot of the page and not in remote endnotes that require readers to oscillate between text and notes. To Peter Dougherty at Princeton University Press, my thanks for pressing to have the English version appear in print. And to my percipient friend and Renaissance scholar, James L. Shulman, I am happily indebted for his knowing Introduction. Jennifer Lee and Maritsa Poros provided faithful research assistance in the slow beginnings of the Afterword, while Elizabeth Needham Waddell helped bring it to a close through her indomitable research and preparation of the Indexes and the Select Bibliography. Other, specific bibliographic aid came from Nicole Radmore and Michelle McKowen, librarians of the Russell Sage Foundation, and from Susanne Pichler, librarian of the Andrew W. Mellon Foundation. To the Eugene Garfield Foundation, my thanks for a continuing grant-in-aid of research.

I close with words of boundless gratitude to my wife and collaborator, Harriet Zuckerman, for her thoughtful vetting of the Afterword as for her vetting of my other writings over the years.

Publisher's Note

Written originally in English, *The Travels and Adventures of Serendipity* was initially published in Italian, by Il Mulino, in 2002. The Princeton University Press edition of this book marks its first appearance in English. The main text goes back untouched to its original form as written by the late Robert K. Merton and Elinor Barber in 1958. Editing required the reconstruction of several passages lost in time-worn pages, the back-translation of several other difficult-to-read passages from the Italian edition, and the translation of several new footnotes added to the Italian translation by Robert K. Merton.

Professor Merton died in February 2003, just after learning that Princeton would publish this edition but before editing of this edition commenced. Hence, the publishers did not have the benefit of his wisdom in completing the editing of the manuscript (nor that of Dr. Barber, deceased since 1999). However, the editors benefited greatly from the contribution of Professor Merton's wife, Dr. Harriet Zuckerman of the Andrew W. Mellon Foundation, and Dr. Elizabeth Needham Waddell of Columbia University in resolving editorial issues concerning the published version. We, the publishers, are in their debt.

Peter J. Dougherty
Group Publisher for the Social Sciences
Princeton University Press

Introduction

J A M E S   L .   S H U L M A N

*I*n a footnote in chapter 47 of *On the Shoulders of Giants* (1965),
Robert K. Merton alluded to "a carefully unpublished" manuscript titled
*The Travels and Adventures of Serendipity: A Study in Historical Seman-
tics and the Sociology of Science*. In this footnote, the author confesses
that, though tempted, he cannot quote "the supremely pertinent passages from that manuscript" because of "the warning of its authors that it
is 'not to be quoted, abstracted, or reproduced without specific permis-
sion.'" Years went by and some of the many fans of *OTSOG* (the acronym by which *On the Shoulders of Giants* became known to its admirers)
would occasionally write to Merton to inquire about *Serendipity*. Still, no
book appeared.

By 1987, another footnote (number 15 in his "Unanticipated Consequences and Kindred Sociological Ideas: A Personal Gloss"[1]) mentions
the work as "a still unpublished monograph." The polysemous "still"
reminds us that the document both continued to be unpublished and
had not moved or changed shape. Captive the book remained. Were
these footnotes the confessions of a guilty conscience or were they ran-
som notes? What is the story of the *Serendipity* book and why does the
work come to light only now, forty-five years after its composition? Like
the three princes of Serendip (in the story that launched the word), we
must act as detectives, examining physical objects for vestigial clues to
this mystery.

The primary physical evidence consists of the now-delivered book it-

---

[1] R. K. Merton, "Unanticipated Consequences and Kindred Sociological Ideas: A Personal Gloss," in *L'Opera di R. K. Merton e la sociologia contemporanea*, ed. C. Mongardini
and Simonetta Tabboni (Genoa: Edizione Culturale Internazionale Genova, 1989), pp.
307–329.

self. It traces serendipity—the word and the concept—from its birth in 1754 through the then-contemporary days of 1958. Derived from the ancient name for Sri Lanka featured in the *Peregrinaggio de tre giovani figliuoli del re di serendippo*, the word serendipity appeared to Horace Walpole out of the East—reminiscent of other exotic prizes, such as Boiardo's Angelica (the captivating daughter of the emperor of Cathay) who would eventually drive the hero mad in Ariosto's *Orlando Furioso*. Defining the word is a task that I must leave to the authors, for the word's resistance to precise interpretation represents a crucial aspect of its tale. Serendipity can be about finding something of value while seeking something entirely different or it can be about finding a sought-after object in a place or a manner where it was not at all expected. The word is always about discovery and always about what Walpole called "happy accident," but the exact mixture of wisdom and luck—and the homiletic gloss on each tale of serendipitous discovery—varies as the word is employed in different contexts.

Today, English speakers have embraced the word serendipity; so much so that one may sense an excessive richness in the word's current popularity. For we see serendipity everywhere—in the names of cruise lines, software companies, scholarships, bookstores, marketing firms, a ranch in Australia, and a nudist camp outside of Atlanta, Georgia. But in the late 1950s, writers employing the word serendipity attached generous parentheses full of explanatory definitions. Although the content of these parentheses varied, the authors of this book describe the history of how various audiences encountered the alluring idea of a process hovering ambiguously between the clever and the clairvoyant, the incisive mind and the wheel of fortune.

We know that sometime in the mid-1950s, with his colleague Elinor Barber, Merton wrote the enclosed travel guide. Beginning with the invention of the word, the authors chart where it went, how long it stayed, with whom it dined and resided, and they note whether it was loved or despised. The book is sophisticated enough to make all stops on the itinerary. Some people would like to believe that a word appears (or is invented) and history and the writers of dictionaries automatically throw open the gates; not so. A word has to fight for its life. To survive, a word must claim its place and convince an audience.

The authors follow the word across oceans and across academic specialties, tracking its use and abuse. They chase it down surreal burrows (for example, when they consult a linguistic wise man to inquire about the pleasing sound of the word) and into the illustrious halls of the Harvard School of Medicine. Thankfully for our investigation of the mystery of the book's having remained unpublished until now, the physical evidence lay undisturbed. This book does not take up the chase beyond the chronological boundaries of its original composition; in fact, except for the filling in of bibliographical citations, it is untouched. Unlike many tardy academic manuscripts, it has spent its forty-five-year seclusion not in endless cycles of revision, but rather in suspended animation. Not a word has changed from the shrouded manuscript alluded to in *OTSOG*, chapter 47.

In addition to the absence of never-written chapters on the post-1950s travels of serendipity, we find other signs of the book's unfinished state. As it is chased by Merton and Barber across its early years, the elusive serendipity and its pursuers can be found meandering; characters (such as Dr. Walter Cannon of the Harvard medical school) appear briefly before disappearing and reappearing four chapters later. Vivid and marvelously instructive passages—such as the authors' explanation of how the door-to-door salesman faces different species of uncertainty than the railway switch operator—beg for expansion and a place of their own (as the scientists have earned). Some crucial philosophical frameworks—such as the discussion of fate and the varieties of luck—remain offstage until later than might be expected. Meanwhile, the authors' second close reading of the Walpolean moment of word-coinage pops up only far down the line, when the reader, thinking the eighteenth century long gone, is more than ready to peek into the serendipitous birth of penicillin.

Nevertheless, the unfinished text remains as it was then, when serendipity was first exposed to the public spotlight. Although one can read it as the tale that it was originally meant to be, it also can be read as a historical document, intended for a 1950s audience but stolen away and transported to a strange land some forty-five years later. And, significantly, it is as a preserved remnant from a time capsule that the book speaks with its more powerful—prophetic—voice. Like Ariosto's Merlin, who "never refused an answer to those who sought word from him of things past or future,"[2] the pages of *The Travels and Adventures of Serendipity* foresaw the now-past future.

The sections of the book that deal with scientific discovery presciently hint at the future history now lived. The authors understood that as the word and its class-bound meaning were becoming transformed from a plaything of the idle class, serendipity found itself at the juncture of epistemological debates and administrative logistics: "It is the obligation of the applied scientists to discover useful items, and the question of how discoveries are made is asked not for its own sake, but as matter of practical strategy" (Chapter 7). In that it emblematizes the Eureka moment in

---

[2] Ludovico Ariosto, *Orlando Furioso*, ed. Lanfranco Caretti (Milan: Ricciardi, 1963), 3:11. English translation is from *Orlando Furioso*, trans. Guido Waldman (Oxford: Oxford University Press, 1983).

its distilled form, serendipity can be read as a synecdoche for all of the ambiguities of the process of discovery more broadly considered. We can see in this crystallized microcosm the seeds of debates that have since taken place both within the scientific community and in interdisciplinary skirmishes over the nature of scientific truth claims.

Within science the ensuing years have seen growing pressure for useful results from those who sponsor the research. As the cost of doing science has increased, debates simmer and occasionally rage over the allocation of resources between basic research and the development process that converts the gains of basic research into useful products. Serendipity magically collapses the time frame of this process: In the narrative of serendipitous discovery, the search (or research) and the *objet trouvée* appear immediately linked. And yet, the elements of the narrative—the far-ranging freedom of basic research, the search for applicable findings, the role of chance, and the role of the mind—all have their costs. As sponsoring institutions—private-sector laboratories, universities, government agencies—evolve, the practical strategists see serendipity either as an alchemist's stone, a shortcut to useful answers, or as fool's gold. And the stakes continue to get higher as research and development in the United States and other industrial nations competes for resources with the ever more enticing practical applications. In such an atmosphere, debates over which research will bear the most fruit are inescapable. And as the financial burden shifts from the government to the private sector, the harvesting of the crops is entrusted to those whose charge it is to feed today's market and not necessarily to watch over the long-term health of the orchard.

The debates over spending priorities coincide with philosophical differences over what sort of truth the various branches of science champion. The physicists' search for a unifying theory postulates that, given data enough and time, every phenomenon from the smallest subatomic particle to the universe itself could be concatenated and understood. At the other end, biologists who trace the complex emergence of life and its many forms maintain that no one theory can reduce the endlessly rich weave of life and the cosmos to a formula. At one end the creativity championed is that of the aperspectival view of the scientist looking over the great expanse of the universe; for the others, the creativity resides in the artful discovering of how complexity makes the universe worth inhabiting. Either side can find in its opposite a claim to truth that menaces the vital purpose of its own quest. In this light, debates about whether the scientific act relies more upon the mind of the seeker or the multiplicity of contingent and unpredictable data transform serendipity into a microcosmic version of the more acute epistemological duels being played out as we begin to explore the new millennium.

Moreover, from outside the walls of science come voices that ask why there should be walls at all. In these cross-disciplinary skirmishes, the specter of an even broader reductionism looms. Some humanists and social scientists, having experienced the *jouissance* of postmodernist skepticism, relativism, and subjectivism, seek to pull the institution of science from its privileged place, asking whether empirical findings in the laboratory are subject to the same slippery truth claims as poetic sentiments or political promises. In the process, such extreme questioning of the epistemological claims of the scientific community implicates a much larger group of sensible thinkers who also seek to make connections across the walls of difference that surround science. In loathing the radical agenda of the extremists, some within the scientific community retreat within their walls and, as is likely in time of siege, mistake every ambassador for an enemy. A great deal of fruitful exchange between the scientific and humanistic communities is lost when, faced with the attacks of radicals, some scientists view all external examination of the scientific enterprise under the same heading and close ranks accordingly.

The days of the middle ground of innocent serendipity—when Dr. Harold Brown urged his Columbia medical students to be open to factors "quite extraneous to the procedures of scientific investigation"— seem lost. Although there are, of course, benefits to be gained from the narrowing of research focus, specialization limits our ability to reach outside of what we practice on a daily basis. In addition, practical pressures (such as the need to establish reputations) lead students of science and medicine to limit their focus at progressively earlier stages of their careers.

But in the 1950s these debates were still young. The *Serendipity* manuscript was composed just before C. P. Snow's now famous 1959 Rede Lecture in which he argued that scientists and humanists were segregating themselves into two increasingly separate cultures. And sociology— the logical field to turn toward the question of what differentiates science as a structured field of inquiry—had begun only in the late 1930s and 1940s to turn its attention toward the historical and sociological structures that play a role in shaping the process of discovery.

What the untouched remnant of *Serendipity* shows is how these tensions were foreseen. In citing the writings of George Burch (of the Tulane University medical school), the authors capture his anger at the association between serendipity and luck, luck and casual random efforts, and such random efforts and "fashionable" discoveries that are important only for their public relations value (Chapter 9). Moreover, the authors show how serendipity has subsequently spiraled far beyond its origin as a mere Walpolean trifle into the administration of science and of industrial invention. The director of the U.S. Navy's Department of Research perhaps best plotted the recent history of what must be thought of as "serendipity management": "The real skill, of course, lies in the ability to forgo the unpromising occurrence, however interesting it may be, and relentlessly pursue that which, found but unsought, may pay great dividends in progress" (Chapter 10). The metaphor of "paying great dividends" foreshadows four decades of constant battles over research priorities, the dividends and the costs of which must be "sold" to the newspapers, the impatient congressman, and in some cases even the corporate CEO whose bonus is tied to quarterly profit margins.

But alas, these well-preserved insights and foresights provide few clues to the book's suppression. The satisfaction of *Serendipity*'s prescience confirmed is small consolation for a forty-five-year wait. Despite its pristine state, the book yields few clues to its slumber. We must turn to our magnifying glass in the hope that an examination of the concept itself might resolve the mystery of its delayed chronicle.

The concept describes a process that is quintessentially ambiguous. How much of serendipity is luck? How much is due to the prowess of the searcher? Its roots are correspondingly ambiguous. Walpole (as the authors note in Chapter 1) must have recognized that the tale of the three princes—who deduced, for example, from the state of the grass along the road that a camel with a bad right eye had been wandering and eating according to what he saw best—was all about sagacity and not at all about the "happy accident" that he also wished to invest in the word. While Walpole tells of its origin in the detective story, he also tells us (in the letter to Mann in which he coins our word) of his habit of playing a *sortes Walpolianae* "by which I find everything I want *à point nommée*, wherever I dip for it." Walpole's habit was a variation of the *sortes Virgilianae*—a popular means of divine prophecy in the Middle Ages and Renaissance played by asking a question, opening the *Aeneid* to a random page, putting a finger on a random line, and interpolating an answer from the randomly chosen verse.[3] The prophetic game, more than Walpole's source text, encapsulates the ambiguity of the term: Is it God who leads the seeker to the key word or line? Is it the author of the text who was divinely inspired? Should all credit be given to the prepared mind of the interpreter who assigns meaning to the randomly chosen verse? Or is

---

[3] Vergil (mostly on the basis of the Christian allegoresis of the *Fourth Eclogue*) was read as a pre-Christian prophet of Christianity and his text as a medium of inspired guidance. A famous example of the related *sortes Biblicae* can be found in book 8 of Augustine's *Confessions*. Having struggled endlessly to find faith, the tormented Augustine hears children singing what he understands as "tollo legge" ("take up and read"). Remembering the conversion of St. Anthony (who was converted by a random reading of Matthew 19:21, which he had happened upon), Augustine plays the *sortes* and lands on Romans 13:13; upon reading it, he is converted.

the discovery truly a linking of past and present, a séance in which the natural and the supernatural blur?

The spirit of the word seems to have elusive roots that may have escaped even the penetrating eyes of Merton and Barber. Other scholars, pursuing very different mysteries than our own, have suggested that the original Italian source (from which Walpole had access through a French translation) may contain even more levels of significant ambiguity. Published (as the authors note) by Michele Tramezzino in the free-thinking Venice of 1557, the work is ostensibly an Italian translation of Persian tales by M. Christoforo Armeno. Some literary scholars have argued that Tramezzino himself invented the supposed Armenian Christoforo and gathered the stories not from Persia but from popular folk tales and assorted classical stories (including Plautus), "packaging" them as an exotic import.[4] This theory suggests the possibility that even in its charming magical infancy, the source of serendipity was already being mined and manipulated for its commercial potential. Rather than arising out of the East on a magic carpet, the grandfather of Walpole's brainchild may have been the product of shrewd marketing and impression management. Ironically, despite Merton's recognition in the 1993 postface to *OTSOG* that authors' appeals to the authority of preceding authors might represent a strategic use of what Merton's teacher George Sarton referred to as "ghost-writing in reverse"[5] and despite the tenacity that allowed Merton to crash through deceptively authoritative but erroneous citations of the *OTSOG* aphorism, the potential manipulation of serendipity's source may have eluded him.

Was *The Three Princes of Serendip* an Italian import that underwent a progressive series of editorial conversions before presenting itself to Walpole's wandering eye? Or was it an Italian creation, born of clever fancy and planted knowingly before a gullible audience? Either way, one can find in the later appropriation of the word a transmigration of souls. As the word headed westward (thanks to the either insightful or duped Walpole), the word's ambiguous roots were amplified.

And although the ultimate source of the fairy tale may eventually be traced to Persia or elsewhere, we are still left with the mystery of why the 1950s book *The Travels and Adventures of Serendipity* comes to us only

---

[4] Jocelyn Penny Small, "Plautus and the Three Princes of Serendip," *Renaissance Quarterly* 29, no. 2 (summer 1976): 192.

[5] In his 1993 "postface" to *On the Shoulders of Giants: The Post-Italianate Edition* (Chicago: University of Chicago Press, 1993), p. 313, Merton cites Jacques Le Goff on why writers might falsely attribute their inventions to others: "In the twelfth century, Arab writers were so fashionable that Adelard of Bath slyly remarked that he had attributed many of his own thoughts to the Arabs so that they would be more willingly accepted by his readers." Le Goff, *Medieval Civilization 400–1500* (Oxford: Basil Blackwell, 1988).

now. Because we know the impresario who both summoned the word from its hiding place a half-century ago and now by an act of prestidigitation has made it reappear, the next set of clues is likely to come when we check out Merton's alibi.

While one story begins with Walpole's ambiguous coining of the word from the ambiguous source story, the other begins in 1924 with fourteen-year old Robert Merton learning the rudiments of prestidigitation from Charles Hopkins, a Philadelphia neighbor who would eventually marry Merton's sister and became "a surrogate father" to Merton himself. Acquiring a wand, a cape, and a top hat, the young Merton learned to dazzle audiences with magic tricks and "enchanting mysteries"; his talents were surely not insubstantial, as he was able to help pay for college with the proceeds of his performances.[6] The subsequent capture of the magician by the only slightly less arcane field of sociology proved complex. Neither the specter of sprawling social structures nor the minutiae of academic specialization could contain him. Rigid chains of a singular methodology, a single topic, or a single ideological banner that strangle most researchers through long periods of their careers were either burst or subtly unlocked. The magician won out. Over the course of his career, he summoned from his sociological top hat some two dozen books, countless articles, and almost single-handedly the subfield now established as the sociology of science.[7]

The movement toward *Serendipity* can be recreated through a cursory examination of Merton's collected writings. The beginnings of the tale are clear. In the 1930s at Harvard, Merton first ventured to explore how and why intentional social actions often have unintended consequences. His dissertation, for example, described how the theology of ascetic Protestantism unwittingly contributed to the rise of modern scientific thought. Over the course of his career, this theme—the unanticipated consequences of social action—crossed disciplinary boundaries as if they simply did not exist. It helped to explain mysteries of urban planning at the same time that it gave clues to the failure of financially strong banks.[8] It led also, since Merton inevitably returned to the study of science as a

[6] See Robert K. Merton, "A Life of Learning," from the Charles Homer Haskins Lecture presented at the annul meeting of the American Council of Learned Societies, Philadelphia, April 1994. First published as Occasional Paper 25 (New York: American Council of Learned Societies, 1994), it was reprinted in R. K. Merton, *On Social Structure and Science* (Chicago: University of Chicago Press, 1996), pp. 339–358.

[7] His influence in sociology, the study of science, and the history of ideas goes far beyond such a casual description. For the story of his life's work and influence, the reader should seek out Piotr Sztompka's introduction to *On Social Structure and Science*.

[8] These examples are from "The Self-Fulfilling Prophesy" (1948), reprinted in *On Social Structure and Science*, pp. 183–201, an exceedingly powerful formulation of one variety of unanticipated consequences.

social institution, to an interest in the role of the unanticipated in scientific discovery—both the "by-products" of intended research and, correspondingly, the main products of unintended research.

Sometime in or before 1945, while looking for the definition of some now-forgotten word in volume 9 of the *Oxford English Dictionary* (*OED*), Merton's eye "happened upon the strange looking but euphonious word 'serendipity.'"[9] Just as Walpole wrote of his habit of playing a *sortes Walpolianae*, a random flip of the page led Merton to serendipity. It was, when Merton originally stumbled upon it, a strange beast pacing restlessly within the confines of a few learned vocabularies. Had he not chosen to spend a significant portion of his third-year graduate student stipend on the then twelve massive volumes of the *OED*, he might not have ever stumbled on the word. Had he heeded the call of whatever his pledged mission was that day—learning about *sequestration* or *seraphim* or *sepulcher*—this sociological tale of the wanderings of serendipity would have been stalled, ensnared in the maze of the dictionary, imprisoned from further adventures until some other wandering eye might find it and send it on its way.

The strange object gathered by bibliomancy—the playing of the *sortes OEDinae*—appeared to Merton and was quickly put to good use. Although the word first found its way into Merton's published work via—no surprise—a footnote (in his 1945 article "Sociological Theory"), it was in his talk to the American Sociological Society in March 1946 that Merton unveiled the concept of "the serendipity pattern" in empirical research. In it he discerned a pattern "of observing an unanticipated, anomalous, and strategic datum which becomes the occasion for developing a new theory or for extending an existing theory." All three elements are crucial—the datum is not what was expected; it is surprising "either because it seems inconsistent with prevailing theory or with other established facts." In order to fulfill the pattern, the datum must also be of strategic importance in setting the course of a new theory. From this point in the mid-1940s, the word entered the social sciences. As Merton returned to his study of scientists in and out of the lab, serendipity loomed larger as a topic. The intertwining of Merton's study of the unanticipated in general and his original sociological petri dish—science—made an investigation of serendipity predestined.

Merton and Barber demonstrate that, in a sense, the word's history enacts its essence. All three of the crucial attributes (unanticipated, anomalous, and strategic) required to fulfill the serendipity pattern in research have also been met for the word's absorption into the lexicon. Coined words not only need to be striking, they must also "earn their keep" by providing a name for something that is worth naming. Seren-

---

[9] Merton, "Unanticipated Consequences and Kindred Sociological Ideas," p. 325.

dipity had to pass the same market test as any other "invention"; it may have been "unanticipated" and "anomalous" to the early Victorian historian and critic Thomas Babington Macaulay, who despised Walpole, but, to Macaulay, anything Walpole created or contributed was the antithesis of "strategic." Later in its history, when the word reappeared and when the prepared cultural context saw it, the dictionaries swerved to greet it.

So why did the telling of its tale not excite Merton enough to take the small final steps necessary to publish the chronicle decades earlier? A diversion might provide a clue. It is well known that Merton disliked the oversimplifying aura that envelops published arguments. As early as 1938 he observed that the telling of a scientific finding depends upon creative storytelling: Work is "presented in a rigorously logical and 'scientific' fashion (in accordance with the rules of evidence current at the time) and *not* in the order in which the theory or law was derived."[10] Later, as the implications of the story-telling process became even clearer to Merton (both in the subject of his research and in the process of his own work), he would write that "the etiquette governing the writing of scientific (or scholarly) papers requires them to be works of vast expurgation, stripping the complex events that culminated in the published reports of everything except their delimited cognitive substance."[11] In short, the audience demands a well-formulated argument that retrospectively imposes logical form on the romance of investigation. It is, however, a well-known fact that daemons pay little heed to the dicta of the marketplace.

Soon after the 1950s composition and deliberate nonpublication of *Serendipity*—the logical compilation of the true history of the travels of the word and the concept—the prepared mind found its datum. Merton chose to chase an image that he couldn't let go or, as he would later acknowledge, "it chose me." Merton's important 1957 presidential address to the American Sociological Society—written contemporaneously with the *Serendipity* manuscript—had an unanticipated consequence far removed from forging the link between the study of the reward system in scientific research and the ethos of science; reading the published version of the address also led Merton's friend, the eminent Harvard historian Bernard Bailyn, to inquire about the source of Merton's quotation of the Newtonian colorful aphorism, "If I have seen further, it is by standing on ye shoulders of Giants." Upon sitting down to write a response, Merton found what he had not anticipated: "I soon found myself engulfed in a Shandean stream of consciousness. And once caught up in it, I could not escape. The daemon had taken over. Except for occasional catnaps and

---

[10] R. K. Merton, *Science, Technology and Society in Seventeenth-Century England* (New York: Howard Fertig, [1938] 1970), p. 220.

[11] Merton, "Unanticipated Consequences and Kindred Sociological Ideas," p. 309.

stray excursions to the kitchen, I remained staked out at my desk with its IBM electric typewriter . . . for the next three to four weeks."[12] Sparked unexpectedly, greeted with playful curiosity, and only to be explored by someone with Merton's learning, tenacity, comfort in moving across time and space, and willingness to err, that response, *On the Shoulders of Giants*, fulfills the romance of magical serendipity. The IBM electric typewriter was the hippogriff upon which he took flight. His books were his Faerieland, replete with voices living and dead, castles enchanted and disenchanted, songs sung and trapdoors leading either to barren dungeons or sparkling treasure. Cyclical in nature, it is the story of the quest, wherein the finding of sources and answers does little to lessen the rich texture of the search. It provides, as Stephen Jay Gould noted, a celebration of "the inestimable value of error"[13]—of methods tried and failed, of paths followed fruitlessly, of structure built only to be abandoned when the final brick simply doesn't exist.

And though it was the frenzy of possession that provided the passion, it was the modus operandi first adopted in *Serendipity* that seems to have cleared the way. Despite his frustration with the way in which logical exposition drains away the richness of the research exploration, Merton generally had relented and had published numerous and seminal "traditionally structured" pieces such as those collected in the 1949 *Social Theory and Social Structure*. For the sake of explaining phenomena such as self-fulfilling prophecies, Merton was willing to provide in his work what we as consumers demand, a narrative "bottom line," complete with a beginning, a middle, and an end. But with the pursuit of serendipity came something new. *Serendipity*'s structure—tracing the use and reception of an idea across fields and the centuries, acknowledging the strange and varying reception of the notion by individuals of all ilk, joyfully exploring the most arcane and the most absurd—is eerily reminiscent of the structure that undergirds the ludic energy of *OTSOG*. But whereas serendipity's nonsensical, ahistorical, and impressionable roots enable each inheritor to overrun the word with whatever meaning strikes his fancy, the *OTSOGIAN* image provides a clearer and more distinct baseline against which to gauge the changing contexts that receive it. In this way, the pursuit of the mythic and exotic serendipity can be thought of as a *preparazione OTSOGIA.*[14]

---

[12] Merton, "Triennial Thoughts," *East Hampton Star*, 1 July 1993, section 2, p. 13.

[13] Stephen Jay Gould, "Published Pebbles, Pretty Shells: An Appreciation of *OTSOG*," in *Robert K. Merton: Consensus and Controversy*, ed. J. Clark, C. Modgil, and S. Modgil (New York: Falmer, 1990), p. 43.

[14] This function draws upon two notions: (1) the humanist tradition of rationalizing why the reading of classical pagan texts was really a means of preparing the mind for the truth of scripture, and (2) the Mertonian observation in his 1993 "postface" that the aphor-

Legend has it that daemons can be either good or evil, but they are most likely to be found among the ruins—they may be the spirits of soldiers who lay unburied, of people who died an untimely death (such as Hamlet's father), or of the gods that are left after a city has been ransacked and its inhabitants dragged away. In writing his response to Bud Bailyn, a daemon of an idea overtook Merton; this consuming idea—the image—may have been lurking in the well-built, but recently abandoned, history-of-an-idea structure of the *Serendipity* manuscript.

And as *OTSOG* burst forth, returning to the predecessor investigation must have seemed to Merton like an unnecessary retracing of his steps. There was so much else to do that the shelving of the *Serendipity* manuscript must have evoked, at most, the quickly passing pain of the deleted footnote. In this way the forty-five-year slumber of the book can be seen not as a tragedy, but merely as another casualty of the ongoing romance of investigation.

Late in its life, we return to the word serendipity, which, like Angelica, was whisked from the imaginary East, wrought by Italian hands, and grew tired of being a foil for the passions of others. Today we can see its life story as that of an object overwritten, semantically overcoded by our *innamorato*, our *furioso*, a mystery upon which we imprinted our deep but otherwise unspoken beliefs. Weary of the battles and the marauding heroes, Angelica is last seen in Ariosto's poem heading east, turning toward home. Merton must have known that work of yesteryear had something new to offer: It could, in its intact form, now be read as an early eyewitness depiction of serendipity's journey from mysterious plaything to demystified catchword. And only with the telescopic device of Merton's twenty-first-century Afterword are we treated to an eyewitness account of that post-1958 diffusion.

Although Merton frequently chastised himself both orally and in print for procrastination in failing to publish both this book and other long-unpublished works, he did so in an anti-Mertonesque spirit. For in order to absorb the unanticipated datum, he had to be prepared to spur his horse to a full gallop directly away from his previously intended path. Each work completed, each idea donated to the field, is the cousin of another goal laid aside. In the case of *Serendipity*, the alarm was raised by the paradigmatic 1957 paper on the institutionally distinctive reward system of science and the path to be pursued was *OTSOG*. New adventures presented themselves: In addition to the paper itself, this period spawned

---

ism's acronym (*OTSOG*) "is admirably suited to all the various word formations." *OTSOG* adapts and escapes "the harsh and lonely fate" of obscurity. But in that the acronym never fails to carry within it its alluded-to roots, it functions as a counterpoint to the objectified, marauded, and abused word serendipity.

crucial works such as the second, canonical edition of *Social Theory and Social Structure* and the pioneering study of professional socialization, *The Student-Physician*. Soon to follow were such seminal works as "The Role Set," "Singletons and Multiples in Scientific Discovery," "The Matthew Effect," and "Sociological Ambivalence." In *OTSOG*, the chronicler of serendipity put down his pen, took up his arms, and set out to join in the clash of distant battles.

In some ways, he must have hoped that telling the tale of serendipity would weave together various threads of his work on the unanticipated. But whereas the "serendipity pattern" had provided a highly appropriate—and satisfying—description for a specific tendency in the conduct of research, the word serendipity may have proved too chimerical. Still, something unanticipated and positive resulted; tracking the word's movements led Merton to explore a digressive writing style that resonated with his sense of how the process of research truly occurs. This style would only blossom later when it was granted the freedom to pursue a more historically rooted object. Although the central image of *OTSOG* may have been as elusive as "serendipity," the Newtonian aphorism would prove to be better grounded both by its long heritage of reasonably consistent usage and by its powerful visual analogy. For the scholar tracing its footsteps over the centuries, it would prove less thoroughly exposed to the whims of whichever historical figure possessed it at a given moment. As Merton set off in search of new adventures, the curious and exceedingly impressionable import—serendipity—was left to fly unchecked into our arms.

Mystery—of forty-five years' delayed publication—provisionally solved. And as with all good tales of romance, it ends, happily, as it began: Once upon a time . . .

THE
TRAVELS AND ADVENTURES
OF
Serendipity

Chapter 1

THE ORIGINS
OF SERENDIPITY

The letters that passed between Horace Walpole and Horace Mann form what Wilmarth S. Lewis calls the Andean range of the Walpole correspondence.[1] The two friends, who were also distant cousins, exchanged these letters over a period of forty-six years (1740–1786), although, after Walpole's visit to Florence in 1741, he and Mann, who long remained British minister to the Court of Florence, never saw each other again. Walpole wrote all his many letters for posterity, but these letters to Mann were particularly designed to be a "kind of history,"[2] a chronicle of important political and social events. Inevitably, and as a matter of his characteristic taste, many "unimportant" incidents crept into his letters, too, and one such item came to mean much more to a small and growing segment of posterity than Horace Walpole could possibly have anticipated.

Writing to Mann on January 28, 1754, apropos of the arrival in England of the Vasari portrait of the Grand Duchess Bianca Capello, which Mann had had sent to him, Walpole told of how he made a "critical discovery" about the Capello arms in an old book of Venetian arms:

This discovery I made by a talisman, which Mr. Chute calls the *sortes Walpolianae*, by which I find everything I want, *à pointe nommée* [at the very moment], wherever I dip for it. This discovery, indeed, is almost of that kind which I call *Serendipity*, a very expressive word, which, as I have nothing

[1] Wilmarth S. Lewis's introduction to the Walpole-Mann correspondence, in *The Yale Edition of Horace Walpole's Correspondence*, ed. W. S. Lewis (New Haven, Conn.: Yale University Press, 1937–1983), vol. 17, p. xxiii.

[2] Letter to Mann, 28 January 1754, in *Walpole's Correspondence*, vol. 20, pp. 407–411.

better to tell you, I shall endeavour to explain to you: you will understand it better by the derivation than by the definition. I once read a silly fairy tale, called *the three Princes of Serendip*: as their Highnesses travelled, they were always making discoveries, by accidents and sagacity, of things which they were not in quest of: for instance, one of them discovered that a mule blind of the right eye had travelled the same road lately, because the grass was eaten only on the left side, where it was worse than on the right—now do you understand *Serendipity?* One of the most remarkable instances of this *accidental sagacity* (for you must observe that *no* discovery of a thing you *are* looking for comes under this description) was of my Lord Shaftsbury, who happening to dine at Lord Chancellor Clarendon's, found out the marriage of the Duke of York and Mrs. Hyde, by the respect with which her mother treated her at table.

Since he had "nothing better to tell," therefore, Walpole was reporting to his friend a bit of whimsy, a word he had coined. His attitude toward it was half-pleased (the word is "very expressive"), half-mocking and deprecatory. Had Mann looked into the fairy tale that helped Walpole to mint the word, he might have been confused, for its story line scarcely resembles Walpole's account of it or the allegedly parallel examples he provides. Walpole *was* looking for information about the Capello arms and only happened, by "serendipity," to find it at just the right moment, but the three princes of the fairy tale *found* nothing at all, but merely gave repeated evidence of their powers of observation. Moreover, Lord Shaftesbury actually did make a useful discovery that he had not anticipated, one that he could not have made without considerable "sagacity" about the minutiae of the symbols of respect and deference, just as one now gauges impending changes in the status of Soviet leaders by noting their location in the Kremlin ensemble on public occasions. The complexity of meaning with which Walpole endowed serendipity, carelessly and inadvertently, at its inception, was permanently to enrich and to confuse its semantic history.

The "silly fairy tale" that Walpole referred to was called *The Travels and Adventures of Three Princes of Sarendip*. According to the title page, it was "translated from the Persian into French, and from thence done into English," and printed in London for Will. Chetwode in 1722. As far as Walpole knew it was anonymous, but we shall have more to say later about its authorship and history. The three princes of the title are the sons of Jafer, the philosopher-king of Sarendip (or Serendib, which is the ancient name for Ceylon).[3] King Jafer had seen to it that his three promising sons received the best possible education from the wisest men in the

---

[3] [Nowadays, Ceylon is called Sri Lanka. However, this manuscript was written in the 1950s, so all references to Ceylon remain unchanged.]

kingdom, and now he wished them to travel in order that they might gain in experience to complement their book learning. Above all, he wanted them to learn about the customs of other peoples. There is never any mention of a search for treasure, which has so often been ascribed to them by those who know the tale at second or third hand.

"As their Highnesses travelled" they had various adventures and made certain "discoveries." Their adventures resulted from the use they made, and that other people made, of their keen wits; and their "discoveries," which were of the nature of Sherlock Holmesian insights rather than more conventional "treasures," often proved valuable to those whom they encountered. In two episodes they used their ability to make careful observations and subtle inferences, practicing this skill for the sheer pleasure its exercise afforded. In another episode, they did their host, the Emperor Behram, a valuable service, when, by virtue of their keen observations and their intuitive understanding of human psychology and physiology, they were able to save him from the vengeance of a treacherous minister. At still another court they visited, they passed yet another age-old test of wit, the solution of riddles, both humorous and serious. In all these adventures they conducted themselves with great courtesy and modesty.

Of all these many incidents, the one that seems to have impressed Horace Walpole the most is one of the princes' exploits of observation and inference. (It is, in fact, the first incident that occurs in the course of their travels; perhaps Walpole never got any further in this "silly fairy tale.") As the princes were riding along, they met a camel driver who had lost one of his camels and asked if they had seen it. Since they had seen various clues that might indicate the lost animal, they asked him the following three questions: Was the animal blind in one eye? Was it lacking one tooth? And was it not lame? The driver answered all these questions affirmatively, so they in turn told him that they had passed his animal and that it must have gone quite far by now.[4]

The camel driver searched the road for twenty miles without finding his missing animal, so he returned and again came upon the three youths. He told them that he thought they had merely been teasing him, so they gave him further evidence: that the camel was laden with butter on one side and honey on the other, that it was being ridden by a woman, and that this woman was pregnant. Now the driver was sure that the princes must have stolen the camel, and he had them brought to justice before the Emperor Behram. The princes confessed that they had never really seen the camel and that they had only told the driver of inferences drawn

---

[4] Motifs of this kind were common in the eighteenth century, as we shall see. Voltaire was only one of many to anticipate the techniques of Sherlock Holmes in this way.

from the clues they had observed, which happened to coincide with the facts.

The incident ended happily when the camel was found. The emperor, now vastly impressed, wished to know how the princes had so accurately inferred its characteristics. They explained to him their guess that the camel must be blind in the right eye because the grass had been cropped on the left side of the road, where it was worse than on the right; that they had found bits of chewed grass on the road, of a size indicating that they had fallen out between the animal's teeth where a tooth was missing; that its footprints showed that it was lame and was dragging one foot; that its load of honey and butter could be inferred from the trail of ants on one side of the road, for ants love butter, and of flies on the other, for flies love honey; that at one place they saw footprints that they attributed to a woman rather than a child because they also felt carnal desires there; and finally, that this woman must be pregnant, because they had seen the imprints of her hands on the ground, where, in her heavy state, she had used them to get to her feet again.

It was the "discovery" of the blind right eye that Walpole evidently remembered best and which he used to illustrate the princes' peculiar talents. By the time he was "deriving" serendipity for Mann's benefit, however, his memory had transformed the camel of the original story into a mule. As an Englishman he was certainly more familiar with mules than with camels; perhaps this is why the alien camel was transformed into the more familiar mule. For this story in its essentials is, as we shall see, an old one. As it was told in India, for example, it involved an elephant, while in Palestine and Arabia it generally was the camel, as in the tale of our three princes.[5] In each case the cultural background produced at least this small variation in the protagonists of the story. In like manner, the already complex meaning of Horace Walpole's "very expressive word" was on many future occasions to be slightly or drastically modified by the social context of its use.

These, then, were the immediate occasions of the invention of serendipity: an episode in a story of three princes of Serendip in which they displayed their powers of observation and found certain clues they had not been looking for; Horace Walpole's unexpected discovery of an item missing from his knowledge of heraldry, one among many such accidental discoveries; and, finally, Walpole's letter to Sir Horace Mann, in which he indulges himself by elaborating on the nature of certain aspects of the process of discovery. But all this tells nothing of how it was that Horace Walpole, living in England, in the year 1754 came to merge these partic-

---

[5] See Joseph Schick, *Die Scharfsinnsproben*, vol. 4, part 1 of *Corpus Hamleticum* (Leipzig: Harrassowitz, 1934).

ular ingredients to fill a minute space in the English language by creating this strange new word, serendipity. From all indications, this was the result of two unrelated sets of circumstances: One is the great efflorescence of interest in the Orient in the eighteenth century; the other, Walpole's idiosyncratic propensities, which he brought to the reading of the tale of the three princes of Serendip.

Both England and France had had some contacts with the East and with Oriental history and literature in the sixteenth and seventeenth centuries, but the great upsurge of interest did not come until Antoine Galland translated the *Arabian Nights* into French, between 1704 and 1717. His translation of the *Arabian Nights* was quickly followed by Petis de la Croix's translation of *La histoire de la Sultane de Perse* . . . (1717) and *Les Mille et un Jour* [*sic*] (1710–1712). In France, these tales from the Orient were welcomed for several reasons: they provided an escape from the restrictions from classicism, they were found to be a useful device for social criticism (Montesquieu's *Lettres persanes* and Voltaire's *Zadig* are, perhaps, the most famous examples), and they provided writers such as Crébillon *fils* with a takeoff point for his *contes licencieux*, which satirized the then-popular *contes morals*.[6]

The response in England to the tales from the Orient was in some respects similar to that in France. "The magical atmosphere, the rich variety of dramatic incident, the spirit of adventure, and the brilliant background" of the *Arabian Nights*, the telling of a story for its own sake, and the food these stories provided for peoples' "imagination, their fancy, their emotion" were congenial with the incipient romanticism of the period, in England as in France.[7] In England, the social and literary satire that used oriental tales was, however, far milder than the French: "French satire, more pervasive and more penetrating, expressed—especially when touched by the genius of Voltaire and Montesquieu—something of the deep unrest of France in the eighteenth century, the era before the Revolution. . . . The typical English writer of philosophic oriental tales, on the contrary, dwelt in an imaginary country of pure speculation, and entered the world of fact only for the purpose of moralizing."[8] The moralizing tendency was extremely powerful in England in this period and it stifled the oriental tale. "Too exotic to become easily acclimated, such tales were regarded as entertaining trifles, to be tolerated seriously only when utilized to point a moral."[9] Except for their

[6] Martha P. Conant, *The Oriental Tale in England in the Eighteenth Century* (New York: Columbia University Press, 1908), p. xxv.

[7] Ibid., pp. 12, 241, 245.

[8] Ibid., p. 231.

[9] Ibid., p. 233.

romantic appeal, then, the chief reason for the vogue of the oriental tale in Francophile England was its vogue in France.[10]

Walpole's interest in and familiarity with oriental tales was no greater than might be expected of a literary man of his time. Nor was his mockery of these tales unusual, in the later eighteenth century especially, and it is, in part, his longevity that is responsible for the gamut of his attitudes. Walpole was fond of the *Arabian Nights*, and the contempt he expressed for the *Three Princes* was, as Mancroft suggests, at least partly feigned.[11] Walpole himself, in his *Letter from Xo-Ho* (1757), made use of the oriental tale for satiric ends, commenting on the contemporary scene by means of the pseudoletters of an oriental observer. The *Letter from Xo-Ho* was successful and went through five editions in a fortnight. "It is a brief, witty satire, aimed chiefly at the injustice of the system of political rewards and punishments, as exemplified in Admiral Byng's recent execution. . . . The oriental disguise is extremely thin, but it is cleverly used to point the satire."[12] Nearly thirty years later, in 1785, Walpole mocked the literary worth of the oriental tales in his parody, the *Hieroglyphic Tales*. The preface to these tales, according to Miss Conant, "is rather a clever satire on the pretentious, highly moralistic, and would-be scholarly prefaces to oriental tales. . . . Walpole's tone of supercilious mockery toward the oriental tales was typical of critical opinion generally between the middle of the century and the end of our period (c. 1786)."[13]

Two of the moral themes of eighteenth-century oriental tales are worth isolating here, because they lead us back, more or less directly, to The Word, serendipity. One of these recurring moral themes is that of the hedonistic paradox. In two of Hawkesworth's tales, for example, the heroes find "that the attempt to be happy at any cost ends in greater pain. Both tales represent an idea that was persistent in the philosophy of the eighteenth century, and was to find its most artistic expression in *Rasselas* and *The Vanity of Human Wishes*."[14] Walpole seems to be falling in with this moralistic theme when he stresses the importance of *not* looking for the object of discoveries by serendipity. Yet, it must also be said that the oriental tales are philosophically and morally hostile to the notion of the operation of chance. In Miss Edgeworth's moral tale, "Murad the Unlucky,"[15] modeled on the oriental pattern, ill-luck turns out to

---

[10] Ibid., p. 238.

[11] See Arthur M. Samuel, Baron Mancroft, "Serendipity," in *The Mancroft Essays* (London: J. Cape, 1923).

[12] Conant, *Oriental Tale in England*, pp. 187–188.

[13] Ibid., pp. 220–221.

[14] Ibid., pp. 96–97.

[15] Maria Edgeworth, "Murad the Unlucky," *Popular Tales* 2 (1804): 199–282. This story, although not published until 1804, was similar to many in the earlier period.

be identified with imprudence; and in Voltaire's *Zadig,* one of the most important themes is "the part played in human life by destiny—the apparent supremacy of Chance and the real supremacy of a foreknowing and overruling Providence."[16] Walpole was undoubtedly familiar with this moral and philosophical problem (he had read *Zadig* in the English translation in 1749), but he seems to have rejected the current formulations of the answer. It was not, perhaps, sheer whimsy that made him substitute serendipity for "what Mr. Chute calls *sortes Walpolianae,*" for, whether *sortes* is translated as "luck" or "fate," it lacks the mixture of those two ingredients that Walpole irrevocably included in the complex meaning of serendipity: accident and sagacity. It may be that, of the two, Walpole preferred to accent sagacity rather than accident, and it is certainly true that in the future many users of the word were to try to minimize the accidental component in the meaning of serendipity. But whether he so intended it or not, Walpole's new word has done much to emphasize the role of accident in the process of certain kinds of discovery.

Inevitably, certain of Walpole's personal, often idiosyncratic, traits have crept into our allusions to him as a man of his time. Only the thoroughly second-rate mind would not have certain individualized responses to the intellectual problems of the age. But Walpole's idiosyncrasies amounted more nearly to departures from the norms of his group—what the sociologists call deviance—than to individualism. Many of the things that had value and significance for him were of small account in the lives of his contemporaries. His sensitivity and timidity, his almost effeminate withdrawal from the social and intellectual rough-and-tumble of the time, might have made of him only an ineffectual and ridiculous eccentric had he not also had the unusual strength to turn his weaknesses into virtues. Walpole made the most of his defenses, he cultivated his special tastes and preferences, and since he was a man of considerable talent, much that he created was of great and permanent merit.

Walpole would not enter the hurly-burly of politics in which his father had thrived, and rarely if ever used his seat in the House of Commons to make a contribution to the ongoing debates. Instead, he devoted great energy and effort to the building of Strawberry Hill, to his Strawberry Hill press, to his published writings, and to his letters. In the fields that were congenial to him he was more than a mere dabbler; rather, he was a careful student. This is true even of his gossip. Without doubt, Walpole was a gossip, even a malicious one, but he gossiped with so much perceptivity and thoroughness—even, we might say, with such thorough conscientiousness—that posterity has long since dignified his gossip with the name "social history."

---

[16] Conant, *Oriental Tale in England,* pp. 101, 137.

Walpole enshrined the odd and the quaint, and often made a great deal of what appeared to others to be trivial. But even his most antipathetic critic, Thomas Macaulay, had to admit that his juxtaposition of oddities or incongruities could be unusually fruitful: "He had a strange ingenuity peculiarly his own," Macaulay says, and in another place: "He coins new words, distorts the senses of old words, and twists sentences into forms which make grammarians stare. But all this he does not only with an air of ease, but as he could not help doing it. His wit was, in its essential properties, of the same kind with that of Cowley or Donne. Like theirs, it consists in an exquisite perception of points of analogy and points of contrast too subtle for common observation. Like them, Walpole perpetually startles us by the ease with which he yokes together ideas between which there would seem, at first sight, to be no connexion."[17]

Walpole was rarely able to take seriously anything that did not touch him personally; his interest in politics, for example, was almost always directly related to the fate of his father's reputation or the political fortunes of his friends. But he had the kind of originality that led his idiosyncratic and egocentric interests to be of lasting interest to others. The finding of "new connexions" and "subtle analogies" delighted him; one such connection led him to the neologism serendipity. But Walpole's defensiveness made it impossible for him not to treat the things he valued without self-mockery and self-deprecation. As a result, he himself contributed to delay in their recognition. As Leslie Stephen says: "Walpole was no colossus; but his peevish anxiety to affect even more frivolity than was really natural to him, has blinded his critics to the real power of a remarkably acute, versatile, and original intellect. We cannot regard him with much respect, and still less with much affection; but the more we examine his work, the more we shall admire his extreme cleverness."[18]

It was in this characteristic spirit, half-delighted, half-deprecatory, that Walpole reported to Sir Horace Mann his coinage of serendipity. Once we know something of the eighteenth-century background and, more particularly, of interest in the Orient current at that time, and once we examine more closely Walpole's propensities, that one brief fragment of a letter in which Walpole writes about serendipity stands revealed as an essential product of these ingredients. Here is Walpole, familiar with oriental literature and fond of the *Arabian Nights*, yet seeming to deprecate, and rightly, the "silly fairy tale" he has read; here he is a collector of art and an expert on heraldry, yet one who makes light of a new acquisi-

---

[17] Thomas Babington Macaulay, "Walpole's Letters to Sir Horace Mann," *Edinburgh Review* 58 (1833): 227–258.

[18] Leslie Stephen, *Hours in a Library* (London: Smith, Elder, and Co., 1876), vol. 2, p. 197.

tion and of new knowledge; and here he reports a word coinage that delights him, but half apologizes for telling his friend about it, as he has "nothing better to tell."

Walpole was an inveterate maker of words. Macaulay testifies to Walpole's fondness for neologisms, and so also does Odell Shepard in his novel *Jenkins' Ear*, which purports to be a "narrative attributed to Horace Walpole, Esq." Shepard has the fictitious editor of Walpole's fictitious manuscript say, "I soon found myself shortening interminable sentences, inserting marks of quotation where I knew he was using the words of other men, adding necessary punctuation, and changing neologisms such as 'smuggle' and 'serendipity' and 'womanagement' into words that can be found in Dr. Johnson's Dictionary."[19] As befalls every creator of words, many of Horace Walpole's creations were short-lived and never resurrected. Judging by *womanagement* (one of the early portmanteau words in a tradition to be developed to its peak by Lewis Carroll and vulgarized by *Time* magazine), it would be rash to say that this was because they dealt with unimportant problems; but it is certainly not too much to say that in the world of "discoverers" of all kinds, Walpole's neologism serendipity has an aptness—to say nothing, for the present, of its other qualities—which has greatly facilitated its eventual diffusion. We shall see that those engaged, vocationally or avocationally, in looking for "finds," whether they are antiquarians, book collectors, or scientists, seem to have a special proclivity for "accidental sagacity"; and as a consequence, they have much enjoyed making use of Walpole's single expressive word to designate their experience.

Serendipity occurs only once in all of Walpole's writings, although in that instance, in his letter to Mann, Walpole speaks of his discovery as "of a kind which *I call* [our italics] *Serendipity*." He certainly implies frequent usage here. But if he did use it often, it could only have been in conversation. On March 2, 1754, Walpole wrote to Richard Bentley of "a new instance of the *Sortes Walpolianae*,"[20] which enabled him to identify a certain portrait. But though, in the letter to Mann, he claimed to prefer serendipity to *sortes Walpolianae*, he gives no sign of the preference here, just a month after he had reported it to Mann. Perhaps he hesitated to run it into the ground by an excess of repetition. Many years later, in a letter to Miss Mary Berry, February 2, 1789, he was to write: "It is a misfortune that words are become so much the current coin of society, that, like King William's shillings, they have no impression left; they are so smooth, that they mark no more to whom they first belonged than to

---

[19] Odell Shepard and Willard Shepard, *Jenkins' Ear: A Narrative Attributed to Horace Walpole, Esq.* (New York: Macmillan, 1951), p. vi.

[20] *Walpole's Correspondence*, vol. 35, pp. 161–164.

whom they do belong, and are not worth even the twelvepence into which they may be changed: but if they mean too little, they may seem to mean too much too, especially when an old man (who is often synonymous for a miser) parts with them."[21]

The eighteenth century has been described as the greatest age of conversation. This only emphasizes the comparatively little "oral history" of that century. Until the very recent development of "oral history," there are relatively few records of the spoken word, and the great bulk of conversations of the past are of course lost beyond recall. The few notable exceptions are such records as Luther's or Goethe's table talk, or such a gem as that handed down to us by Boswell. Even an experienced anthologist, James R. Sutherland, could find no instances of eighteenth-century "polite conversation" for his *Oxford Book of English Talk*, and so we can only speculate rather vaguely about this kind of conversation.[22]

In quality, eighteenth-century conversation must surely have attained considerable distinction, for in the salons of Paris and London a great premium was placed both on the substance of what was said and on the wit with which it was conveyed.[23] In Paris, especially, the art of conversation was enhanced by the rivalry among the formally organized salons. As far as the quantity of conversation is concerned, Bernard Berenson maintains, perhaps rashly, that in this respect too the eighteenth century was par excellence the age of conversation and deserves more recognition as such: "And conversation should have the same privilege that is granted—reluctantly enough—to the other fine arts, the privilege from utilitarian purpose. The result may be of little consequence, as eighteenth century conversation doubtless was; the more so as in that least unhappy of centuries a larger number of people were enjoying talk than at any previous moment in history, even if we include the Athens of that greatest of all conversationalists, Plato's Socrates."[24]

Whether the use of superlatives is justified in discussing the eighteenth-century conversation is of no great importance; many of the more educated and intellectual men and women of the time did find in conversation a form of recreation, and, indeed, of competitive recreation. To excel in conversation it was desirable that one be knowledgeable without being pedantic, or, to put it another way, that one be both well-informed and witty. Serious subjects were best concealed in humorous disguise, since, for good or ill, people tended to become bored by the serious or affected

[21] Ibid., vol. 11, p. 3.

[22] James R. Sutherland, *Oxford Book of English Talk* (Oxford: Clarendon Press, 1953).

[23] M. Glotz, *Salons du XVIIIe siècle* (Paris: Hachette, 1945); Valerian H. Tornius, *Salons: Pictures of Society through Five Centuries* (New York: Cosmopolitan Book Corporation, 1929).

[24] Bernard Berenson, *Sketch for a Self-Portrait* (New York: Pantheon, 1949), p. 33.

to be so, and to be boring was the ultimate failure. Subtly interwoven with this battle of wits there was another kind of battle, a battle for social prestige. In an age when the aristocracy was the undisputed highest class in the society, the canons of aristocratic conduct were widely acknowledged and the greatest social prize was acceptance by the aristocratic elite. Many a drawing room conversationalist was fighting not merely for the laurels of that art but for the social rewards his wits might bring him.

We can, again, only speculate about the compatibility of these two goals of success in the realms of conversation and of "Society." Clues for such speculation are to be found in the relationship between class status and language, in the stratification of language. As Otto Jespersen noted, some time before the voguish interest in *U-speech*,[25] we may "speak of an 'upper class' language and a 'lower class' language: 'the classes and the masses' are distinguished by their speech as much as by their clothes and their ways of thinking."[26] To what extent then was conversational brilliance in the eighteenth century compatible with the use of "upper-class language"? And, more specifically, to what extent did (or does) upper-class language tolerate the use of neologisms?

Opinions vary on this interesting question, and there is not much evidence. Alexander Pope, in his *Essay on Criticism*, held that

> In words, as fashions, the same rule will hold,
> Alike fantastic if too new or old;
> Be not the first by whom the new is tried,
> Nor yet the last to lay the old aside.[27]

Since Pope was a member of the kind of society in which Walpole moved, his opinion holds special interest for us. Pope's dictum has its ambiguities. For one, it is not clear whether he is addressing himself to all his readers, including those who were already among the "best people," who must be assumed to know how to conduct themselves in these matters. His advice, then, might have been meant for those who wished to emulate the best people and so win acceptance into their circle. In that

[25] [The expressions "u-speech" and "non-u speech" became fashionable in England and in America in the mid-1950s, after the publication of an essay by Nancy Mitford, herself a member of that English upper class she satirically portrayed in her novels. She derived both terminology and concept from "Linguistic Class Indicators in Present-Day English," an article by Professor Alan S. C. Ross that was first printed in a learned Finnish journal, *The Bulletin of the Neo-philological Society of Helsinki* (1954), and reprinted as "U and non-U: An Essay in Sociological Linguistics" in Alan S. C. Ross et al., *Noblesse Oblige: An Enquiry into the Identifiable Characteristics of the English Aristocracy*, ed. Nancy Mitford (London: Hamish Hamilton, 1956)].

[26] Otto Jespersen, *Mankind, Nation, and Individual from a Linguistic Point of View* (Cambridge, Mass.: Harvard University Press, 1925), pp. 141–142.

[27] Alexander Pope, *An Essay on Criticism* (1711), part 2, verses 131–135.

case, there still remain two possible interpretations of his dictum: He might be suggesting that the social elite does, in fact, behave in the way he has described, or that such behavior is necessary only for those striving to win acceptance by this group. The latter appears more likely, for if all followed his dictate, how could anything new ever arise? If Pope is reporting fact in the form of a homily for climbers, we would infer that the use of the would-be telling neologism was familiar to the established upper class in the eighteenth century.

More than a century later, Thorstein Veblen dealt with the same matter, but unlike Pope, he tried to analyze the differential use of words by different social classes rather than to advocate particular use. At first sight, Veblen appears to come to a conclusion different from the one we finally attributed to Pope; he appears to maintain that neologisms are not a part of upper-class speech: "A discriminating avoidance of neologisms is honorific, not only because it argues that time has been wasted in acquiring the obsolescent habit of speech, but also as showing that the speaker has from infancy habitually associated with persons who have been familiar with the obsolescent idiom. It thereby goes to show his leisure class antecedents. Great purity of speech is presumptive evidence of several successive lives spent in other than vulgarly useful occupations."[28] Behind Veblen's declaratory statements, however, are the same optative and manipulative implications as in Pope's dictate: Veblen can also be read as saying that for those who would wish to be accepted as members of the upper class, regardless of their actual antecedents, great purity of speech is particularly essential.

Two examples come to mind here, examples of the use of a kind of private language, composed both of new words and of "distortions of the sense of old words" (see Macaulay on Walpole's language), and in each case the users came from families that had definitely arrived in Society. The first is "Glynnese," created by William E. Gladstone's in-laws, the Glynnes. Writing of Glynnese, Gladstone's biographer Philip Magnus says:

> The Lytteltons and the Gladstones [Lord Lyttelton and Gladstone married the Glynne sisters] were so numerous and devoted, so quick, eager and vital, that for many purposes they felt themselves to be self-sufficient. They invented a kind of language for themselves, which was formally embodied by Lord Lyttelton in 1857 in a glossary which was privately printed. It was entitled *Contributions Toward a Glossary of the Glynne Language by a Student (George William, Lord Lyttelton)*. . . . Gladstone, who loved to hear Glyn-

---

[28] Thorstein Veblen, *The Theory of the Leisure Class* (London: Macmillan, 1899), pp. 303–304.

nese spoken, did not often use the language himself. But Mrs. Gladstone used it on every possible occasion.[29]

Now, Gladstone was never really able to win full acceptance in that Whig society of which his wife was by birth a charter member; Gladstone was felt by the old Whigs to be "Oxford on the surface, Liverpool underneath"; and the Hon. Emily Eden said of him: "there was an element of parvenuism about him, as there was about Sir Robert Peel. . . . In short, he is not frivolous enough."[30] Gladstone, the middle-class outsider, might perhaps have felt that it was unsuitable for him to use the Glynne language, symbol of the pretense he avoided of belonging in the society of his wife's family.

The other family with a language of its own creation was the Barings, and they too were well established as members of the social elite. Sir Edward Marsh, a great friend of Maurice Baring, describes the language in his autobiography: "I have mentioned the Baring language, or to speak more idiomatically, 'The Expressions.' It was started, I believe by Maurice's mother and her sister, Lady Ponsonby, when they were little girls, and in the course of two generations it had developed a vocabulary of surprising range and subtlety, putting everyday things in a new light, conveying in nutshells complex situations or states of feeling, cutting at the roots of circumlocution. Those who had mastered the idiom found it almost indispensable."[31] Among those who had mastered it were high officials in the Foreign Office, members of the literary elite such as Desmond MacCarthy, and many others. In this circle, the "Expressions," far from being frowned upon, were used as a symbol of unquestioned membership and helped mark off the boundaries of the group.

The foregoing is only an apparent digression, for it leads us back to the nature of conversation in the social world in which Horace Walpole moved, and to the part he played in this conversation. Although Walpole was not of the oldest or highest aristocracy, his acceptance in the highest social circles of his day was unquestioned. In matters of social class, at least, Walpole gives no evidence of insecurity, though his timidity might have made him withdraw from certain more lusty kinds of social gatherings. It is at least plausible that he did not hesitate to entertain and amuse his friends with his latest word coinages. It is reasonable to assume that here, as elsewhere, he made the most of his somewhat eccentric talents, and that if he was able to embellish his conversation with a pretty new word, he probably did so. In this way, Our Word, serendipity, may well

[29] Philip Magnus, *Gladstone: A Biography* (New York: Dutton, 1954), pp. 125–127.

[30] Ibid., pp. 141–142.

[31] Edward Marsh, *A Number of People: A Book of Reminiscences* (New York: Harper and Brothers, 1939), p. 72.

have found its way into many a drawing room, there to enhance Walpole's reputation as a witty conversationalist.

Having gone so far with our speculations, perhaps we may carry them yet one step farther. In the world of fashions of all kinds, there a "trickle-down" process has been identified. In the course of this process, items introduced as "fashionable" at the very highest level of society gradually come to be used in ever wider circles, which model their behavior on that of the elite. As the popularity (in the strictest sense of the word) of these items increases, they become cheapened, and they are, consequently, discarded by the elite in favor of a new item. Though this process is most often associated with fashions in clothes,[32] it operates in the same manner with regard to other articles of consumption, including words—exclusive usage enhances their value, while popular usage diminishes it. So appealing a word as serendipity (we shall see much evidence later of its appeal) might have become popular enough in Walpole's conversational circles for its author to feel compelled to abandon his word-child. It would be of interest to know whether the boom serendipity was to have in the middle of the twentieth century is a repetition of an earlier boom, however brief, in the mid-eighteenth century.

Horace Walpole was not the only writer of the eighteenth century to be particularly taken with the story of the three princes and the camel (or mule). Voltaire seized upon the same theme and incorporated it in one of the episodes of his novel *Zadig* (1748).[33] Indeed, it is a theme that has fascinated people through the ages.

The basic "plot" of the tale is the demonstration of skill in detection, which, in turn, is proof of general quick-wittedness. Tales of detection of this kind—tales of *Scharfsinnsproben*, as German scholars have called them—had their origin in antiquity in the Far East, in India and China, and in the Semitic countries of the Near East. They have many themes: the discovery of paternity and bastardy, the distribution of an inheritance, and, finally, the description either of an unseen object or of the provenance of a known object from various clues and traces. It is this last theme that is involved in the adventure of the three princes of Serendip that attracted both Walpole's and Voltaire's attention. The three princes "describe" a camel that they have never seen, as well as its rider and its load. (In a later episode, they detect that the wine they are drinking came

---

[32] See Bernard Barber and Lyle S. Lobel, "Fashion in Women's Clothes and the American Social System," *Social Forces* 31 (1952): 124–131.

[33] François Marie Arouet de Voltaire, *Zadig: Histoire orientale* (London-Amsterdam, 1747).

originally from a vineyard in a cemetery, that the lamb they are eating was once nursed by a bitch, etc.) Although, as we mentioned earlier, the cultural context in which the story occurs may cause some variation in its exact content, many of the significant details and the grounds of inference vary only slightly from story to story, and those in the story of the princes of Serendip overlap considerably with the common core. Similar stories can be found in the Babylonian Talmud, in the Jewish Midrasch Ekāh, and much later, in the eighteenth century, in the *Arabian Nights* story of the "Sultan of Yemen and His Three Sons." René Basset, in his *Contes Populaires d'Afrique*, recounts yet another similar story.[34]

European interest in the literature of the East came as a by-product of commercial contacts and political involvements. The highly educated commercial aristocracy of Venice, especially, found much of interest in Oriental culture, and Venetian ambassadors to Constantinople and points east learned much of the language and customs of these countries. Our tale of detection, more specifically our camel story, appears in Italian literature for the first time in the writings of one of Boccaccio's students, Giovanni Sercambi (1344–1424), as a tale called "De Sapientia." More important, there appeared in 1557 in Venice the *Peregrinaggio di tre giovani figlivoli del Re di Serendippo*, by Christoforo Armeno, "dalla Persiana nell'Italiana lingua trapportato."

This Christoforo was, indeed, an Armenian, who in the middle of the sixteenth century spent three years in Venice and there wrote that loosely connected series of tales, the *Peregrinaggio*. Christoforo's immediate model for the episode of the princes and the camel, which serves to get the long and rambling story going, as well as for other episodes in the *Peregrinaggio*, was the *Hast Bihist* of the Persian writer of the late thirteenth and early fourteenth century Amir Khosrau.[35]

There is considerable evidence for the popularity of Christoforo's work. Four new editions in Italian appeared within less than a century: in 1584, 1611 (this one is in the Harvard College Library), 1622, and 1628. Further, there were numerous translations. It was translated by Johann Wetzel into German and published in Basel in 1583, and this edition was republished in 1599, and again, in a reworked form, in 1630. In French, there appeared one translation by Francois Beroalde in 1610, a very free translation by Simon Gueulette in 1712 (which Voltaire used for *Zadig*), and a more accurate one by de Mailly in 1719. From de Mailly's translation three further translations were made, into English in

---

[34] René Basset, *Contes populaires d'Afrique* (Paris: E. Guilmoto, 1903).

[35] This is the opinion of Joseph Schick, whose study is the "last word" on the subject. Earlier authorities had different opinions, which we shall discuss later in connection with scholarly interest in the subject of these tales.

1722 (*The Travels and Adventures of Three Princes of Serendip*, which
Walpole read), into German in 1723, and into Dutch in 1766. Although
much of the interest in this work was only a part of the more general
interest in everything oriental, some part of it may have been generated
by the particular theme of the story.

By the early nineteenth century, the long tradition of genuinely scholarly interest in folklore and mythology converged with interest in the
Orient, and from that time on, we find recurring in scholarly literature
discussions of tales of detective skill in general and of our camel story in
particular. The first such scholarly treatment (apart from editions of the
Arabian Nights, etc.) was probably that of J. C. Dunlop. Dunlop retells
the camel story when he discusses the sources of the episodes of *Zadig*,
and he traces it from Voltaire back to Gueulette, from Gueulette to
Christoforo, and thence to an Arabic work of the thirteenth century titled
Nighiaristan.[36] (According to Schick, our final authority, the *Nighiaristan* was itself a copy of the work of Amir Khosrau.) The next orientalist
to pay attention to the tale was Joseph von Hammer, who translated the
Nighiaristan version in his *Geschichte der Schönen Redekünste Persiens*
(1818). He also is familiar with the connection between this story and
Zadig, and says that "unless he is mistaken," Voltaire found the story in
d'Herbelot's collection of oriental tales.[37]

Again, the camel story appears in one of the manuscripts collected by
Col. Colin Mackenzie and edited by H. H. Wilson, (1828), and Wilson
draws attention to it as "illustrative of the oriental origin of part of
Zadig."[38]

More and more was added to scholarly knowledge about the camel
story in general and the history of Christoforo's story in particular. In the
notes to his edition of the *Arabian Nights*, A. Loiseleur-Deslongchamps
points out the similarity of the story of the "Three Sons of the Sultan of
Yemen" to the episode in *Zadig* of "Le chien et le cheval," and he suggests that Voltaire could have based this episode either on Gueulette's or
de Mailly's translation of Christoforo. He refers also to Persian and Indian stories of this kind, and to one by the early Danish writer Saxo
Grammaticus.[39] In Germany in the later nineteenth century, a scholar by
the name of Theodor Benfey took a great interest in these stories of keen

---

[36] J. C. Dunlop, *The History of Fiction* (Philadelphia: Carey and Hart, 1842), pp. 329–
330.

[37] Joseph von Hammer, *Geschichte der Schönen Redekünste Persiens* (Vienna: Heubner
und Volke, 1818), pp. 307–309.

[38] Colin Mackenzie, comp., *The Mackenzie Collection*, ed. H. H. Wilson (1828; reprinted Calcutta, Madras: Higgenbotham and Co., 1882), p. 222.

[39] Notes to his edition of *Arabian Nights*, ed. A. Loiseleur-Deslongchamps (Paris,
1841), pp. 690–691.

observation and inference, and in 1864 he published a fragment of his translation of the *Peregrinaggio* with an introduction.[40] In this introduction he proposes new possibilities as to the origin of the *Peregrinaggio*, and he, too, describes many similar stories with only minor variations on the basic theme.

From here on, scholarly discussions become more and more intricate and refined, and it is only worth mentioning some of the leading scholars and the main direction of their thought. In an article written in 1885, Israel Levi points out numerous Jewish prototypes of our tale.[41] Georg Huth, in 1889, points to many Indian and Arabic versions, but leaves the question of priority open.[42] But in the next year, in the same publication, Siegmund Fraenkel claims that the stories could have originated only in Arabia.

As far as the *Peregrinaggio* specifically is concerned, it was reprinted in 1891 in *Erlanger Beiträge zur Englischen Philologie* by Heinrich Gassner, with a brief introduction by Heinrich Varnhagen. Johann Wetzel's early (1583) translation into German was published with extensive editorial notes by Hermann Fischer and Johannes Bolte in the Bibliothek des Literarischen Vereins.[43] Fischer and Bolte present for the first time the detailed genealogy of translations from Christoforo. They again mention numerous oriental versions of the story: Arabic, Jewish, Turkish, and Indian. In 1932, to commemorate the fiftieth anniversary of Benfey's death, Richard Fick and Alfons Hilka published his entire translation of the *Peregrinaggio* as *Die Reise der drei Söhne des Königs von Serendippo*, with a long introduction. Here, they put forward the novel theory that Christoforo never existed at all and that the "translation from the Persian" is a literary fiction. They believe the *Peregrinaggio* was compiled by an Italian, possibly by the publisher of the first edition, Michele Tramezzino.

The last authority, in every sense, is Joseph Schick's *Die Scharfsinnsproben*.[44] The objective of Schick's monumental work appears to be the tracking down of the themes of the Hamlet legend (sad and needless to say, Schick never completed his work), and one of these themes is the *Scharfsinnsprobe*, or proof of skill in observation and inference. In the

---

[40] Theodor Benfey, fragment of his translation of the *Peregrinaggio* (with introduction), *Orient und Occident* 3 (1864): 256–288.

[41] Israel Levi, in *Revue des études juives* 11 (1885).

[42] Georg Huth, in *Zeitschrift für vergleichende Literaturgeschichte* N.F. 2 (1889): 404–414; Sigmund Fraenkel, in *Zeitschrift für vergleichende Literaturgeschichte* 3(1890): 303–330.

[43] Christoforo Armeno, *Die Reise der Söhne Giaffers*, trans. Johann Wetzel, ed. Hermann Fischer and Johannes Bolte, vol. 208 of Bibliothek des literarischen Vereins in Stuttgart (Tübingen: Literarischer Verein in Stuttgart, 1895).

[44] Schick, *Scharfsinnsproben*.

early version of the Hamlet legend of Saxo Grammaticus, Hamlet gives evidence of great skill of this kind. While he is visiting the court of the king of England to obtain the hand of his daughter in marriage, Hamlet disdains a meal that had been prepared for him. He is overheard saying that he cannot eat it because there is blood in the bread, the water tastes of iron, and the meat smells of corpses. Further, he remarks that the king has the eyes of a lowly servant, and the queen has thrice prostituted herself. The king proceeds to make inquiries, and Hamlet turns out to have been correct in each of his observations and inferences. The king is so impressed with the keenness of his intelligence that he gives Hamlet his daughter's hand.

The history of the story of the description of the unseen camel (or elephant, or ass) is, therefore, a *Scharfsinnsprobe* of great importance in Schick's work. His work appears to comprehend everything that has been written on the subject before and to add a great deal to this body of scholarship. As far as the *Peregrinaggio* is concerned, Schick resurrects Christoforo after Fick and Hilka discredited his existence: He was, it seems, the Armenian Chachatur, who Italianized his name. And as we have already mentioned, he points to the work of Amir Khosrau as the undoubted source of the *Peregrinaggio*. (As for *Zadig*, Schick maintains that Voltaire's source of inspiration was certainly Gueulette.) Schick's research on stories of the detection of unseen animals, a specific case in the generic pattern of *Scharfsinnsproben*, places their origin in early Indian literature, where they involved the description of an elephant. The camel story as such probably occurred first in Arabia and was transmitted from there.

Although the Hamlet in the legend of Saxo Grammaticus does indeed perform intellectual feats very similar to those of our three princes of Serendip, in Shakespeare's version of the Hamlet story this particular incident has disappeared. There, Hamlet is not so much the keen observer who happens to notice an odd or incongruous detail and draws a useful inference from it, but the experimentalist who *contrives* a situation in which it may be possible to make significant observations to support or discredit an existing hypothesis. He sets up the play within the play to test and study his uncle's reactions, to find proof of his guilt or innocence. The element of contrivance makes the play within a play a planned experiment in the reconstruction of what is or was, from traces observed here and now. Hamlet's procedure is almost like that followed by experimental scientists, who frame a hypothesis and then set up an experiment to "see what happens." Less often, the scientist will share the experience of the three princes of Serendip: He will *happen* to notice "something" he had not expected, and his inferential reconstruction of how it came to be may put him on the track of a discovery. In this manner, for instance, Wilhelm Roentgen happened to notice that certain of his photographic plates had become unexpectedly clouded, and he inferred the action of X-rays on these plates.

Directly or indirectly, the story of the three princes and the camel has led us to two important patterns of scientific thought. Another analogy of the patterns of thought of the three princes of Serendip can be found in the world of science. This analogy was observed not immediately in connection with the tale that delighted Horace Walpole, but with that which has aroused so much more interest by virtue of its author's eminence: *Zadig* by Voltaire. In the episode titled "Le chien et le cheval," Zadig demonstrates that same skill in observation and inference as the three princes of Serendip, only in his case his skill enables him to describe a lost royal bitch and royal horse. The cynical and satirical overtones are Voltaire's own contribution: Zadig's superior ability gets him into trouble with the authorities, and he is not, ultimately, rewarded. So he vows in future to keep his observations to himself, but this only involved him in further trouble.[45]

In an article titled "The Method of Zadig," T. H. Huxley shows how important an element this "method" is in scientific thinking, and especially in certain of the sciences. "What, in fact, lay at the foundation of all Zadig's arguments," says Huxley, "but the coarse, commonplace assumption, upon which every act of our daily lives is based, that we may conclude from an effect to the preexistence of a cause competent to produce that effect?" And he goes on to say: "the rigorous application of Zadig's logic to the results of accurate and long-continued observation has founded all those sciences which have been termed historical or palaetiological, because they are retrospectively prophetic and strive toward the reconstruction in human imagination of events which have vanished and ceased to be."[46] In effect, the method of Voltaire's Zadig must be used in science when the experimental method of Shakespeare's Hamlet *cannot* be used.

---

[45] As we saw, Voltaire probably used Gueulette's translation of Christoforo as the basis of his tale. In any case, Fréron's charge, in *L'année litteraire* (1767), that Voltaire plagiarized de Mailly seems to be inspired chiefly by malice, since it can scarcely be said that a story that has gone through so many versions can be plagiarized.

[46] T. H. Huxley, "On the Method of Zadig: Retrospective Prophecy as a Function of Science," *Nineteenth Century* 7 (1880): 929–940; reprinted in T. H. Huxley, *Science and Hebrew Tradition* (New York: D. Appleton and Company, 1896), quotations from pp. 7, 9. [The extremely accurate but alas obsolescent term *palaetiology* was coined by the historian and philosopher of science William Whewell, "to describe those hypotheses which . . . refer to actual past events, but try to explain them by causation laws" (as in geology). See W. Whewell, *History of the Inductive Sciences* (London, 1837), vol. 3, p. 481. Not only Huxley, but also other Victorian scientists relied on Whewell's ingenuity in inventing scientific terms. Incidentally, it was Whewell who coined the English word *scientist*, anonymously in 1834 and explicitly in 1840. See Robert K. Merton, "De-Gendering 'Man of Science': The Genesis and Epicene Character of the Word *Scientist*," in *Sociological Visions*, ed. Kai Erikson (New Haven, Conn.: Yale University Press, 1997), pp. 225–253.]

Unlike experimentation and retrospective prophecy—"retroduction" in the language of Isaiah Berlin and others—which cannot, by Huxley's definition of the latter, occur simultaneously, it may depend on the interest of the observer whether one and the same discovery is described as a retrospective prophecy or a discovery by serendipity, or happy accident. If we compare the different descriptions of the discoveries Heinrich Schliemann made in the process of his excavations, we can see how two different observers made different abstractions from the same events, or, to put it another way, how they described these events with varying emphasis. C. W. Ceram, in his book *Gods, Graves and Scholars*,[47] mentions some unexpected finds made by Schliemann, but for him these are of negligible importance compared to the staggering amount of material that Schliemann found that he *had expected* or prophesied; Hendrik Van Loon, on the contrary, in *The Arts*,[48] makes Schliemann's discoveries the very exemplification of serendipity, and stresses how much of value he stumbled on in the course of his excavations, over and beyond any anticipation. We shall have repeated occasion to see later that it is such different emphases in description, sometimes ideologically conditioned, sometimes not, that play a considerable part in the receptivity of different people to serendipity, both as a pattern of behavior and as a word.

Horace Walpole's somewhat confusing "derivation" of serendipity came about, in all probability, by just this kind of discriminating use of emphasis. It was not so much that he misunderstood the import of the fairy tale he had read, but that he highlighted those aspects of it that were significant to him and obscured those of lesser interest. What appealed to him in the story was the *unplanned, accidental* factor in the making of the discovery, and the "sagacity" necessary to make it. The three princes had not set out to find a lost camel, he himself had not set out to find the Capello arms, and Lord Shaftesbury had not planned to make any discovery about Anne Hyde's marital affairs when he accepted her father's invitation to dinner. But without the princes' keen powers of observation, without his own know-how in the field of heraldry, and without Shaftesbury's profound knowledge of etiquette, none of these "discoveries" could have occurred. What Walpole obscured in his explanation of serendipity was the nature of the object discovered: whether it was a known quantity or an unknown quantity; whether it was something that might have been expected (retrospectively prophesied) or not; and, finally, whether it was of any significance or not. It is in the latitude that these obscurities give to individual interpretation that the complexity of

---

[47] C. W. Ceram, *Gods, Graves, and Scholars: The Story of Archeology*, trans. E. B. Garside (New York: Knopf, 1951).

[48] Hendrik Van Loon, *The Arts* (New York: Simon and Schuster, 1937).

meaning of serendipity has its origin. Even had Walpole stated positively that serendipity had to do only with accidental discovery, his ambiguity about the finder's foreknowledge of the object of discovery meant that discoveries by serendipity came to be regarded as more or less accidental. But these initial ambiguities became compounded as the word serendipity acquired a variety of meanings in the course of its diffusion to varied social groups.

Chapter 2

EARLY DIFFUSION OF SERENDIPITY

*F*or seventy-nine years after Horace Walpole explained to his friend Sir Horace Mann, in that letter of January 28, 1754, the nature of certain discoveries that are "of that kind which I call *Serendipity*," there is no record of any further appearance of this "very expressive word" in writing. During all those years, serendipity remained dormant in that peculiar limbo of history, the world of unpublished documents. Facts and words lead a passive existence in this limbo: Frozen rather than fruitful, they can be brought back to active life only through publication. In his will, Walpole consigned his correspondence with Mann to this unpublished state for a certain period of time, though he intended that ultimately it should emerge into public view. He bequeathed these letters to John James, sixth earl of Waldergrave, the son of a niece, when the earl became twenty-five years of age in 1810. Waldergrave turned them over to Lord Holland, who edited Walpole's *Memoirs of the Reign of George II*, and he, in turn, asked Lord Dover to edit the letters to Mann. In 1833, this edition of the Mann correspondence, including the crucial letter, was published in three volumes by Richard Bentley of London.[1]

Patterns of Critical Reaction to Walpole in Nineteenth- and Early Twentieth-Century England

It is a common occurrence in the world of letters that the works of an author go through alternating periods of great popularity and great ne-

[1] Lord Dover, ed., *Letters of Horace Walpole, Earl of Oxford, to Sir Horace Mann, British Envoy at the Court of Tuscany*, 3 vols. (London: Richard Bentley, 1833).

glect. While the author enjoys a "boom," his works receive frequent and sympathetic critical attention, but before or after such a period of popularity they may be elaborately despised or, worse, wholly ignored. Often it is possible to relate such shifts in literary taste to changes in more general social and moral attitudes, to climates of opinion; and, contrariwise, these shifting literary tastes may be used as clues to the prevailing climate of opinion, for preferences in literature tend to be compatible with other values or preferences. As Carl Becker pointed out,[2] there is a kinship between certain historical epochs and a kind of alienation between others; some speak the same basic language, others do not.

When we examine the patterns of critical reaction toward Walpole in the nineteenth and early twentieth centuries in England, we find that both the man and his works were generally rejected by the early Victorians, but the later Victorians and Edwardians received them with increasing sympathy and respect. This shift in literary taste, and the changing climates of opinion that it reflects, may, in turn, have affected the receptivity of the readers of Walpole not merely to his interests such as his collection of antiques and of objets d'art, but also to his word coinages—specifically, to serendipity.

Walpole's letters to Mann were not the first of his letters to be published in the nineteenth century. In 1818 one volume of Walpole's letters to George Montagu appeared as did another of his letters to William Cole, both published by Rodwell and Martin. In 1820, the *Private Correspondence of Horace Walpole* was published in four volumes, but this edition did not include Walpole's letter to Mann in which he explained serendipity. Walpole's letters to the earl of Hertford, brother of his great friend Henry Conway, and those to the Reverend Henry Zouch were edited by John Wilson Croker and published in one volume in 1825. A certain pattern of critical reaction to Walpole's writings had, therefore, been established by 1833, when the Mann correspondence appeared, and reviews of the letters to Mann may be considered in relation to earlier criticism of Walpole.

The fact that, on their appearance in 1818, the letters to George Montagu received attention from two distinguished literary figures makes these reviews all the more interesting: William Hazlitt wrote a piece about them in the *Edinburgh Review*, and an article in the *Quarterly Review* has been attributed to Croker.[3] Hazlitt was pleased and amused by the letters, but he had little respect for their author. "Horace Walpole

[2] Carl Becker, *The Heavenly City of the Eighteenth-Century Philosophers* (New Haven, Conn.: Yale University Press, 1932).

[3] William Hazlitt in *Edinburgh Review* 31 (1818): 80–93; [John Wilson Croker] in *Quarterly Review* 19 (1818): 118–131.

was by no means a venerable or lofty character:—But he has here left us another[?] volume of gay and graceful letters, which, though they indicate no peculiar originality of mind, or depth of thought, and are continually at variance with good taste and right feeling, still give a lively and amusing view of the time in which he lived." Although he was annoyed by Walpole's taste and sentiments, Hazlitt described Walpole's personality without any great animus: He thought Walpole to be a hypocritical, shallow, quarrelsome fellow, self-important and narcissistic, incapable of affection. "The key [to his social relationships] is that Walpole was a miser. He loved the arts, after a fashion, but his avarice pinched his affections."

Hazlitt was to be neither the first nor the last to think of Walpole as a man preoccupied with trifles. Already two of Walpole's famous contemporaries had so stigmatized him, for Dr. Johnson said of him that he "got together a great many curious little things and told them in an elegant manner," and Gibbon is said to have referred to him as "that ingenious trifler." Neither Johnson nor Gibbon, evidently, was able to deny Walpole's skill in what he did with these trifles. Hazlitt, too, branded him as a little man concerned with petty things:

> His mind as well as his house was piled up with Dresden china and illuminated through painted glass; and we look upon his heart to have been little better than a case full of enamels, painted eggs, ambers, lapis-lazuli, cameos, vases, and rock-crystals. This may in some degree account for his odd and quaint manner of thinking, and his utter poverty of feeling: He could not get a plain thought out of that cabinet of curiosities, his mind;—and he had no room for feeling,—no place to plant it in, or leisure to cultivate it. He was at all times a slave of elegant trifles, and could no more screw himself up into a decided and solid personage, than he could divest himself of petty jealousies and miniature animosities. In one word, everything about him was little, and the smaller the object and the less its importance, the higher did his estimation and his praise of it ascend. He piled trifles to a colossal height—and made a pyramid of nothing "most marvellous to see."

Hazlitt seemed to grant Walpole reluctant admiration for his perverse virtuosity in the domain of the curious, the odd, and the trivial, but felt it to be a sterile virtuosity. He implied that Walpole's "odd and quaint manner of thinking" was as barren of valuable ideas as his arid little soul was incapable of warm emotions. For Hazlitt, "plain thoughts" by "solid personages" contributed to progress in this world of serious affairs.

Since we shall later in this study devote considerable attention to the problem of what constitutes "triviality" and what "significance" in science, it is interesting to note, in anticipation, the attitudes expressed by the reviewers of Walpole's works toward the worth of his interests. In the sciences, theoretical or methodological advances may lead to the fruitful reexamination of what were once considered to be trivia and certain common "trivial" occurrences may appear anomalous or "curious" to the scientist's trained eye and lead him to new discoveries. Sophisticated humanists also realize that the value of many cultural artifacts is not fixed but shifts with prevailing tastes. The sturdy provincialism of the early Victorians limited their perspective and led them to ridicule Strawberry Hill and all it stood for; the later Victorians and Edwardians, on the other hand, somewhat weakened their standards of discrimination and became susceptible to the charms of the curiosity shop. It remained for the scientists, both natural and social, to put the "odd" and the "trivial" in a new light: neither despised nor exalted, but as a vital element in the process of accidental discovery.

In Croker's review of Walpole's letters to Montagu, pleasure in the literary quality of the letters weighed somewhat more heavily in the balance than annoyance with Walpole's character and values, but both elements entered into Croker's appraisal.

> We have here another[?] volume of Letters, from an author who may decidedly claim pre-eminence for ease and liveliness of expression, terseness of remark and felicity of narration above almost all the letterwriters of Britain. The peculiarities and even the foibles of Horace Walpole's character were such as led to excellence in the style of his composition; and although his correspondence has not always taught us to respect the man, the writer seldom fails to amuse us.

Croker realized that another kind of man could not have had the insights or the particular skill in description that Walpole did. But if Croker was charmed by the literary quality of Walpole's letters, Walpole's "peculiarities" and "foibles" did not appeal to him, and, like most nineteenth-century critics, he was alienated rather than disarmed by Walpole's mock modesty and seeming deprecation of what was important to him. Croker resented Walpole's constant "affectation of contempt for what he really valued." Affectation and hypocrisy were charged frequently, and without charity, against Walpole in this period.

If Walpole, because of his great and not altogether unrealistic fear of ridicule, affected far less concern for his interests and hobbies than he really felt, his "affectation" (in twentieth-century American slang it would be called "an act") was at least in some measure successful. Many critics saw through his affectation, but some of his judgments were taken partly at face value. Hazlitt considered Walpole self-important, but if this is so, it was a covert form of self-importance. Self-importance in the grand manner might have won more respect from posterity for his collection at Strawberry Hill, which was dispersed by auction in 1842: Croker ad-

mitted (in 1844, in his review of the second part of the Mann correspondence)[4] that "the aggregate would have been well worth keeping
together in some royal or national repository, as illustrative of the arts and
of the variations in habits and manners." Similarly, his half-apologetic
attitude toward such a word coinage as serendipity (he read a "silly fairy
tale" and made up a word, which he will tell about for lack of more
important news) has affected later attitudes toward that word. If it is
often not taken as seriously as its aptness of meaning, if not its derivation,
warrants, this is, in part at least, the result of the forced carelessness with
which Walpole tossed it off.

Much of Walpole's affected dilettantism, Croker felt, could be laid to
his profound identification with all that was aristocratic. He disdained all
important subjects and affected a lack of real competence in any area
because "he was too much of a gentleman to take the trouble of it." And
the result, said Croker, was that "most of our readers will regret to see a
man of real genius frittering away his time in trifles 'to astonish the natives.'" There is much truth in Croker's contention that Walpole disclaimed literary competence because of his identification with the aristocracy. Literary distinction was somewhat suspect among noblemen in this
period, for genuine distinction in this field could be attained only by hard
work and was becoming, in the eighteenth century, more closely associated with pecuniary gain. In the Veblenian view, the hallmark of the upper or leisure class is its abstention from "productive" labor. "Government, war, sports, and devout observances" are the proper occupations
for this class, for they are at most incidentally and indirectly productive of
wealth, but for none of these occupations did Walpole have either incli-
nation or aptitude. So to indicate his aristocratic status, he had to find
other ways of conspicuously advertising his leisure: He accumulated an
abundance of what Veblen described as " 'immaterial' goods."

Veblen caught the flavor of such an aristocratic existence as Walpole's
when he described the life of leisure:

> The criteria of a past performance of leisure . . . commonly take the form of
> "immaterial" goods. Such immaterial evidence of past leisure are quasi-schol-
> arly or quasi-artistic accomplishments and a knowledge of processes and inci-
> dents which do not conduce directly to the furtherance of human life. So, for
> instance, in our time there is the knowledge of the dead languages and the
> occult sciences; of correct spelling; of syntax and prosody; of the various
> forms of domestic music and other household arts; of the latest proprieties of

---

[4] John Wilson Croker, "Horace Walpole (Part II)," *Quarterly Review* 74 (October
1844): 395–416.

dress, furniture, and equipage; of games, sports, and fancy-bred animals, such as dogs and race-horses.[5]

It was in keeping with the attitudes of such a dyed-in-the-wool aristocrat as Walpole to have quasi-scholarly knowledge of heraldry rather than of science. The objects in his collection of art were certainly displayed by him as "immaterial goods," not all accumulated for their generally acknowledged value. Walpole went one step further than Veblen's man of leisure: He disclaimed even the unpractical learning he had, and affected unmitigated frivolity and idleness. Macaulay recounts in his 1833 review of the Mann correspondence (of which more later):

> Mann had complimented [Walpole] on the learning which appeared in the Catalogue of Royal and Noble Authors; and it is curious to see how impatiently Walpole bore the imputation of having attended to anything so unfashionable as the improvement of his mind. "I know nothing. How should I? I who have always lived in the big busy world; who lie abed all the morning, calling it morning as long as you please; who sup in company; who have played at faro half my life, and now at loo until two and three in the morning; who have always loved pleasure, haunted actions. . . . How I laughed when some of the magazines have called me the learned gentleman! Pray, don't be like the magazines."[6]

Ambivalently, Walpole played down as best he could his serious literary activities, but also dignified them and reassured himself about their legitimacy by preparing a *Catalogue of Royal and Noble Authors*.[7] With such a catalogue, he associated himself with excellent company and bent his critical judgment to suit his aristocratic prejudices. Macaulay appraises his judgment without mercy:

> His judgment of literature,—of contemporary literature especially,—was altogether perverted by his aristocratic feelings. No writer surely was ever guilty of so much false and absurd criticism . . . Who, then, were the first writers of England in the year 1753? Walpole has told us in a note. Our readers will probably guess that Hume, Fielding, Smollett, Richardson, Johnson, Warburton, Collins, Akenside, Gray, Dyer, Young, Wharton, Mason, or some of those distinguished men were in the list. Not one of them. Our first writers, it seems, were Lord Chesterfield, Lord Bath, Mr. W. Withed, Sir Charles Williams, Mr. Soame Jenyns, Mr. Cambridge, Mr. Cov-

---

[5] Thorstein Veblen, *The Theory of the Leisure Class* (London: Macmillan, 1899), pp. 38–39.

[6] Thomas Babington Macaulay, "Walpole's Letters to Sir Horace Mann," *Edinburgh Review* 58 (1833): 227–258.

[7] Printed in Strawberry Hill, 1758.

entry. Of these seven gentlemen, Whithed was the lowest in station, but was the most accomplished tuft-hunter of his time. Coventry was of a noble family. The other five had among them two peerages, two seats in the House of Commons, three seats in the Privy Council, a baronetcy, a blue riband, a red riband, about a hundred thousand pounds a-year, and not ten pages that are worth reading.

With such an attitude toward learning and literary merit, it is no surprise that, as Austin Dobson says, Walpole's own place in English letters is on the small scale. "If masterpieces could have been dashed off at a hand-gallop; if antiquarian studies could have been made of permanent value by the exercise of mere elegant facility; if a dramatic reputation could have been secured by the simple accumulation of horrors upon Horror's head, his might have been a great literary name. But it is not thus the severer Muses are cultivated; and Walpole's mood was too variable, his industry too intermittent, his fine-gentlemanly self-consciousness too inveterate to admit of his producing anything that (as one critic has said) deserves a higher title than '*opuscula.*'"[8] It was fortunate for Walpole that he lived in an age when letter writing was a perfectly acceptable activity. His undeniable talent for writing thus found the ideal outlet, one that caused him no embarrassment and was to win for him the admiration of posterity.

Both Hazlitt and Croker admired the letters for their style if not their substance. A more generous review than theirs, discussing Walpole's letters both to Montagu and to Cole, appeared anonymously in the *Monthly Review.*[9] This reviewer felt that although Walpole was never so happy as "when he was picking up an old portrait, a bit of painted glass, or some antique whimsicality" for the decoration of Strawberry Hill, "he was nevertheless perfectly conscious of the frivolity of his own favourite pursuits and amusements, in collecting the *nugae antiquae,* the rarities, the nicknackatories of every virtuoso within reach." Rather clumsily, perhaps, the reviewer was trying to extenuate Walpole's frivolity by pointing out that Walpole knew what he was doing. The reviewer even tried to vindicate the very triviality of the subject matter of Walpole's letters, for "he said 'the first thing that came uppermost'" and "filled his sheets with the chapter of accidents,—and an amusing chapter it always is." The reviewer was evidently a kindred and sympathetic spirit, and, deliberately or unwittingly, touched on that component in Walpole's thought which moved him to coin such a word as serendipity: his appreciation of the

[8] Austin Dobson, *Horace Walpole: A Memoir with an Appendix of Books* (London: Osgood McIlvaine and Co., 1893), pp. 291–292.

[9] *Monthly Review* 90 (September 1899): 1–14.

twists given to experience by the unplanned, the unanticipated, the accidental.

At last, in 1833, Lord Dover's edition of Walpole's correspondence with Horace Mann was published, and a distinguished reviewer indeed went to work on it—Thomas Babington Macaulay. We have already quoted a passage from Macaulay's review in connection with the remarks made earlier by Croker, in 1818, about Horace Walpole's aristocratic self-consciousness. Macaulay's opinion on that subject was the same as Croker's, only Macaulay felt much more strongly about the fact that Walpole was, as he saw it, a terrible snob. And so it is with the whole of Macaulay's review: It is similar in substance to previous critical opinion but far more strongly worded. Macaulay wrote about his review to his sister Hannah (October 14, 1833):

> I have just finished my article on Horace Walpole. This is one of the happy moments of my life: a stupid task performed; a weight off my mind. I should be quite joyous if I had only you to read it to. . . . I was up at four this morning to put the last touch to it. I often differ with the majority about other people's writings, and still oftener about my own; and therefore I may very likely be mistaken; but I think that this article will be a hit. We shall see. Nothing ever costs me more pains than the first half; I never wrote anything so flowingly as the latter half; and I like the latter half the best. I have laid it on Walpole so unsparingly that I shall not be surprised if Miss Berry should cut me.[10]

From this it seems that Macaulay was somewhat self-conscious about this review, which was something of an outburst, though he says nothing of the basis for his being provoked. Many years later, Norman Pearson was to write of this antagonism: "Walpole's indolent, fastidious, effeminate temperament was intensely distasteful to the strenuous, combative, and essentially masculine vigor of Macaulay. To him it bespoke nothing but 'an unhealthy and disorganized mind': and had the two been contemporaries, Walpole, who detested the very atmosphere diffused by strong and assertive natures—such as Johnson's, for instance—would have returned the dislike with interest."[11] It was this animosity of Macaulay's, coupled with his rigorous and immensely colorful writing, that made this review a hit, indeed; it became one portrait of Horace Walpole that was never to be forgotten. As with several other such portraits by Macaulay, no future biographer was ever able to remain unaffected by his interpretation, whether he was moved to applaud the author or to defend the victim.

---

[10] George Otto Trevelyan, *The Life and Letters of Lord Macaulay* (New York: Harper and Brothers, 1909), vol. 1, p. 305.

[11] Norman Pearson, "Neglected Aspects of Horace Walpole," *Fortnightly Review* 92 (1909): 482–494.

To Macaulay, it seemed that Walpole had such unerring skill in the inversion of the true value of things that he invariably exalted the trivial and mocked the important. He felt that Walpole was "drawn by some strange attraction from the great to the little, and from the useful to the odd." Walpole was the devoted trifler par excellence: "The conformation of his mind was such that whatever was little seemed to him great, and whatever was great seemed to him little. Serious business was a trifle, and trifles were his serious business." That was not all. An even worse perversity on Walpole's part was his inability to admit, in a manly and forthright fashion, to having those inverted values and tastes; instead, he affected contempt for what interested him most, and this "affectation" (which we discussed earlier) completed Macaulay's alienation. Walpole's ultimate affectation was his denial of seriousness and competence even in his chosen field of trivia, and his dislike of being considered, as Macaulay put it, "that vulgar thing, a learned gentleman."

Much as he might have liked to, however, Macaulay was unable to deny the excellent quality of Walpole's writing. He appreciated this quality all the more keenly, perhaps, since, in his animosity to Walpole, he could see in that literary excellence only another symptom of Walpole's moral and intellectual perversity and felt compelled to probe carefully the sources of the strange appeal of this monster. The appeal was certainly there: "Walpole's writings have real merit, and merit of a very rare, though not of a very high, kind. . . . And we own that we expect to see fresh Humes and fresh Burkes before we again fall in with that peculiar combination of moral and intellectual qualities to which the writings of Walpole owe their extraordinary popularity." But Macaulay found it hard to overcome the feeling that the merit of such a literary product was tainted by the character of its author: "as the *pâté de foie gras* owes its excellence to the diseases of the wretched animal which furnishes it, and would be good for nothing if it were not made of livers preternaturally swollen, so none but an unhealthy and disorganized mind could have produced such literary luxuries as the works of Walpole." Plainly, such literary luxuries would ultimately be found to be as indigestible as pâté de foie gras.

If Macaulay noticed that particular literary delicacy, serendipity (which in that period could only have been a tidbit), appearing for the first time in print in the Walpole letters under review, he made no mention of it. But Macaulay did highlight many of those characteristics of Walpole's personality and of his style that contributed to the coinage of serendipity, among other such neologisms. Macaulay saw Walpole as an indefatigable collector of objects of "trifling intrinsic value," as a man with a passion for the curious for its own sake, and, in the end, as a man who could produce much that was pleasant, if not useful, by delving in his storehouse of trivia. "What then," Macaulay asked, "is the charm, the irresistible charm, of Walpole's writings? It consists, we think, in the art of amusing without exciting. . . . [With] the Sublime and the Beautiful Walpole has nothing to do; but . . . the third province, the Odd, was his peculiar domain." It was in this province of the odd that Walpole evidently found the new words that Macaulay remarked: "He coins new words, distorts the senses of old words, and twists sentences into forms which make grammarians stare." He goes on (this bears quoting again) to compare Walpole to Cowley and Donne: "Like theirs, [his wit] consisted in an exquisite perception of points of analogy and points of contrast too subtle for common observation. Like them, Walpole perpetually startles us by the ease with which he yokes together ideas between which there would seem at first sight to be no connection." To Macaulay, this subtle yoking together of ideas and this habit of word coinage were Walpole's peculiar foibles, at best amusing, at worst symptomatic of an "unhealthy and disorganized mind."

As Macaulay saw it, there was no serious substance in Walpole's subtle wit, no moral and philosophical insights that might help Macaulay and his contemporaries the better to order their lives. Walpole seemed to him to be clever but not wise, acute without being profound. Walpole's letters made entertaining "light reading," bedtime reading at best, providing a dubious kind of relaxation after the serious work of the day was done. In the ordinary way, Macaulay would never quote Walpole; his whole tone suggests that he would have been chagrined had his pleasure in Walpole's literary excellence been mistaken for identification on his part with Walpole's frivolous interests and perverted values.

Only the early letters from Walpole to Mann appeared in 1833, those written between 1741 and 1760. The later letters of this long correspondence, those from the period 1760 to 1785, did not appear until 1843–1844. Their review in the *Quarterly Review*[12] is again attributed to J. W. Croker, who is presumed to have reviewed the letters to Montagu back in 1818. Although Croker had no great admiration for Walpole, he did not find it as difficult as Macaulay to admit the merits in his work. (Croker was, after all, the editor of Walpole's letters to Lord Hertford; Macaulay would never have permitted himself any such close association with Walpole.) Croker had become slightly mellower in his judgment of Walpole since 1818. He found Walpole's taste questionable, and said of him that he was "an excellent judge of everything up to mediocrity" (thus setting the mold for what Dorothy Parker was one day to say of Katherine Hepburn, that she ran the gamut of emotions from A to B). But he did not grudge Walpole due acknowledgement of the valuable

---

[12] *Quarterly Review* 72 (1843): 516–552.

results of his petty preoccupations: "He [Walpole] will probably be for ages remembered as the creator of a new style of domestic architecture. . . . Great discoveries are sometimes made from small circumstances, and the repairs of a little citizen's box at the corner of two high roads revealed to Walpole the great secret of the combined beauty, convenience and grandeur which a revival of our old English architecture was capable of producing." It is hard to imagine Macaulay agreeing wholeheartedly that one man's trivia might be another man's inspiration.

Croker, like Macaulay, admired what he called Walpole's wit and what Walpole himself might have preferred to describe as "sagacity." Croker went so far as to compare Walpole's wit with Voltaire's:

> It generally fulfils very exactly the first part of Locke's definition of wit—as "lying in the assemblage of ideas, wherein can be found any resemblance or congruity and putting these together with quickness and variety"; but it frequently fails in the second, for it is not very fertile in "pleasant pictures or agreeable visions." It is a wit that never makes one laugh—seldom even smile—but on the other hand, is wonderfully terse, forcible and descriptive. It tells more in one or two words than any one else would say in many sentences.

Croker must have been a humorless gentleman indeed, if Walpole's wit could hardly make him smile, or perhaps it was only that he lived in an agelastic period when smiles were generally suppressed. If it is true that state funerals in early Victorian times were taken with total seriousness, it would have been hard for Croker to laugh with Horace Walpole at the antics of the aristocracy at the funeral of George II.

Whether we read the reviews of Walpole's letters by Hazlitt or Macaulay or Croker, we receive a strong impression that these men looked upon Walpole as a creature from another world, stimulating to their curiosity about the gossip of the past century, amusing, perhaps even delightful to their literary sensibilities, but rarely one to be taken seriously. Since they were all eminently serious men, they could not without difficulty find a place in their world for that strange creature that was Horace Walpole. Unreceptive as they were to his interests, ideas, and values, they—Macaulay especially, but also Hazlitt and Croker—and all those for whom they spoke, would scarcely have adopted any of Walpole's strange and presumably frivolous word coinages into their vocabularies.

A generation or so after Croker wrote his last review of Walpole's letters in 1843, the reactions of the literary critics to Walpole had changed perceptibly. Where the early Victorians had found Walpole quite alien to their own temper and had dealt with him more or less unsparingly, the later critics tried increasingly to rehabilitate the merit of Walpole's attitudes and the dignity of his values as well as to praise his stylistic brilliance. Implicitly or in so many words, these critics were battling the Macaulay "version" of Walpole.

Leslie Stephen's article on Walpole in 1872[13] is, perhaps, a kind of bridge between the early Victorian antipathy and the Edwardian admiration of Walpole. As Noel Annan said in his biography of Stephen, he "would hardly be likely to find the skinless, spiteful, quarrelsome, frivolous intriguer, Horace Walpole, a congenial spirit; yet he took pleasure in exhibiting Walpole's tolerance, scepticism and shrewdness, appreciated his hatred of dullards and boors, and placed him above the mass of commonplace writers by virtue of his literary power." Stephen deplored the "hard measure" Walpole received from Macaulay, and laid bare the basis of what he called Macaulay's "fervour of rebuke": "There is something which hurts our best feelings in the success of a man whom we heartily despise."[14]

The early Victorians had done little more than describe Horace Walpole's character and his style, but Leslie Stephen tried to explain the sources of Walpole's interests and to appraise his position in English arts and letters. Rather sympathetically, he traced Walpole's eccentric tastes to his inability to live up to the requirements for manly sociability of the age in which he lived. "The men of that age may be divided by a line which, to the philosophic eye, is of far more importance than that which separated Jacobites from loyal Whigs or Dissenters from High Churchmen. It separated the men who could drink two bottles of port after dinner from the men who could not." Since Walpole was one of those who could not, he "retired to Strawberry Hill, and made toys of Gothic architecture, or heraldry, or dilettante antiquarianism. . . . When you can only join in male society on pain of drinking yourself under the table, the safest place is to retire to the tea-tables and small talk."

Unlike Macaulay, Stephen tried to understand Walpole in terms of the social and cultural setting of his life. Stephen felt very strongly that "men must be manly and women womanly; and the slightest androgynous taint must be condemned or satisfactorily explained."[15] He did not condemn Walpole, partly, perhaps, because he could explain his effeminate traits, and partly, also, perhaps, because he realized that although Walpole was sensitive and timid, he was not altogether spineless; he stuck to his eccentric guns. Again strikingly unlike Macaulay, Stephen "practically always

[13] Leslie Stephen, "Horace Walpole," *Cornhill Magazine* 25 (June 1872): 718–735.

[14] Noel Annan, *Leslie Stephen: His Thought and Character in Relation to His Time* (London: MacGibbon and Kee, 1951), p. 233.

[15] Ibid., p. 224. But of course Stephen would have none of that genteel temper unforgettably expressed in such a work as Lady Gough's *Etiquette* (1863): "The perfect hostess will see to it that the works of male and female authors be properly separated on her bookshelves. Their proximity unless they happen to be married should not be tolerated."

overcame any natural repugnance he may have had for certain kinds of mental and moral attributes." Even in the case of Coleridge, for example, Stephen made an effort "to make a serious judgment of Coleridge's life and work. . . . He ridicules his own utilitarian outlook, deplores attempts to judge Coleridge by rule of thumb, and declines to answer whether the author of the *Ancient Mariner* is entitled to neglect his children and break the Ten Commandments."[16] And so with Walpole.

Stephen had far more charity for Walpole's pretensions and affectations than did the earlier critics, for example, his disavowal of literary ambition and learning. He tried to explain why a man like Walpole might want to dissociate himself from the professional writers of the time; he sought Walpole's acceptable motives as well as his disreputable ones: "There is too much truth in his disavowals to allow us to write them down as mere mock-modesty; but doubtless his principal motive was dislike for entering the arena of open criticism. . . . The anxiety of men in that day to dis-avow the character of professional authors must be taken with the fact that professional authors were then an unscrupulous, scurrilous and venal race. Walpole feared collisions with them as he feared collisions with the 'mountains of roast beef.'" A manly fellow like Johnson might be able to compete in that arena, but Stephen could not blame Walpole for running for cover.

As for Walpole's place in the world of letters, Stephen saw the unique contribution such a "trifler" as Walpole could make. Stephen is worth quoting at some length:

> He was an arrant trifler, it is true; too delicately constituted for real work in literature and politics, and inclined to take a cynical view of his contempor-aries generally, he turned for amusement to antiquarianism, and was the first to set modern art and literature masquerading in the antique dress. . . . In truth, Walpole has no pretensions whatever to be regarded as a great original creator, or even as one of the few infallible critics. . . . But he was infinitely superior to the great mass of commonplace writers, who attain a kind of bastard infallibility by always accepting the average verdict of the time. . . . There is an intermediate class of men who are useful as sensitive barometers to foretell coming changes of opinion. Their intellects are mobile if shallow; and perhaps their want of serious interest in contemporary intellects renders them more accessible to the earliest symptoms of superficial shifting of taste. They are anxious to be at the head of fashions in thought as well as in dress, and pure love of novelty serves them to some extent in place of genuine originality. Amongst such men Walpole deserves a high place; and it is not

---

[16] Annan, *Leslie Stephen*, p. 233.

easy to obtain a high place even among such men. The people who succeed best at trifles are those who are capable of something better.

Stephen went quite a long way toward redeeming Walpole as a human being and also redeemed him as a writer not merely of stylistic brilliance but of some social importance. Trifler that he was, incapable of "real" (that is, serious) work, a shallow lover of novelty rather than an original creator, Walpole, as Stephen saw him, nevertheless had that perceptivity and even courage that a leader of fashion, whether in clothes or ideas, must have. Such leaders of fashion may have little influence on fundamental values and attitudes, but they mold the more superficial tastes of their society. Walpole's works must, therefore, be read not merely as light entertainment, but as a reflection of real, if ephemeral, shifts of taste. If for Stephen and his contemporaries such a shallow, trifling dilettante as Walpole could produce works of value, perhaps, also, theirs was a climate of opinion prepared to accept such a creation of Walpole's as serendipity on its semantic merits, while recognizing the frivolous circumstances of its origin.

The first full-length treatment of Walpole was by that great eighteenth-century scholar Austin Dobson. His *Horace Walpole: A Memoir*, published in 1890, was an objective, balanced appraisal, willing to borrow, as Dobson himself said, both from Macaulay's "unsparing" portrait of Walpole, and from Miss Berry's adulatory one.[17] "Walpole's character," wrote Dobson, "may be considered in a fourfold aspect, as a man, a virtuoso, a politician, and an author. The first is the least easy to describe." Dobson agreed with Macaulay, that as a man, Walpole was an aristocrat first, and that this accounted for his unsympathetic attitude to the great literati of his day. Dobson also maintained that Walpole was "very much in bondage to public opinion, and morbidly sensitive to ridicule." Walpole was not only an aristocrat, then, much concerned to preserve the dignity of his class position, but a sensitive aristocrat at that, and hence doubly fearful lest his literary activities expose him to criticism. "To the friends of his own class he was constant and considerate, and he seems to have cherished a genuine affection for Conway, George Montagu, and Sir Horace Mann. . . . But his closest friends were women. In them . . . he found just that atmosphere of sunshine and *insouciance*—those conversational 'lilacs and nightingales'—in which his soul delighted, and which were most congenial to his restless intelligence and easily fatigued temperament."

Such an atmosphere of "sunshine and *insouciance*" rarely produces

---

[17] Dobson, *Horace Walpole: A Memoir*, p. 275.

works of great depth, and if Dobson considered it to be Walpole's most congenial milieu, we can appreciate why Dobson did not take Walpole's "serious" literary efforts with equal seriousness. These works, he felt, were mere *opuscula*. But as a letter writer, Walpole is without peer: "It matters nothing whether he wrote easily or with difficulty; whether he did, or did not, make minutes of apt illustrations or descriptive incidents: the result is delightful. For diversity of interest and perpetual entertainment, for the constant surprises of an unique species of wit, for happy and unexpected turns of phrase, for graphic characterization and clever anecdote, for playfulness, pungency, irony, persiflage, there is nothing in English like his correspondence" (p. 294). Unique wit, playfulness, pungency, happy and unexpected turns of phrase—these are the elements out of which serendipity came to be created, and Dobson fully appreciated these qualities in Walpole's writing.[18] Beyond all this, Dobson recognizes Walpole's letters as a valuable social chronicle "of a specially picturesque epoch by one of the most picturesque of picturesque chroniclers."

The emphasis in triplicate is one bit of testimony that the picturesque had won a place for itself in late Victorian taste. Witness Dobson's judicious and ultimately sympathetic estimate of Walpole's collection at Strawberry Hill:

> He was an indiscriminate rather than an eclectic collector; and there was considerable truth in that "strange attraction from the great to the little, and from the useful to the odd" which Macaulay has noted. Many of the marvels at Strawberry Hill would never have found a place in the treasure houses— say of Beckford or Samuel Rogers. . . . At the same time, it should be remembered that several of the most trivial or least defensible objects were presents which possibly reflected rather the charity of the recipient than the good taste of the giver. All the articles over which Macaulay lingers, Wolsey's hat, Van Tromp's pipe case, and King William's spurs were obtained in this way; and (with a laughter) Horace Walpole, who laughed a good deal himself, would have made as merry as the most mirth-loving spectator could have desired. . . . In any case, however, it was a memorable curiosity shop, and in this modern era of *bric-a-brac* would probably attract far more serious attention than it did in those practical and pre-aesthetic days of 1842, when it fell under the hammer of George Robin.[19]

---

[18] Dobson may have known the word serendipity. He contributed a chapter to Andrew Lang's book, *The Library* (London: Macmillan, 1881), in which, as we shall see presently, Lang used the word in his preface.

[19] Dobson, *Horace Walpole: A Memoir*, pp. 287–288. Dobson also included "A Day at Strawberry Hill" in the first series of his *Eighteenth-Century Vignettes*. Properly enough, the copy in Butler Library (in Special Collections) bears the bookplate of Col. W. F. Prideaux, who in 1903 sent a note about serendipity to *Notes and Queries*.

The later Victorians had, perhaps, more humor and irony as well as more "aestheticism" than their ancestors, and so they could understand Horace Walpole's half-serious, half-mocking fondness for "curiosities." This was not long before, in the first decade of the twentieth century, Everard Meynell was to establish in London the Serendipity Shop, a bookshop where bibliophiles found many treasures among the literary bric-a-brac.

By the Edwardian period, Walpole was given a serious and friendly appreciation he had never before had. Literary critics now emphasized his strengths rather than his weaknesses, when they did not try to conceal his weaknesses altogether. In an article that appeared in the *Edinburgh Review* in 1904,[20] the anonymous author refused to consider Walpole a "mere dilettante"; on the contrary, Walpole was described as having had one serious aim in life, which by dint of hard work he had realized, "to obtain fame as a writer of letters." He credited Walpole with being "at once sagacious and fortunate in perceiving the form of expression best suited to his mental equipment", namely, the writing of letters. Where earlier critics saw Walpole as a creature of utter frivolity, this anonymous critic has transformed him into a dedicated writer. Where the early Victorian critics had thought Walpole's form of expression only peripheral to literature, and the aptness of that expression was for them its only redeeming feature, now the form was at least partly attributed to Walpole's wisdom of choice, and the aptness was only one of its virtues.

The writer in the *Edinburgh Review* went on to give Walpole the accolade of being modern—sophisticated, trenchant, realistic—especially in his use of words:

> He has the quickness of a later age than that in which he lived, its capacity for being bored, its preference for art, its appreciation of medievalism. Gray on one occasion took him to task for using the words "tinker up." He did not say in so many words that Walpole was using what in modern phrase would be called "slang"; he says something about want of dignity of style. But Walpole actually has the temerity to defend his words as being the most telling and suitable for the purpose. . . . [H]e defends the expression for its vividness and truth, and expression reproduced from actual life and applied to literary uses.

In this more favorable climate of opinion a new Walpole emerged: the fearless judge of apt expression rather than the craven dabbler in literature who happened to have a flair for description; the uncompromising writer concerned, moreover, for the expressiveness rather than the dig-

[20] *Edinburgh Review* 199 (1904): 432–456.

nity of his style, and, it is almost implied, little concerned with dignity in general.

Evidently, Walpole's kind of expressiveness, as well as his values and tastes, found more resonance among the Edwardians than among Macaulay's contemporaries. The article in the *Edinburgh Review* explicitly defended Walpole against Macaulay's assault on his taste. Walpole, the writer felt, may have been carried away by his enthusiasm for Gothic architecture, but "Macaulay's attack on Walpole's rather indiscriminate mass of curios and objects of art shows how unable was the Early Victorian to understand the value and interest of such a collection." Norman Pearson, too, in the article we referred to earlier, tried to vindicate Walpole against Macaulay's charge that he inverted the true value of everything: "This . . . is an extravagant misrepresentation. . . . [Walpole] was by no means indifferent to things great, and it would be truer to say that he could always see the greatness of things little. For it is just the trifling details on which he so skilfully seized which give to his descriptions not only their charm but their vitality."[21]

The failure of the early Victorians to appreciate the "true worth" of the so-called trifles that preoccupied Walpole has by the Edwardian period become a kind of refrain in the appraisals of Walpole both as writer and as collector: as a theme, it was introduced without emphasis by Croker, who regretted the dispersing of the Strawberry Hill collection; it was accented somewhat more by Dobson, who referred to the "pre-aesthetic days of 1842"; and it was stressed strongly by Pearson and his contemporaries, who saw in Walpole's preoccupation with small things the very strength of his writing. Pearson even attributed to Walpole one of the characteristics of the truly scientific mind, which is able to see the potential significance of the apparently trivial.

The reactions of literary critics to Walpole changed quite markedly between the beginning of the nineteenth century and the beginning of the twentieth. Throughout the first half of the nineteenth century, the critics saw a man who glorified the trivial, who was preposterously "affected," and who made a cult of frivolity; half a century later, Walpole appeared as a writer who possessed the genius of seeing "the greatness of little things," who disarmingly laid bare his own failings, and who had a serious purpose, even a sense of dedication, in the world of letters. A possible context for these changing attitudes toward one author, Walpole, is a detectable change in the English climate of opinion generally. Climates of opinion are of course difficult to characterize. The "portrait" of any age so very quickly tends to become a jumble of inconsistencies and paradoxes, of "on the one hand this" and "on the other hand that"; the

---

[21] Pearson, "Neglected Aspects of Horace Walpole," pp. 482–494.

historian all too soon loses sight of the salient features as he seeks to encompass the concrete variety of details. But paradoxical and contradictory as every age appears when examined closely, certain dominant features stand out if we adopt the necessary perspective.

In Victorian England the moral and intellectual doctrines that dominated the climate of opinion were Utilitarianism and Evangelicalism.[22] Utilitarianism was profoundly individualistic and rationalist, it "abhorred muddle and waste. . . . It stood for the self-supporting, self-reliant individual."[23] The individual was held in large measure accountable for his actions, and he tried to control as nearly as possible the consequences of those actions. At the same time, the influence of Evangelicalism endowed the Victorians with a deep concern for the seriousness and righteousness of all they did. As Buckley points out, even the free-thinkers of the age, such as George Eliot and Charles Bradlaugh, were filled with moral fervor. Their moral zeal filled the Victorians with zest and energy in their activities; they were very sure of themselves. William Gladstone's lifelong, devoted work for the rescue and rehabilitation of prostitutes epitomizes this aspect of the "Victorian temper."[24]

By contrast, the later Victorians and Edwardians would appear to have been less solidly convinced that they knew what they were doing and why they were doing it. Doubt and self-doubt, moral uncertainty and tension had arisen; a Leslie Stephen or a John Stuart Mill, to say nothing of a Matthew Arnold or a Samuel Butler, experienced moral crises unknown to their fathers. Compare the moods in which Tennyson wrote "Locksley Hall" and "Locksley Hall Sixty Years After"—the recuperative optimism in the former and the despair in the latter. At the turn of the twentieth century, thoughtful men were frequently pessimistic, sometimes even cynical, about man's nature and destiny, as many intellectuals in the eighteenth century had been.

It would be rash to claim any direct connection between these changing climates of opinion and changes in the patterns of critical reaction to Walpole. At most, this relationship can be expressed in terms of more and less compatibility. But it can be suggested that the salience of the doctrines of Utilitarianism and Evangelicalism might have reduced the likelihood of sympathy on the part of the early Victorian critics for Walpole, to whom he must have appeared as a man who capriciously and wastefully frittered away his time, and who lacked even the courage of his frivolity.

[22] See Jerome H. Buckley, *The Victorian Temper: A Study in Literary Culture* (Cambridge, Mass.: Harvard University Press, 1951).

[23] David Churchill Somervell, *The Victorian Age* (London: Published for the Historical Society by G. Bell and Sons, 1937), p. 81.

[24] See Roy Jenkins, *Gladstone: A Biography* (New York: Random House, 1997).

And since serendipity was a characteristic creation of Walpole's, any attitude toward Walpole was apt to encompass his word-child. The serious early Victorians were not likely to pick up serendipity, except perhaps to point to it as a piece of frivolous whimsy. Such an etymological oddity of a word was not congenial to them, nor did its meaning recommend itself as particularly congruent with their thinking. Although the Victorians, and especially Victorian scientists, were familiar with the part played by accident in the process of discovery, they were likely neither to highlight that factor nor to clothe the phenomenon of accidental discovery in so lighthearted a word as serendipity. By the same token, at the end of the century, some men not only took pleasure in Walpole's literary style (this had never ceased to delight), but felt at home in his brittle, troubled world. Some of them, too, be it coincidence or consequence, adopted serendipity into their vocabularies.

Chapter 3

ACCIDENTAL DISCOVERY IN SCIENCE: VICTORIAN OPINION

*T*he social world is not of a piece: It is differentiated into groups, classes, by educational strata, and so on through the entire range of statuses. Among these groups, there are two in particular that are generically concerned with searching and finding: collectors (be they bibliophiles, antique collectors, or antiquarians in general) and scientists. In the later nineteenth and early twentieth centuries, some bibliophiles and antiquarians were among the first to appreciate the aptness of the word serendipity to describe their experience of making happy accidental discoveries, whether of desired books or of needed scraps of historical or literary information. Like Walpole, who was no longer despised, they realized that chance played a part in the success of their various collections.

The scientists in the nineteenth century, too, were aware of the role played by chance or accident in the process of research and discovery. Indeed, in the 1940s and '50s it has been the scientists who have been especially taken with the "expressiveness" of serendipity to describe the familiar phenomenon of accidental discovery. And among these scientists, it has been those who have had "broader interests" in the humanities who started the serendipity boom and introduced their less learned colleagues to the word. In nineteenth-century England, presumably, many more scientists had, by virtue of their education and the still relatively unspecialized nature of their work, those broader interests that today are more exceptional. It would not be surprising to find that Victorian scientists read and enjoyed Walpole's letters in their leisure time. And yet, unlike the collectors, none of the writers dealing with science appears to have made use of Walpole's apt term in their discussions of accidental discovery.

There were ample occasions that could have called for the use of the word serendipity. Numerous well-known accidental discoveries occurred in the early and mid-nineteenth century, the period when serendipity came into the market of words. In 1833, the year in which Walpole's letters to Mann were first published, there occurred one of the more spectacular examples of what was later to be cited as a case of serendipity, for in that year William Beaumont published his report on the new discoveries he had made about the digestive process, which resulted from his study of a gastric fistula that had accidentally come to his attention. Shortly before, in 1828, Friedrich Woehler accidentally obtained the compound urea in his laboratory, by treating potassium cyanate with ammonium sulphate, and the synthesis of this substance, held to be identical with urine, laid the foundation of the science of organic chemistry. Again, in 1839, Goodyear accidentally discovered the process for vulcanization of rubber, and in 1856, William Henry Perkin by accident synthesized the first coal tar dye, "mauve." These discoveries inevitably received considerable public attention, but, unlike, say, Fleming's discovery of penicillin, about a century later, they were not described as cases of serendipity.

The problem of the role of accident in such discoveries as these seems to have preoccupied Victorian writers not a little. The great authority on the history and philosophy of science William Whewell engaged in bitter controversy on the subject with his anonymous reviewer in the *Edinburgh Review* (the reviewer gives the impression of himself being a man of considerable stature). The dispute began when Whewell's *History of the Inductive Sciences* was published in 1837. This was an important and influential although inadequate piece of work, as George Sarton testifies:

> The first modern history [of science] is the history of the inductive sciences by the Reverend William Whewell . . . a book which maintained the dignity of a classic in English libraries and colleges during the whole Victorian age and even beyond. . . . Whewell's work was not historically up-to-date at the time of its first publication; it is at present entirely out-of-date. It is a dangerous book for young students of the history of science, but it has itself become a document of great value enabling us to recapture the scientific outlook of a hundred years ago. Nothing illustrates better the backwardness of our studies than the fact that Whewell's book was still commanding the respect of many thoughtful readers at the beginning of [the twentieth] century.[1]

[1] George Sarton, *A Guide to the History of Science* (Waltham, Mass.: Chronica Botanica, 1952), pp. 49–50.

Whewell went a long way in this book toward denying any importance to the factor of accident in discovery. Writing of Pythagoras's discovery in geometry, he said that "this, like all other fundamental discoveries, required a *distinct and well-pondered idea* as its condition . . . as in all cases of supposed accidental discoveries in science, it will be found that it was exactly the possession of such an idea which made the accident possible." Whewell was, in effect, arguing that it was not so much the role of circumstances, nor even the appearance of a man of great gifts or genius that was the decisive factor in scientific discovery, but rather the availability to the scientist of a theory, of an "idea" as he called it. Whewell's reviewer disagreed strongly:

> It cannot, we think, be questioned that many of the finest discoveries in science have been the result of pure accident. By the accident of placing a rhomb of calcareous spar on a book or a line, Bartolinus discovered the property of Double Refraction. [Then follow other examples: Huygens's discovery of polarization, Malus's discovery of polarization by deflection, etc.] Now all these are fundamental discoveries in which no appropriate idea had any share, and which, though they were made by men of the most distinguished talents, and pursued with the most consummate dexterity, might, nevertheless, have been made by the most ordinary observer.[2]

The reviewer implied that in those frequent and important discoveries that were "the result of pure accident," the particular person making the discovery might have been replaced by any one or more of numerous others, without preventing the occurrence of the discovery: Given the right circumstances, the discovery was bound to be made. (Later we shall see that this conception of accidental discovery is related to the inferences that were beginning to be drawn at this time from simultaneous independent discoveries: namely, that at a given stage of scientific development certain discoveries occur almost of necessity, and that there is no need for outstandingly qualified scientists to bring them about.) Whewell's reviewer seemed to suggest, however, that the follow-up of these accidental discoveries might not have been made with such "consummate dexterity" by men of lesser talent than the actual discoverers. Perhaps he had in mind the fact that in accidental scientific discoveries (unlike accidental discoveries made by collectors), the accidental observation constitutes only the beginning of a scientific advance.

In his next book, the *Philosophy of the Inductive Sciences*, Whewell resumed the battle:

> *No scientific discovery* can, with any justice, be considered *due to accident*. . . . The common love of the marvellous and the vulgar desire to bring down the greatest achievements of genius to our own level, may lead men to

[2] *Edinburgh Review* 74 (1837): 265–306.

ascribe such results to any casual circumstances which accompany them; but no one who fairly considers the real nature of great discoveries, and the intellectual processes which they involve, can seriously hold the opinion of their being the effect of accident. . . . *Such accidents never happen to common men.* Thousands of men, even the most inquiring and speculative, have seen bodies fall; but who, except Newton, ever followed the accident to such consequences?[3]

Whewell was defending the great scientists against what he took to be the belittling of their genius as well as against the deprecation of their "intellectual processes." He adumbrated a theory of interacting components in the process of accidental discovery, of chance events interacting with the prepared mind—not merely the inquiring and speculative mind, but the "uncommon" mind of the great scientist. Whewell now had his antagonist on the defensive. Reviewing Whewell's new book, the anonymous reviewer modified his argument appreciably; only certain kinds of discoveries depended on accident, he wrote (he explained this elaborately in terms of the psychology and physiology of perception), and he maintained that he had never intended to imply that great discoverers were not prepared for such accidents. Each of the two disputants had, in the course of this controversy, done some violence both to his own argument and to that of his opponent for the sake of emphatic exaggeration, but in joining the issue they had brought into clear view those points that in many future discussions of accidental discovery were to have the greatest salience.

The latter half of the nineteenth century continued to see discussions of the role of accident in discovery, some written for the educated layman, some for that same public for which Whewell wrote, which was assumed to have considerable sophistication about the nature of science and of its development. The layman who read magazines such as *Macmillan's Magazine* or *Chambers's Journal* could receive some enlightenment not only about the fact of accidental discovery in science but also about some of its alleged causes. One J. Coryton, writing on "Accidental Invention" in *Macmillan's Magazine*,[4] introduced the subject with what he felt to be necessary cautions: "It is needless, we hope, to deprecate insinuations as to our being the unqualified eulogists of Lucky Accident, or encouraging 'loafery' by the instances we are going to adduce of Idleness and Scampishness succeeding where Philosophy failed. . . . One element there must be in common to all invention, be the immediate causes what they may. There must be Genius—that particular species of Genius

[3] William Whewell, *Philosophy of the Inductive Sciences* (London: Frank Cass and Co. [1847] 1967), vol. 2, p. 23.
[4] J. Coryton, "Accidental Invention," *Macmillan's Magazine* 4 (May 1861): 75–85.

which Dr. Johnson termed as *'knowing the use of tools.'* " Like Whewell, Coryton believed that the inventor must be in some way prepared to take advantage of an accident; and like many who were in the future to be concerned with serendipity, Coryton felt it incumbent on him to defend the beneficiaries of lucky accidents against the charge that they were getting more than they deserved. Coryton's article contained a whole catalogue of accidental inventions or chance insights, enough thoroughly to familiarize the reader with the role that accident has played in the history of science.

Another article, this one anonymous, appeared in *Chambers's Journal*.[5] This author explained much more clearly for the benefit of the layman the interaction of chance with the prepared mind:

> Seldom do men sit down with a steady resolve, a determined purpose, to discover some new principle or invent some new process. When they do so, there is a lurking idea of the thing they want, a dim perception of the direction in which success may reasonably be sought. Generally speaking, something is concerned which, for want of a better term we call "accident." . . . But the important point to notice is that the value of the accident depends on the kind of man, on the kind of mind, by whom or by which it is first observed. If the soil is not sufficiently prepared, the seed will not grow. Thousands of men had seen light reflected from distant windows, and variations in the light according to the angle of reflection, but a well-prepared mind on one occasion suddenly drew from this phenomenon an idea which established the beautiful science of the Polarization of Light.

Again many familiar and less familiar examples, both legendary and true ones, follow: Newton's discovery of gravitation, the discoveries of the glazing of pottery and the printing of calico, William Lee's stocking frame, Hargreaves's spinning jenny, and others.

The more academic discussions of the role of accident in the development of science were, in the second half of the century, refined far beyond the rather crude assaults that Whewell and his reviewer had made on the problem. William Stanley Jevons undertook in 1874 to specify more precisely the conditions under which accidental discoveries were more and less likely to occur.[6] He maintained that accidents were more important at the inception of a branch of science, and that "the greater the tact and industry with which a physicist applies himself to the study of nature," the more likely it was that he "will meet with fortunate accidents, and will turn them to good account." (Many years later, in the

[5] *Chambers's Journal* 53 (1876): 245–247.
[6] William Stanley Jevons, *The Principles of Science* (London: Macmillan, 1874), pp. 164–165.

1930s, two teachers at the Harvard Medical School, Drs. W. B. Cannon and M. J. Rosenau, will speak instead of the cultivation of serendipity.) Another attempt to specify the conditions of accidental discovery was made in 1878 by G. Gore:[7] "Different discoveries occur with very different degrees of unexpectedness; great ones rarely come unawares. The quantitative relations of known scientific truths are also rarely found by accident, because definite researches are specially made to find them. The discovery of new qualitative facts is usually the most unpredictable, and the most unexpected discoveries in physics and chemistry are generally those of isolated phenomena of an entirely novel and peculiar kind." Again, in 1945, a sociologist examining the activities of scientists will speak, instead, of the serendipity pattern in research, which involves the observation of surprising, anomalous data. In the nineteenth century, scholars and laymen alike were evidently interested in the occurrence of accidental discoveries; although they sought acceptable explanations for these discoveries and came to sense the connotations on the moral level implied in the factor of accident, they did not use the apt term coined by Walpole to describe accidental discovery.

The Use of Neologisms by Scientists in Nineteenth-Century England

It is in the nature of science that new concepts, facts, and instruments constantly emerge, and there is a continual concomitant need for new terms to designate them. With the accelerated pace of scientific development in the nineteenth century, the need for new terms was frequently felt and as frequently met by the construction of neologisms. Scientists had no antipathy to new words as such: hundreds and then thousands were being coined; for examples, Berzelius in chemistry introduced "isomerism," Pelletier and Caventou in 1817 isolated the green pigment of plants and called it "chlorophyll," and Helmholtz invented the "ophthalmoscope." It is known, moreover, that at least two great English scientists, Sir Charles Lyell and Michael Faraday, were genuinely concerned about the linguistic quality of the words they were inventing to designate new facts and concepts in geology and physics, respectively.

Lyell and Faraday turned for advice on this subject to William Whewell, who was sensitized to the role of technical language in science and devoted much attention to the search for apt words and phrases. Whewell realized "how powerful technical [sic] language is for the per-

---

[7] G. Gore, *The Arts of Scientific Discovery* (London: Longmans, Green, 1878).

petuation either of truth or error,"[8] and he gave careful considerations to Lyell's and Faraday's suggestions and inquiries. He weighed their sound qualities, their aptness, and their etymology. Thus, he wrote to Lyell that the "termination *synchronous* seems to me to be long, harsh and inappropriate." He complimented Lyell on "your introduction of ἠώς though it is somewhat poetical, for it will be well remembered: but I wonder you never thought on the same principle of calling the tertiaries 'nychto-synchronous' which is a very pretty looking word and quite new." Whewell did not like words with hybrid derivations: "*Cenary* or *Coenary* is a bad word: the termination is a Latin one, and it will not be scholarly and wise to stick it to a Greek root." Or, as Whewell wrote in a more general vein to Faraday:

> I had the pleasure of being at the R.S. at the reading of your paper, in which you introduced some of the terms which you mention, and was rejoiced to hear them, for I saw, or thought I saw, that these novelties had been forced upon you by the novelty of extent and the new relations of your views. In cases where such causes operate, new terms inevitably arise, and it is very fortunate when those, upon whom the introduction of them devolves, look forwards as carefully as you do to the general bearing and future prospects of the subject; and it is an additional advantage when they humour the philologists so far as to avoid gross incongruities of language. I was well satisfied with most of the terms that you mention; and shall be glad and gratified to assist in freeing them from false assumptions and implications, as well as from philological monstrosities.[9]

Whewell was evidently well aware of the circumstances in which the need for new words arose, as well as of the dangers of ambiguity and of etymological impurity in the coinage of such words. The conservative H. W. Fowler[10] considered even such coinages by Lyell as "eocene, miocene, pliocene" to be "regrettable Barbarisms," though Whewell had approved of them. But such a "philological monstrosity" as serendipity would certainly not have recommended itself to Whewell: No measure of aptness or even poetical quality could have redeemed its etymological bastardy.

The early and mid-Victorians, then, whether they were preoccupied with arts and letters or with science, would appear to have had little disposition to welcome or adopt Walpole's word coinage serendipity. In humanistic circles, the moral and aesthetic temper was unsympathetic to

[8] William Whewell, *History of the Inductive Sciences* (London: Frank Cass & Co., [1837] 1967), vol. 1, p. 60.

[9] Isaac Todhunter, *William Whewell: An Account of His Writings* (London, 1876), vol. 2, p. 99.

[10] H. W. Fowler, *A Dictionary of Modern English Usage* (Oxford: Oxford University Press, 1926), p. 440.

Walpole: He presented himself as a frivolous, affected, shallow man, whose tastes were for the trivial and the odd, for things of little intrinsic beauty or value. And by the same token, his neologism serendipity was likely to seem alien and undesirable, the product of a frivolous whim. If the scientists read Walpole, his description of discoveries by accident and sagacity would have had a familiar ring: Many of them were aware of the role of accident in the development of science, and some were interested in it. Many were aware that scientific advances often required the coinage of words.[11] But, once again, the seeming whimsy of Walpole's word coinage made it unlikely to find any resonance among the scientists: it would, in all probability, not be in keeping with the seriousness, and perhaps humorlessness, with which they set about their work. In both social worlds that were potentially interested in accidental discovery—that of humanistic collectors on the one hand, and of scientists on the other—serendipity was fairly certain to be victim of the Victorians' aversion to frivolity and whimsicality.

Serendipity Enters the Literary Circle

Forty-two years after the first publication of Walpole's letters to Horace Mann in 1833, and almost twenty years after the reprinting of those letters in Cunningham's edition of the Walpole correspondence in 1857, serendipity was for the first time used in print by another writer. This first appearance of the word was in the periodical *Notes and Queries* (over the signature of Edward Solly, of whom more presently), and serendipity was therewith launched into literary channels. The word was to remain in these channels for a long time. For more than fifty years, serendipity was to be used almost exclusively by people who were most particularly concerned with the writing, reading, and collecting of books.

Notes and Queries is a periodical that commends itself to people of diversified humanistic inquisitiveness—not merely people interested in literature, but a more varied group, including historians, classicists, lexicographers and bibliographers, antiquarians, and general "erudites." The periodical was founded in 1849 by William John Thoms, a learned bibliophile. Earlier, in 1846, Thoms had approached the editor of the *Athenaeum*, Charles Wentworth Dilke, with the idea of starting a column in that periodical that would contain articles and correspondence on "the manners, customs, observances, superstitions, ballads, proverbs, etc. of

---

[11] Lavoisier, for example, in his classic *Traité élémentaire de chimie*, had argued back in 1789 that the language used in science served as an analytical tool, as his own revolutionary system of chemical nomenclature aptly demonstrated.

olden time."[12] Dilke approved of the scheme, and "on the 22nd of August [1846] the first article appeared, Mr. Thoms writing under the pseudonym of 'Ambrose Merton' and giving his investigations the title of 'Folk-Lore.' In the number published on the 4th of September, 1847, Mr. Thoms revealed himself to be 'Ambrose Merton,' and at the same time claimed the honour of introducing the expression 'Folk-Lore,' 'as Isaac D'Israeli does of introducing "Fatherland" into the literature of the country.' *The Athenaeum* of the same date states that 'in less than twelve months the word 'Folk-lore' has almost attained the dignity of a household word.' "[13]

By 1849, the volume of articles and correspondence in "Folk-lore" had become too extensive for the *Athenaeum* to handle, and so, with Dilke's blessing, the periodical *Notes and Queries* was begun. Among the distinguished literary figures associated with it in its early days were several whom we have encountered as a result of their direct or peripheral interest in Horace Walpole: Peter Cunningham, who brought out the 1857 edition of Walpole's letters; John Wilson Croker, who edited Walpole's letters to Lord Hertford in 1825 and reviewed several other volumes of his letters; and John Doran, who became editor of *Notes and Queries* in 1878, and who edited the letters of Walpole to Sir Horace Mann under the title *Mann and Manners at the Court of Florence* (1876). The participation of these and other serious writers and scholars is explained by Thoms's purpose in founding the periodical: It was not to be simply a collection of literary gossip and chitchat; rather, he hoped "to reach the learning which lies scattered not only throughout every part of our own country, but all over the literary world, and to bring it all to bear upon the pursuits of the scholar; to enable, in short, men of letters all over the world to give a helping hand to one another."[14]

By the time the hundredth number of *Notes and Queries* had been issued, Thoms was able to report his pleasure that he had received letters from all parts of the world, "letters expressive of the pleasure which the writers (many of them obviously scholars 'ripe and good,' though far removed from the busy world of letters) derive from the perusal of *Notes and Queries*." Thoms's readership evidently came to encompass a wide range of well-educated men, some directly engaged in scholarly pursuits, but also many others who had been separated from the world of letters by their professions—ministers, lawyers, civil servants, perhaps even scientists and army officers—and who were happy to pursue their continuing avocational interests in literary matters through a medium such as

[12] See John C. Francis, *Notes by the Way* (London: T. F. Unwin, 1909), p. 36.
[13] Ibid.
[14] Introduction to number 52 of *Notes and Queries* 10, no. 2 (1904).

Notes and Queries. Frequently, these readers came to take advantage of the diverse audience of *Notes and Queries* to seek information on obscure minutiae, points that might be of interest to only a very few people but that were presumably saved from the stigma of "mere" triviality by the serious purpose of the inquirer. Founded as it was by an antiquarian and bibliophile, by a successful coiner of words and inveterate collector of books, who in the course of his "bookstalling" was delighted to pick up, by happy accident, the odd and the curious as well as the valuable,[15] *Notes and Queries* was committed to a catholicity of interests. Without difficulty or embarrassment its readers might one day correspond through its pages about that "philological monstrosity" serendipity.

This correspondence about serendipity began in 1875 and continued sporadically for nearly sixty years. The item that initiated it was an inquiry by one M.N.S., one of the many anonymous contributors to *Notes and Queries*. In the issue of February 27, 1875, M.N.S. inquired: "Where in his admirable letters does Horace Walpole refer to the story of the Princess of Serendip, and where is the story itself to be found?"[16] With that first query about an alleged "Princess" began the process of distortion and of garbling that was to accompany so many future references to Horace Walpole's "silly fairy tale." M.N.S.'s query suggests that he had read Walpole's letters and his transformation of the princes of Serendip into a princess appears to be the result of a slip of memory. In later years, when serendipity came to be used by less well-read people, the distortions of the title of the fairy tale, of its authorship, or of its plot, had other sources: In part, these errors were due to pretentious ignorance; but in part, also, they were due to the something less than scholarly efforts on the part of those encountering the word serendipity to establish its original source.

M.N.S. had not long to wait for a reply to his query; only a few weeks later, in the issue of April 17, 1875, Edward Solly sent in a reply.[17] Solly, as we noted earlier, was originally a chemist by profession, a respected one, who was associated with Faraday when the latter was at the Royal Institution, and who was made a Fellow of the Royal Society. In 1849, however, at the age of thirty, Solly quit the field of chemistry, and for the rest of his life worked for the Gresham Life Assurance Society. By avocation, Solly was well known as an antiquarian and bibliophile. Grenville Arthur James Cole wrote in his sketch of Solly for the *Dictionary of National Biography*: "Solly collected a large library, which was particularly

---

[15] See the article by William John Thoms, "Gossip of an Old Bookworm," *Nineteenth Century* 10 (July 1881): 63–79.

[16] *Notes and Queries* series 5, vol. 3 (1875): 169.

[17] Ibid., p. 316.

rich in eighteenth-century literature, and his wide genealogical and literary knowledge was always at the service of *Notes and Queries*, the *Bibliographer*, *Antiquary*, and other periodicals of a similar nature. In 1879, he edited *Hereditary Titles of Honour* for the Index Society, of which he was treasurer."[18] The articles that Solly contributed to these periodicals were on enormously diverse subjects; they were all concise and informative, but if they reveal the breadth of his learning, they give no clue as to the kind of man he was.

With what was apparently his usual readiness to provide information, Solly wrote to *Notes and Queries* in response to the query by M.N.S.: "Horace Walpole used the word *serendipity* to express a particular kind of natural cleverness, and in his letter to Sir Horace Mann, CCLI, 28th January 1754, he thus described it:

> I once read a silly fairy tale, called *The Three Princes of Serendip*: as their Highnesses travelled, they were always making discoveries, by accidents and sagacity, of things which they were not in quest of: for instance, one of them discovered that a *mule* blind of the right eye had travelled the same road lately, because the grass was eaten only on the left side, where it was worse than on the right—now do you understand *Serendipity?*

I presume it is the story of these Princes, and not of the Princess, that your correspondent is inquiring after."

If Solly had complied strictly with M.N.S.'s request for information (i.e., where in his letters did Horace Walpole refer to the story . . . ?), he need never have mentioned serendipity at all. But Solly expanded his answer somewhat and took the opportunity to report Walpole's new-minted word that had not before circulated, at least in print. More than that, he added a brief explanation of his own to Walpole's "derivation" of the word: Instead of quoting Walpole's remark about his "talisman . . . by which I find everything I want, *à pointe nommée*, wherever I dip for it," Solly defined serendipity as "a particular kind of natural cleverness," and by this rephrasing stressed Walpole's seeming implication that serendipity was a kind of innate gift or trait. Thus began that subtle process of selective definition and redefinition that was over the long run to effect startling changes in the meaning of serendipity. Each user was likely to select those elements from Walpole's complex concept that adapted its meaning to his special interests and experience, and, as we shall see presently, it was probably the collector in Solly that made him give the meaning of serendipity the particular twist he did. But Solly's first note about serendipity also demonstrated the existence of a certain curb on this pro-

[18] Grenville Arthur James Cole, "Edward Solly," in *The Dictionary of National Biography* (London: Oxford University Press, 1917), vol. 18, p. 622.

cess of redefinition—the quotation of Walpole's original letter, in which he himself discussed his word coinage. Although even this letter comes to be quoted selectively, its repeatedly quoted text does help to keep definitions partly within the scope of Walpole's intended meaning.

It was not so much serendipity as Serendip that excited further comment by the readers of *Notes and Queries* in the following months. On May 22, 1875, R. S. Charnock of Gray's Inn sought to explain the origin of Serendip as "an Arabic corruption of *Sinhala—devipa* (island of lions), now corrupted down to *Ceylon*."[19] He was corrected, on June 26, 1875, by R. C. Childers, who maintained that *Sinhaladvipa* does not mean "island of lions" but "island of the Sinhalese people," and that *Ceylon* is a corruption of *Sinhala* only.[20] These two readers, at least, probably noticed Solly's remarks on serendipity.

Three years later, another anonymous reader, C., made this inquiry of *Notes and Queries*: " 'Serendipity'—A word coined by Horace Walpole to express the luck of a person who sooner or later obtained what he desired. Can anyone suggest any history to the word? It does not appear to have any root or etymology. Did it not more probably arise from some mere passing table talk? The word has been quoted in some recent monthly."[21] C. evidently learned of the word in the "recent monthly," which we have been unable to identify. He thought he knew what serendipity meant, but the reference in the recent monthly must have been fragmentary, for he was ignorant of Walpole's "derivation." The etymological oddity of the word was apparent to C., as it was likely to be to any educated man of the time; even fifty years later it would not always be immediately noticed. And C. sensed that serendipity was a fugitive, whimsical creation.

It was again Edward Solly who immediately provided an answer (August 3, 1878):

"Serendipity," as the word was used by Walpole, meant the discovery of things which the finder was not in search of. I have recently noted at the end of the *Impartial History of the Life and Actions of Peter Alexowitz, Czar of Muscovy*, London, 8vo, 1723, an advertisement of the book which Horace Walpole referred to, entitled: *Travels and Adventures of Three Princes of Serendip.* . . . Walpole was about five years old when this "silly fairy tale," as he calls it, was published. It is now scarce, for I have not been able to meet with a copy in the last few years, during which I have been hoping by "serendipity" to find one, whilst looking for other things.[22]

[19] *Notes and Queries* series 5, vol. 3 (1875): 417.
[20] Ibid., p. 517.
[21] *Notes and Queries* series 5, vol. 10 (1878): 68.
[22] Ibid., p. 98.

Solly was concerned to set C. straight about the meaning of serendipity: The finder, he emphasized, was *not* in search of the object of discovery by serendipity; and he gave an apposite illustration to stress what he felt was the proper definition. Solly did not explain to C. the curious etymology of serendipity, but, as is the case with all articles on the same subject in *Notes and Queries*, his note in reply to C. carried references to all previous discussions both of serendipity and of Serendip in that periodical. If C., or any other reader, was sufficiently interested in the subject, he could now easily look it up.

As in the case of Solly's first note on serendipity, in 1875, this last one again stimulated comments from two readers on Serendip rather than serendipity. Alfred Curwen of Harrington Rectory, Carlisle, wrote on November 2, 1878: "My grandfather, Mr. Henry Curwen, has left behind him, at Workington Hall, Cumberland, a very large number of notebooks, extending from 1805 to 1860. They are almost illegible. . . . I constantly meet the word 'Serendip' and was not a little puzzled by it until I found its origin in your paper."[23] Mr. Curwen was evidently one of those many "scholars 'ripe and good' though far removed from the busy world of letters," who read with pleasure and profit the esoteric literary and historical capsules that made up *Notes and Queries*, and who felt a little less out of touch with the centers of scholarly activity when they published even fragmentary evidence of their own continued intellectual pursuits.[24]

It was Edward Solly who, having given an example in his note to C. of how serendipity might operate, provides us with the first actual "case" of a discovery that is described as one of serendipity. In 1880, he published *An Index of Hereditary English, Scottish and Irish Titles of Honour*, to give it its full title. In the preface he informed the reader of the uses of such an index, and he pointed out that the frequent changing and merging of titles makes it very difficult to trace them without the aid of an index. Here is his illustration of that point:

> Some years ago the compiler of this Index was asked by a friend, who had just taken "Amyand" House, to inform him who Sir   Amyand, Baronet, was? Reference to the current Baronetages of England, Scotland, and Ireland, showed that no such title was in existence; and reference to the extinct Baronetages did not show that the title was extinct. The inquirer was at fault, and it was not till some weeks later, when, by the aid of *Serendipity*, as Horace Walpole called it—that is, looking for one thing, and finding an-

---

[23] Ibid., p. 358.

[24] The second reader we mentioned was R. W. Ellis, who inquired (November 30, 1878) whether Mr. Henry Curwen's notes on *Serendip* made any reference to the town of Swa-Nabha, in which he was interested.

other—that the explanation was accidentally found, that the Baronetcy was
conferred by George III, on the son of his grandfather's skilful medical attendant, Claude Amyand, in 1764, for very good reasons, been changed into
Conewall, by the second Baronet, in 1771, under which title, therefore, the
Baronetcy is now to be found.[25]

The use Solly made of serendipity in this context was to be more lasting
than he could anticipate, for his usage was to be quoted years later in the
definition of serendipity in the *Oxford English Dictionary* (1913).

In this preface to *Titles of Honour*, Solly again tried to keep Walpole's
definition of serendipity intact, according to his lights. He stressed that
the discovery he made was of something he was *not* looking for—*at that
time*. He did presume it to exist and to be discoverable, and he seems to
have felt that an *active search* for something else was not unlikely to turn
up the desired object eventually. Solly would have doubted that anyone
not actively concerned, say, with genealogy or bibliophilia would turn up
either the explanation to a genealogical puzzle or a rare book. Of all the
ingredients in Walpole's complex "derivation" of serendipity, Solly was
most firmly committed to that of "looking for one thing and finding
another." But Solly's interpretation of serendipity differed sharply from
that which the scientists were to give to the word when they adopted it
into their vocabulary: Solly knew just what it was he hoped to find by
serendipity, and once he had found it, the "discovery" was complete; the
scientists can only hope by happy accident to notice some fact or relationship whose significance was *previously undefined*, and to use this discovery as the first step in an advance of scientific thought. (The circumstances of accidental discovery in science will be discussed at length later.)

Solly's interpretation of serendipity cannot, however, be taken as representative of the interpretation current in the literary world generally.
There were men of letters who gave quite a different reading to Walpole's
explanation of serendipity, and, unlike Solly, instead of emphasizing the
discovery of something not immediately sought, defined as serendipity
those experiences they had had of coming upon the *very item* they
wanted *at a given time*, although they could by no means expect to
discover it in so timely a fashion. This interpretation can, of course, be
legitimately referred to Walpole's explanation of serendipity, for Walpole
claimed, after all, "to find everything I want, *à pointe nommée*, wherever I
dip for it." One scholarly gentleman who used serendipity in this manner
wrote to *Notes and Queries* under the pseudonym "Vebna" (October 8,
1881), and rather apologetically recounted this experience:

---

[25] Edward Solly, *An Index of Hereditary English, Scottish and Irish Titles of Honour*
(London: Longmans, Green, 1880), pp. v–vi.

Several references have been made to Horace Walpole's use of this word, [serendipity], which he explains in his letter to Mann of January 28, 1754. Such a singular instance of this gift, or good fortune, or whatever else it may be called, occurred to myself in 1877, that I venture to think, that, trifling as it is, it may be worth recording. Reading Taylor's *Holy Dying* in Pickering's octavo edition . . . I wanted to verify a quotation from Cicero. I looked in the index of my copy of the Barbou edition, in fourteen volumes, but could not find it. While I searched I came upon a cancel leaf of another volume which had got sewn up in the index. The first line of the leaf was the passage I sought.[26]

This was a happy accident, indeed: The item discovered was the very one that was sought, but the circumstances of its discovery did not fit into the expected pattern of search. Rather, the discovery resulted from a coincidence, a trivial coincidence to be sure, and "Vebna," groping for a cause (was it "this gift, or good fortune, or whatever else it may be called"?), felt, perhaps, that the strange word serendipity best described the strange experience.[27] The word, moreover, permitted him to circumvent the question of whether it has some inner quality or merely outward circumstances that led to his "discovery," and enabled him to hint at mysterious forces. His discovery does seem to have appeared somewhat mysterious to him, and he was only the first to seek in serendipity a semimystical, final explanation for those strange coincidences, those "mysterious" workings of the universe, that other men take as only the starting point for rational, scientific investigation. So, the word created in a context of the trivial begins its peregrinations through a series of trivial finds.

"Vebna's" identity is concealed by his pseudonym: The episode he tells reveals him only as a man interested in religion and in the classics. We know far more, however, about another writer who used the word serendipity in print in the same year as "Vebna," and in much the same sense—Andrew Lang. Lang was an unusually versatile man: He distinguished himself as a scholar, a poet, and a journalist, and he also tried his hand, less successfully, at writing novels. As a scholar, he was not concerned with a single field, but did important work in anthropology, where he wrote on folklore, mythology, and primitive religion; in classics, where his work on Homer won praise from Gilbert Murray; and in history, more particularly the history of his native Scotland. His contribution to the science of social anthropology was, probably, his most significant, and it was this work that he himself valued most highly.

[26] *Notes and Queries* series 6, vol. 4 (1881): 294.

[27] We shall see that some twenty-five years later a Mrs. Emma Carlton perceived that serendipity in such a context as this was "coincidental sagacity" rather than "accidental sagacity." *New York Times Saturday Review of Books*, 29 April 1905.

By profession, Lang was a journalist rather than a scholar, excelling in the writing of light essays. G. S. Gordon writes of him in *The Dictionary of National Biography*, "In the lighter play of the essay as in some of the daintier forms of verse, in the short causerie falling just between literature and gossip, Lang had no rival." For this kind of work he was "unusually qualified," "combining [as he did] with a lively scholarship and wit a remarkable range of miscellaneous and immediately applicable knowledge." Although Lang's anthropological work has won recognition,[28] it suffered to a certain extent not only from Lang's other than scientific interest in the occult, but also, more superficially, from certain qualities of his style. He was a shy, proud man with high ideals, who designated himself as a "belletristic trifler" and "had a horror of persons who persist in taking themselves seriously in and out of season."[29] The effect of this orientation on his writings was sympathetically appraised by Salomon Reinach:

> Poor Lang! He suffered from one great fault; he was too witty. Had he been a German professor, heaping up ponderous materials wrapped in obscure language, rewarding his readers by the meritorious discovery of involved truths, we should not have been told in many obituary notices, after words of praise on his poetry, novels and Greek scholarship, that he also dealt with folk-lore and mythology. All his books on mythology sold well; several of them have been translated; but there is something about them that deters the reader from taking them quite seriously. That "something" is more and worse than wit: it is a certain misuse of it. Lang was not only witty but jocular; he found it difficult to abstain from fooling, from affecting "naughtiness."[30]

Lang appears to have had in common with Walpole, though in far smaller degree, the need for compulsive self-deprecation and for the affectation of mockery and self-mockery. Even the diffuseness of his interests may have been a form of self-protection, which permitted him to claim he was not *really* a classicist or a historian (he wrote, incidentally, historical detective novels—what might be considered history in a lighter vein) or an anthropologist. Self-confessed triflers must surely have significant common characteristics.

In addition to his many other activities, Lang was also a collector of books, and it was in connection with his description of the pleasures of the book hunter that he had occasion to use the word serendipity. In his

[28] See R. R. Marett, *The Raw Material of Religion* (London: Oxford University Press, 1929).

[29] John Hepburn Millar, "Andrew Lang," *Quarterly Review* 218 (April 1913): 322–329.

[30] Ibid., p. 218.

usage, Lang made much more explicit the reading, or interpretation, of Walpole's coinage that "Vebna" had conveyed only implicitly. We resort once again to quoting Lang. Writing in one of his bibliophilic books, *The Library*, on the ways of book hunters, he said: "There is a faculty which Horace Walpole named 'serendipity'—the luck of falling on just the literary document which one wants at the moment. All collectors of out of the way books know the pleasure of the exercise of serendipity, but they enjoy it in different ways. One man will go home hugging a volume of sermons, another with a bulky collection of catalogues. . . . But however various the tastes of collectors of books, they are all agreed on one point—the love of printed paper."[31] Unlike Edward Solly, Lang paid no attention to that part of Walpole's "derivation" of serendipity that restricted the meaning of the word to the discovery of things *not* sought, to the desirable but unexpected by-products of other activities. On the contrary, Lang identified serendipity with the pleasure of the unexpected, in the sense of unpredictable, discovery of the very item that the collector *did* hope to find at the moment, much as Walpole found the Capello arms he was curious about when he had just received the portrait of Duchess Bianca Capello: "This discovery, indeed, is of that kind which I call *Serendipity*." The pleasure these discoveries give derives as much from their timeliness as from the value of the object discovered: Here, it is the good timing of the discovery that constitutes the happy accident. But if we compare Lang's conception of serendipity with, on the one hand, Walpole's "most remarkable instance" of it, the case of Lord Shaftesbury stumbling on the secret marriage of Anne Hyde to the duke of York, and on the other, Solly's discovery of the genealogy of the owners of Amyand House, Lang's kind of discovery seems to be far more like Solly's in the sense that the nature of the object to be discovered was clearly defined. What was unexpected was the "how" or "when" of the discovery, not the "what," the object of discovery itself.

These variations in meaning do not, however, exhaust the different interpretations given to the word serendipity even by the first three or four people to use it. If we focus attention on the qualities attributed to the *discoverer*, rather than on the object, or on the process of accidental discoveries, the complexity of the meaning of serendipity increases yet further. Walpole spoke of having a "talisman" and "accidental sagacity"; Solly described serendipity as "natural cleverness"; "Vebna" wavered between "a gift," or "good fortune," or "whatever else it may be called"; and Lang started out with "a faculty," which he identified with "luck" in the next phrase. This brief list foreshadows the basic variations in future interpretations. The many apparent ambiguities in the list reflect, in fact,

---

[31] Andrew Lang, *The Library* (London: Macmillan, 1881), pp. 2–3.

a single fundamental tension in the concept of accidental discovery: a tension between the attribution of credit for an unexpected discovery to the discoverer on the one hand, and to auspicious external circumstances on the other. In general, modesty demands that the writer understate the factor of "genius" or special "gift" in his own accidental discoveries, and that he stress the contribution made by external circumstances, whether the state of scientific knowledge, mystical forces, or just "luck"; in others, the writer is at liberty to explain serendipity in terms of the unusual endowments of the discoverer that permit him to "take advantage" of circumstances, or to "dip" successfully into the very place where treasure may be found.

In one respect, though, there was agreement among all the users of serendipity, from Walpole in 1754 until that time in the 1930s when the scientists first began to use the term: The accidental discovery was conceived as complete in itself. Among those with humanistic interests, that diverse group mentioned before, composed of bibliophiles, lexicographers, humorists, antiquarians, historians, and the like, the fact of accidental discovery was accepted as the beginning and the end of an experience. Whether it is due primarily to chance, to sagacity, to a "sixth sense," or to providence, was not, ultimately, of very great concern to them. "Accidents" and "coincidences" might generate amazement, even awe, but not research into their genesis. For scientists, an accidental discovery is only the initial step, stimulating them to seek explanations for the unexpected or anomalous finding. Nothing intrinsic to science is ultimately accepted as accidental, and insofar as serendipity contributed to scientific advance, it came to be seen as an integral mechanism in the process of research. Both the experience of accidental discovery (serendipity) and the fruits of accidental discovery had to be brought within the orbit of scientific rationality.

The foregoing examination of the first uses of the word serendipity by men in small literary circles sets the stage both for the variation of meanings attached to serendipity in the future, and for the limits to that variation within each social circle of users. Solly and Lang varied widely indeed in their interpretation of Walpole's explanation of serendipity, but the different meanings they ascribed to serendipity arise out of their common experience in the search for those objects valued in the literary world, whether the stuff of erudition or the treasures of bibliophilia. The great changes in the meaning of serendipity were not to come until it entered the vocabulary of the scientific circle and became adapted to the conceptual framework of scientific research.

For even a fragmentary understanding of the English literary world in the latter half of the nineteenth century, the periodical *Notes and Queries* is of considerable usefulness. It would seem to be more than mere historical accident that *Notes and Queries* was founded in 1849, when the early Victorian period had ended, for the character of the periodical was not, perhaps, consonant with the early Victorian temper. Such men as Macaulay, who felt a strong antipathy to Walpole and all he stood for, particularly to his collection of bric-a-brac at Strawberry Hill, were not unlikely to frown on a periodical whose content was so very heterogeneous. It was more than heterogeneous, for though *Notes and Queries* unquestionably had a serious purpose, it often dealt seriously with "small points," with matters that were of serious interest to only a few people, and a preoccupation with such matters was likely to offend the early Victorians who rarely let their attention be diverted from such Big Issues as morality and politics. Underlying the seriousness of *Notes and Queries*, too, was a certain note of humorous detachment, a certain awareness that what appeared important to some might nevertheless appear trivial to others. No early Victorian would have written the following little poem, which Austin Dobson (one of Walpole's first admirers in the nineteenth century) contributed to the Christmas 1882 number of *Notes and Queries*:

> In "N. & Q." we meet to weigh
> The Hannibals of yesterday;
> We trace, thro' all its moss o'ergrown,
> The script upon Time's oldest stone,
> Nor scorn his latest waif and stray
> Letters and Folk-lore, Art, the Play
> Whate'er, in short, men think or say,
> We make our theme—we make our own,—
> Stranger, whoe'er you may be, who may
> From China to Peru survey,
> Aghast the waste of things unknown,
> Take heart of grace, you're not alone;
> And all who will may find their way
> In "N. & Q."[32]

The playful tone of Dobson's poem only occasionally finds its way to the surface in *Notes and Queries*, but even when it was unexpressed, it made the editors and the readers of *Notes and Queries* refrain from scorning such matters as the history of the word serendipity.

The existence of *Notes and Queries*, then, may be connected with the shift to that more tolerant moral and intellectual climate of opinion in which Walpole and his works came to be first accepted and later admired. Walpole's word-child serendipity came, at about the same time, into at

---

[32] *Notes and Queries* series 6, vol. 6 (1882): 501.

least very limited circulation, witness the correspondence on the subject in *Notes and Queries*, the allusion to its occurrence in a "recent monthly" in 1878, and Andrew Lang's use of it in 1881. Although neither curiosity nor information about Serendip or serendipity can be considered representative of the interests of the readers of *Notes and Queries*, their give and take of such information is a fair token of the catholicity of these interests: At the very least, the readers were tolerant of such "trivialities," and sometimes they were explicitly grateful for a "note" on such an oddity as serendipity.

Chapter 4
STOCK RESPONSES
TO SERENDIPITY

*T*he cultural diffusion of a word, more particularly the word serendipity, must be traced through the social channels through which this
diffusion takes place; the changes of meaning the word undergoes as it
diffuses; and the relationship of these changes of meaning to the patterns
of thought of the different segments of the intellectual world in which
the word spreads. An essential element in the process of diffusion is the
initial response to the cultural product that offers itself, as it were, for
consumption. Such responses vary in character from out-and-out rejection of the word, to passive recognition of its existence, to active interest
in its history and continuing usage, and, finally, to the taking for granted
of both its meaning and usage. For the present, we shall deal only briefly
and incidentally with the reasons people have for rejecting or accepting
the word serendipity; the problem of receptivity will be discussed intensively later in this study.

For the first seventy-nine years after Horace Walpole told Horace Mann,
in 1754, of his coinage of that "very expressive word" serendipity, it lay
dormant. In this period, which lasted until Walpole's letters to Mann
were published in 1833, serendipity lingered in Walpole's unpublished
letter, and, unless Walpole or Mann used the word in conversation, no
one had access to it. This can now be only a matter for speculation,
because Walpole had no Boswell to record his conversations for posterity.
Except for the fragmentary records in a few journals and letters, the conversations of Walpole and his circle are entirely lost to us. But it seems
not unlikely that serendipity was a fugitive creation of Walpole's, which,
at best and for a brief period, may have entertained his friends around the
tea table.

In 1833, the letter finally appeared in print in Lord Dover's edition of the Mann correspondence, and it was printed again, in 1857, in Peter Cunningham's edition of what was then considered to be the complete correspondence of Walpole. When serendipity thus first turned up in print, there was, so far as the evidence shows, no assimilation or spread of the word: It was not "picked up" by the reviewers of Walpole's letters, nor is there any record of its having registered on any of the readers of the letters. It will be remembered that the reviewers, at least, had little use for Walpole in the early and mid-Victorian era. It was Walpole's literary style alone that met with their admiration, while his interests and values, the general turn of mind that led him to such an invention as serendipity, rather excited antipathy—serendipity was not likely, therefore, to have much resonance for such men as these when they encountered it. The sheer existence of the word serendipity in print was evidently not enough to give it the momentum necessary for diffusion, the momentum generated by some degree, at least, of positive response to it.

When the word serendipity first took hold in the soil of the English vocabulary, it won its first slight hold, in 1878, among literary people. But it may not be unfair to characterize these people as something less than the robust literary men of action of the earlier period, who turned down Horace Walpole and his coinage of serendipity along with him. This later generation of litterateurs was far more tolerant of Walpole, in some cases even devoted to him, and had more in common with him. Edward Solly and Andrew Lang and the readers of *Notes and Queries* had considerable affinity with Walpole, in that they were men with diverse and generally unrelated interests, and in that many of their concerns, for all the erudition with which they were pursued, were mere "trifles." The word serendipity fitted rather easily into their intellectual world: Its genesis constituted a nice tidbit of esoteric information, and its meaning aptly described many of their adventures in the search of such esoteric morsels.

These amateur scholars—whose avocational interests kept them on the fringes of established academic disciplines—did not come across the word serendipity, or use it, in the course of professional scholarship or research. Rather, as amateurs (by which is not meant as unproficients) in the fields in which they were at work, they had entered a field of inquiry not after rigorous training, but by virtue of their own interests and knowledge. Solly was by profession first in chemistry and then in life insurance, and it was in his avocational pursuits that he encountered serendipity. Lang, too, appears to have come across serendipity and used it in his more marginal activities as a bibliophile. Unlike the scientists of the twentieth century, such men did not discover the word serendipity, or partake of the experience that that word refers to, in their professional roles.

The marginal scholarly activities of the early users of serendipity were bound to arouse some derision, if not outright disapproval, on the part of many in the larger literate public. And since, for a time at least, say until about 1900, the use of serendipity appears to have been restricted to that group of dilettantes, some of the preciosity and "literariness" associated with that circle in which it was used may have rubbed off onto the word itself. Not for a long time was serendipity to shed the stigma of preciosity conferred by its coiner and its early users.

In this connection, it may perhaps be significant that no scholar seriously concerned with Horace Walpole, from Austin Dobson in 1890 to R. W. Ketton-Cremer in 1946, made any reference to serendipity. They must certainly have *seen* the word as they read Walpole's letters, but they neither used it nor referred to its coinage. Only Martha P. Conant refers to Walpole's coinage of serendipity in a footnote, in a study specifically devoted to the influence of Orientalism in English literature.[1] Later, the distinguished scholar-collector Wilmarth S. Lewis uses the word serendipity to describe his own experiences as a collector of Walpoliana.[2]

At the turn of the twentieth century, the ground was broken for the use of serendipity beyond the small circle of literary erudites by Wilfrid Meynell. Meynell was a literary man of action, hearty and energetic, who lived to the age of ninety-six: He was active both as an editor of magazines and as a frequent contributor to them; he was a patriarch with a family of nine children; he was a member of the rather militant Catholic minority in England; and he led the life of a kind of literary country squire at Greatham in Sussex, fostering the social interchange so necessary to the life of the intellectual elite.[3] Serendipity was "a favorite word of Wilfrid Meynell's."[4] To be sure, Meynell was known for his "freakish sense of humour" and the recurrent instance in which Meynell used serendipity in print was in each successive issue of *Who's Who*, under the heading of "recreation" (most unwhimsical Americans in this period listed no recreations at all in their *Who's Who* entries, let alone anything as whimsical as serendipity); nevertheless, Meynell seems to be the first of those whose interest in serendipity was really more than incidental whimsy, but, rather, part of a moral and intellectual outlook. The diffusion that resulted from Meynell's usage of serendipity will be described

---

[1] Martha P. Conant, *The Oriental Tale in England in the Eighteenth Century* (New York: Columbia University Press, 1908), pp. 29–31.

[2] Wilmarth S. Lewis, *Collector's Progress* (New York: Alfred A. Knopf, 1951), p. 241.

[3] Part of the spirit of this portrait of Meynell, if not the actual facts, is derived from D. H. Lawrence's description of him as Godfrey Marshall in the short story "England, My England!"

[4] See Miriam Allen de Ford (Shipley), letter to Mrs. Elizabeth Kingsley, *Saturday Review of Literature* (28 March 1942).

later; for now, it is suggested that his frequent usage not only made the word accessible to a wider circle of literary people, but also that by endowing it with seriousness below its surface frivolity, he made it seem more acceptable to them.

In the period of usage of serendipity that was inaugurated by Meynell (Walpole had been father to the word, and Solly its first guardian), roughly between 1900 and 1935, serendipity was still almost exclusively used in the literary world. It was a far more diversified literary world than the little circle that read *Notes and Queries*, for the word now found acceptance by the makers of dictionaries, by popular essayists, by a bookseller, by a writer of detective fiction, and by others. Increasingly, too, the word was used in connection with the professional activities of the members of this wider circle: as a word that actually "belonged" in the dictionary, as a word with which to point the moral of an essay, as a good name for a bookshop, or as relevant to the discovery of clues to a murder. To a limited extent, at least, the cultural history of the word serendipity entered into a new phase in which it was used for its aptness of meaning, not merely exhibited as an item of esoteric knowledge.

Serendipity has continued to be used in literary circles to the present day, and the nature of its usage in those circles has not changed very significantly. Had the diffusion of the word remained confined to the world of literature, there would seem to have been little impetus for studying the changes of meaning it has undergone, and the relationship of those changes of meaning to the intellectual orbit of its use.

In the 1930s, however, serendipity made a big leap: from the world of letters to the world of science. The key figure in the history of the introduction of serendipity to scientists was Walter B. Cannon, professor of physiology at the Harvard Medical School. Cannon enjoyed using the word frequently, not only to refer to the phenomenon of accidental discovery in science, but also as expressing a whole philosophy of scientific research. Somewhat like Meynell in the earlier period, Cannon's professional stature and his unpretentious authority did much to overcome potential hostility to this exotic and, in its origins, precious word. Unlike bibliophilia and antiquarianism, moreover, science plays an important part in the lives of the vast majority of people, and as a result of the adoption of the word serendipity by scientists, the word became increasingly diffused in more and less popularized writings about science also. From the often rather precious atmosphere of pedantry and literary erudition, serendipity now moved first to the intellectually ascetic climate of the scientific laboratory, and, as we shall see, to the gilt-edged environment of the stockholders' meeting.

Once again, we shall defer any more detailed discussion of this process of increasing diffusion, acceptance, and corresponding modification. But we may hazard at least a conjecture about one final step in the process of the increasing spread of serendipity. As this, or any other, complex word grows in usage, it gathers up many new meanings as it is refracted through the patterns of thought of different social and intellectual circles. These different circles evolve their own connotations of the word and widen its scope of meaning to the point where its original precision of denotation is lost. Thus, with such overuse, a word may become debased, déclassé, and those who first welcomed it for its freshness come to reject it as a cliché. Serendipity started out as an item of precious esoterica; it became, and still is, a useful part of both precise and expressive language. But it may, finally, be "taken up" as fashionable by esotericists, that is, by people who adopt it for fashion's sake alone. Their indiscriminate, often inappropriate, usage of serendipity may lead to its abandonment not only by those who took to the word primarily as a symbol of their esoteric erudition, but also by those whose primary interest lay in its aptness to describe succinctly a complex pattern of experience. This fate has not yet overtaken serendipity, but there are signs that it is waiting offstage.

Over a span of some two hundred years, then, the general response to the word serendipity has changed considerably: After initial rejection, the word was given very limited currency among a few literary erudites; it gradually attained wider currency in more diversified literary circles; and finally it became popular in the world of science and among those journalists who describe the progress of science to the general public. After this introductory overview we shall now examine the kinds of responses— what amounted to stock responses—that contributed to this process of diffusion.

Establishment of Personal Claim on the Word: Praise of Possession, Sense of Discovery

Claims of priority are well known in the field of science, where many inventions and discoveries have been made independently and concurrently by several scientists, and where each of these seeks what credit there is in having made the discovery first. Credit for having effected a significant discovery reassures the scientist of his occasionally doubted worth as an investigator. Although genuine independence in multiple discoveries might seem to reduce the need for claiming first arrival, priority carries with it a certainty of originality that enhances the achievement.[5] The tangible rewards of the inventor are, in part, assured by patent laws, whereas recognition for having originated a new scientific development is

---

[5] This is true, of course, beyond the field of science: However many men may fly the Atlantic alone, or climb Mount Everest, or run a four-minute mile, Charles Lindbergh, Edmund Hilary and Tenzing Norgay, and Roger Bannister will always be remembered for having done so first.

the primary social reward for the scientist. Most scientists are willing, even eager, to communicate the substance of their new knowledge to others and to share what benefits it may confer; so far as they are possessive, this is limited to the credit for "having got there first."[6]

The work of creative writers and of literary scholars rarely receives the same wide publicity as do the achievements of scientists, and, correspondingly, priority disputes in literature and literary scholarship receive less public attention, but this does not mean that anxieties about credit do not also exist in these circles. The creative writer may encounter the problem of plagiarism, an anxiety all the more harrying because plagiarism is often hard either to establish or to deny. As with some simultaneous scientific discoveries, the results of genuinely independent literary efforts may differ only slightly; on other occasions, though, such subtle differences suggest a specious "independence," which is difficult to establish, since the process of literary creation is not public in the same way that scientific work is.

In literary, or historical, scholarship, the problem of priority is again somewhat different, for with the literary detective, as with the polar explorer, it is quite literally a matter of "getting there first," wherever the significant documents may be, and of setting up a flag, as it were—publishing results. Leslie Hotson describes with wit and due measure of feeling the tribulations of the literary detective:[7]

> Suppose you have found your treasure, your Pacific Ocean, your criminal; from that moment, you must guard your find from every curious eye until you have studied it, understood it, clothed it, and are ready to claim it as your own before the world. The period of suspense is often long. For that frantic time you change from the hunter into the hunted. No longer are you the wholehearted detective, pursuing: your apprehensions make you the criminal, pursued. Dangers hedge you in on every side. Each innocent delver among the documents becomes a potential enemy. For mind you, these records are public. Anyone has a right to look at the parchment on which you have made your unannounced discovery.

Hotson tells of his own discovery of documents describing the last hours of the poet Christopher Marlowe, and of his fears that he would be scooped:

---

[6] See Robert K. Merton, "Priorities in Scientific Discovery: A Chapter in the Sociology of Science," *American Sociological Review* 22 (1957): 635–659, which advances the theory that peer recognition of the originality of discovery ("priority") provides the basic social mechanism in the reward system of science. As previously observed, it is this (implicit and explicit) competition for the honor of recognition that is the driving force of the scientific community, just as competition for profit is the driving force of the market.

[7] Leslie Hotson, "Literary Serendipity," *ELH: A Journal of English Literary History* 9 (June 1942): 88ff., n. 2.

It seemed obvious that anyone might discover this priceless document at any moment. What was I to do? Dash off a panicky letter to the London *Times* announcing the discovery, and promising to publish the full details later? Such a course would be certain to bring a raft of searchers down on the document. I took a chance, and modeled my behavior on Uncle Remus's Tar Baby. After all, if the document hadn't been discovered in three hundred years, I might risk a few months more, while I investigated the details and got the story to a publisher. If I had known then what I afterwards found out, namely that there was a keen sleuth on Marlowe's trail at that very moment, perhaps I should not have been so very brave and bold.

On another occasion, when he had discovered some unknown letters of Shelley's, Hotson was not so lucky: His secret leaked out, and only with the aid of pressure from his publisher was he able to publish first, and then hastily. "Such," he says, "are some of the risks and terrors of the serendipitist in literature."

Not only scientific inventions and discoveries, literary works, and literary historical discoveries are claimed as individual achievements, but also new words or phrases. Intellectual life requires new expressions. In the natural sciences, as we have seen in an earlier section on neologisms, new concepts, facts, and instruments require new words to designate them, and scientists are, in fact, continually coining such words. It is easy to justify the need for neologisms in natural sciences—the newly designated element, fact, or uniformity is new in the sense of having been previously unknown, and if the scientist's coinage meets with any objection at all, it is generally the objections of philologists, who disapprove of the etymology of the new word, or of the uninformed laymen who deplore technical terminology as jargon.

In the social sciences and in the humanities, new words and phrases have other functions. The use of a new characterizing word or phrase, the drastic redefinition of an old word, or the resurrection of a term fallen into disuse is an integral part of the development of new perceptions and interpretations. The social scientist must often use new terms to distinguish the systematic abstractions he makes from social behavior from the commonsense abstractions of the layman. Frequently, the social scientist is accused of using "mere jargon," that is, of dressing up commonsense ideas in obscure technical language in order that they might seem deceptively impressive, and, true or false, this charge itself vindicates the sheer importance of words in the social sciences. The humanist who is seeking to reinterpret human experience often uses new words or phrases to reexpress "old truths." Where the social scientist is accused of using jargon, the humanist is accused of preciosity.

The coinage or discovery of an apt new term, then, is often experienced as an achievement, capping the observation that it registers. In these circumstances, the inventors exhibit pride of discovery and possessiveness. So it is with the word serendipity. Walpole himself, however much he might characteristically deprecate the importance of his coinage, could not conceal his satisfaction at having invented a "very expressive word," and those who discovered the word later quite frequently indicated, more and less openly, that they felt that they had come to possess something rather special. Time and again, when they first come upon the word, it is with a sense of discovery and, then, almost of possessiveness.

Serendipity has, by now, diffused widely, but these hints of pride of possession, of early discovery, have a similar ring even when they come from quite different kinds of people. Among those who claimed to have rescued or resurrected the word was, according to Miriam Allen de Ford, Wilfrid Meynell. Meynell did not, as far as we know, write about serendipity himself, but as we just noted, Mrs. de Ford wrote to Mrs. Elizabeth Kingsley, inventor of the Double-Crostic puzzle in the *Saturday Review of Literature*: "It may interest you to know that 'serendipity' was a favorite word of Wilfred [*sic*] Meynell's (Alice Meynell's editor-husband); he claimed to have resurrected it."[8]

More than fifty years after Meynell "resurrected" the word, its "rescue" was being claimed by scientists. An anonymous writer in the trade publication of the Standard Oil Company, *The Lamp*, states that serendipity is "a word that is now being rescued by scientists from the dusty oblivion of unabridged dictionaries." The writer, who appears to identify himself with these prospectors among dusty tomes, adds: "Lately, it has begun to lend its poetic cadence to scientific journals and the published statements of research directors, because it tells how some valuable scientific discoveries have been made." Wilmarth Lewis, indeed, underwrites the claims of the scientists when he says, "The scientific world, which owes so much to Serendipity, has revived the word and given it wide currency, particularly in connection with the discovery of penicillin."[9]

Akin to those who claimed either to have rescued or resurrected the word are those who feel that they knew very early about its existence and early appreciated its worth. Hendrik Van Loon has a chapter in his book *The Arts* devoted to the archeological work of Heinrich Schliemann, which is subtitled "A Short Chapter Which for the Greater Part Is Devoted to an Explanation of the Word 'Serendipity.'"[10] He introduces the word as follows: "I want to say something about my old friend 'serendipity.'" His "old friend" is then amply defined, explained, and illustrated in the remainder of the chapter for the benefit of those many who

---

[8] Miriam Allen de Ford, letter to Mrs. Elizabeth Kingsley, p. 23.
[9] Lewis, *Collector's Progress*, p. 236.
[10] Hendrik Van Loon, *The Arts* (New York: Simon and Schuster, 1937), p. 64.

had not had the privilege of making its acquaintance. Other early initiates appear less forbearing and more aggressive than Van Loon in establishing the superiority of their vocabulary. When David Guralnik, one of the editors of the *Webster's New World Dictionary* mentioned the word in an interview with Harvey Breit of the *New York Times* (March 22, 1953), Mr. Breit reported three weeks later in his column (April 12) that "any number of letters congratulated Mr. Guralnik, with a dash of irony and a tumblerful of sarcasm, on his discovery of the word. Of course, Mr. Guralnik made no such claim; he said 'it's in the air now,' an unmistakable implication that the word existed previously." Such unwarranted accusations, even though they may have been made half in jest, show that people may come to feel that they have a stake in the possession of the word.

As serendipity crossed over from the world of letters to that of science, some of the early users had the genuine, though sometimes mistaken, impression that they were introducing something new into the language of science. Ellice McDonald writes: "While on holiday in Cuba in 1938, I found this word [serendipity] in a detective story by S. S. Van Dine and quoted it in my annual report (*Journal of the Franklin Institute*, January 1939). It seems so pertinent to the vagaries of research that many have used the word since. (*The Way of an Investigator*, Norton, 1945, devotes a whole chapter to it.)"[11] There was no way for Dr. McDonald to know that Dr. Cannon had been familiar with the word long before McDonald encountered it. Again, the word is introduced as if it had never before been heard of among scientists, by the eminent physicist Irving Langmuir, speaking at a Colloquium at the General Electric Research Laboratory (December 12, 1951). Dr. Langmuir attributes the discovery to a colleague: "Dr. Willis Whitney discovered a word which is in the dictionary—serendipity." Dr. Whitney made an "independent discovery," unaware that others, too, were discovering the same thing, and Dr. Langmuir is giving credit where he believes credit is due.

Still other users of the word believe it unlikely that anyone outside of the small group of Walpole devotees is likely to know it. Leslie Hotson, the literary historian, assumes that "lovers of Horace Walpole . . . will recall its meaning with its fanciful derivation."[12] The unwritten corollary to this assumption is that at least the devoted Walpolians, and perhaps only they, will recall its meaning. Similarly, Milton J. Rosenau, director of the Harvard School of Public Health, who himself learned of the word from Dr. Cannon, implies that it is the reading of Walpole that is likely to yield acquaintance with serendipity. He begins his presidential address on

[11] Ellice McDonald, "Notes from the Biochemical Research Foundation—Choice of Research Projects," *Journal of the Franklin Institutes* 247 (1949): p. 421, n. 4.

[12] Hotson, "Literary Serendipity."

"Serendipity" before the Society of American Bacteriologists: "Nowadays, I suppose, no one reads Horace Walpole. . . . If you pick up volume III and turn to p. 204, you will find one of his chatty letters to Horace Mann."[13] He seems to be saying, "Only a few of us old-timers know where to find good things."

Somewhat more recently, we find evidence that as serendipity has diffused, there may develop a new kind of initiate among those who have come to "possess" the word: those who are "devotees of Dr. Cannon," analogous to the earlier devotees of Walpole. A suggestion of this comes in a review of Cannon's book *The Way of an Investigator*, by Carl Binger, an eminent psychiatrist, who calls himself "a former student of Dr. Cannon's, and still a devoted one."[14] Discussing Cannon's chapter on "Serendipity," Binger remarks: "Except to those familiar with Dr. Cannon and his writings, the word serendipity will certainly be a monkey-puzzler. . . . Apparently Horace Walpole coined the word." The word *apparently* appears to have been inserted by Binger to alleviate the reader's sense of ignorance—he implies, generously, that Walpole's coinage was news to him, too. But the word *apparently*'s connotation of skepticism is an unwarranted extension of the so often laudable skepticism of the scholar: There just is no doubt about the fact that Horace Walpole *did* coin serendipity.

Finally, besides those who claim to have rescued or resurrected serendipity, those who used it before it became common, and those who discovered it independently, besides those who are lovers of Walpole, there are also some who have an innocent pride in the possession of a rare word. One of these is Walter Cannon himself, whose pleasure in the discovery of an apt word is evidently enhanced by its rarity. Cannon says of serendipity, in his article "The Role of Chance in Discovery," "The word has not had large usage. It is not commonly found in dictionaries of the English language," and he takes mischievous delight in having a friend guess its meaning. His friend's comical error ("he suggested that it probably designated a mental state combining serenity and stupidity")[15] only served to highlight Cannon's own familiarity with this esoteric word. Another who enjoys the possession of this rare word is Winifred King Rugg, a contributor to the "Home Forum" page of the *Christian Science Monitor*, whose enjoyment becomes independent of the meaning of the word. She says: "For years there has lurked on the outskirts of my imagination, or memory, a word of which I knew neither the meaning nor the

[13] Milton J. Rosenau, "Serendipity," *Journal of Bacteriology* 29 (1935): p. 91, n. 2.

[14] *Saturday Review of Literature* 28 (1945): p. 25, n. 37.

[15] Walter B. Cannon, "The Role of Chance in Discovery," *Scientific Monthly* (19 March 1940): 204.

origin. Many persons, I think, have pet words like that, which they never use, but simply keep, because the sound of them, heard in fancy, is pleasant. This word of mine chimed melodiously in the part of me that hears silent sounds. I could not have made it up, of course, and I intended one day to find out more about it. The word was serendipity."[16] When Mrs. Rugg did find the word elsewhere, in a book of Viola Meynell's and in a bit of light verse (probably by Ogden Nash), she seems slightly disappointed that others know her "pet word," which she "was not averse to letting . . . remain [her] own private mystery." For she "had reached a point where I thought the word actually belonged to me," and now "I saw the word used on the printed page by someone who was so much at home with it as to be able to turn it into dexterous light verse. Evidently the word was not mine alone." And she admits, finally, that serendipity "is not really *my* word" and that "a word that means unexpected discovery is obviously not a word to be miserly about or to try to keep as a shining trinket of one's own, even if one could."

The love of esoterica, which is only one component in the complex pride of possession of many of the users of serendipity, in Mrs. Rugg's case almost entirely excludes everything else. With most of those, however, who seek to establish some kind of claim on serendipity, their pride in esoteric knowledge is combined with their pleasure in producing a single word that recaptures a complex and recurrent experience.

Tracking Down the Meaning and History of the Word

Between the discovery and the use of a "new" word, the discoverer and user-to-be often tries to legitimize his claims to the possession of the word. In most cases, the authority of the dictionary is considered sufficient to establish claims to the proper usage of words, but if the standard dictionaries fail to list a word, knowledge of its history or etymology may be a substitute vindication of its usage. In general, it is assumed that a speaker or writer understands and can explain the words he is using; moreover, in everyday conversation or writing, it is assumed that his audience shares the understanding of his vocabulary. But the less current the words used, the less justified is the user's assumption that his audience will understand them without definition or explanation; and in cases of extreme rarity of words, the burden of proof is on the user to establish the legitimacy of his selection of the rare words, or, in other words, his genuine possession of it.

[16] Winifred King Rugg, "Home Forum," *Christian Science Monitor* 45 (16 November 1953): p. 10, n. 299.

Lately, the word serendipity has had sufficient currency that real or pretended familiarity with its meaning and history is not uncommon, and we shall deal with such familiar usage presently. But until about ten years ago, serendipity was sufficiently rare a word for its use or appearance in print to excite not only active curiosity about its meaning, but also to stimulate efforts on the part of some of those who discovered it to find out how it came to be. In the very early period of the diffusion of the word, accurate information as to its meaning and history was, paradoxically, both more and less easily accessible. It could not, until 1912, be found in any dictionary (the pertinent fascicle of the *Oxford English Dictionary* appeared in that year), but if the discoverer was sufficiently familiar with the resources of the literary world to insert a query in *Notes and Queries*, he could obtain full and accurate information about serendipity. More recently, the first step in the quest of the meaning and history of the word can be made via certain dictionaries, but it is only the first step, for the information in those dictionaries that list it is at best fragmentary and at worst inaccurate. (A later chapter will deal much more fully with the subject of dictionary definitions.)

It will be remembered that the first query about serendipity addressed to *Notes and Queries* came in 1878 (there had been the query about the "Princess of Serendip" in 1875), and the last one, so far as we know, came in 1932. The inquirer, on July 27, 1878, was one "C.," who thought he knew the meaning of the word but was puzzled by its etymology and evidently felt that, without fuller information about it, it was not eligible for use. He wrote: "Serendipity—A word coined by Horace Walpole to express the luck of a person who sooner or later obtained what he desired. Can anyone suggest any history to the word? It does not appear to have any root or etymology. Did it not more probably arise from some mere passing table talk?" C. wished, perhaps, to know whether serendipity was more than a whimsical and fugitive nonce word before he himself took the responsibility of using it. As we saw in Chapter 3, Edward Solly supplied the necessary information, and his reply to C. also carried with it a reference to an earlier note by Solly in which he explained the "Princess of Serendip" to M.N.S. C. now had full information on the root and etymology of serendipity, and should he use the word and be required to explain it, he was safe.

The next query to *Notes and Queries* came on October 31, 1903, and the curious reader was one John Hebb. Hebb had come across a shop bearing the name Serendipity Shop and he sought an explanation of the name: "A shop has recently been opened at No. 118 Westbourne Grove, with the extraordinary name of 'Serendipity Shop.' What is the meaning of 'Serendipity'? I may add that the shop appears to be intended for the sale of rare books, pictures, and what Mrs. Malaprop (was it Mrs. Mal-

aprop?) calls 'articles of bigotry and virtue.'" Hebb was not the only one
to wonder at the name of the shop. In his biography of the poet Francis
Thompson, Everard Meynell quotes a letter to him from Thompson:
"Here is another letter [writes Meynell], written before a visit to the
'Serendipity Shop': 'Dear Ev.,—This is to remind you I shall be at the
shop, whereof the name is mystery which all men seek to look into.'"[17]
Perhaps Hebb did not wish to embarrass the owner of the shop by asking
him for an explanation of its "extraordinary name"; in any case, his allusion to Mrs. Malaprop is apt, for it is by the use of "big words" that she
did not understand that poor Mrs. Malaprop made herself ridiculous, and
Hebb himself, as well as all those who sought to track down the meaning
and history of serendipity, were hoping to avoid just such ridicule. Hebb,
like C. before him, had come to the right place with his query. He received a detailed account from Colonel William F. Prideaux, both of Walpole's coinage and of the nature of the fairy tale that Walpole had read.
Prideaux, an erudite literary scholar and bibliophile (he edited S. T. Coleridge's letters and compiled bibliographies of Robert Louis Stevenson,
Edward FitzGerald, and Coleridge) had succeeded, where Solly had
failed, in obtaining a copy of *The Travels and Adventures of the Three
Princes of Serendip*, and he was happy to give the relevant information
about his find. So Hebb and the readers of *Notes and Queries* were told:
"The old folk-story quoted by Walpole occurs on pp. 9–14. The animal,
however, was not a mule but a camel, and the serendipity of the eldest
prince was displayed in his discovery that it had but one eye. The others
found out, one that the camel was lame, and the other that he had a
tooth less than he should have. They also discovered some other wonderful things about the animal, which are not too easily quotable in *N. and
Q.* They show, however, that Mr. Sherlock Holmes was anticipated by a
couple of centuries."[18] Prideaux knew nothing of the shop that Hebb had
seen—it was, we now know, a book and art shop run by Everard
Meynell, Thompson's biographer and the son of that Wilfrid Meynell
whom we have already met.

   After a long pause, *Notes and Queries* was again searched out as an
authority in 1922. By this time, the word serendipity could be looked up
in two unabridged dictionaries, but for such historical background as one
"F.M." sought, the readership of *Notes and Queries* seemed a likely repository of information. "Serendipity—Is the fairy story 'Three Princes of
Serendip' from which Horace Walpole coined the above word printed in

---

[17] Everard Meynall, *The Life of Francis Thompson* (London: Burns and Oates, 1913), p.
329.
[18] Another inquiry, evidently privately received, was referred by F.G.B. to Prideaux's
note (September 1918, p. 232).

any modern collection? If it bears another title, in Grimm or elsewhere, what is this? Has it been used by any other business except Messrs. Meynell as a trade name?" F.M. was trying to track down the fairy tale, going one step beyond tracking down the word, and his was a difficult task, for there are few copies in existence and it has not been reprinted. F.M. received some help from Edward Bensley, who referred the inquirer to Prideaux's description of his copy of the *Travels and Adventures of the Three Princes of Serendip*, and added that "a shop with the above name was opened in Museum Street, Bloomsbury, a few years ago." Bensley apparently did not know that Museum Street was merely the third home of Everard Meynell's Serendipity Shop. (The second home was in Mayfair, in Shepherd's Market.)

The American bibliophile A. Edward Newton anticipated his readers' curiosity about the Serendipity Shop, which he mentioned in one of his essays in the collection *A Magnificent Farce, and Other Diversions of a Book-Collector.*[19] He evidently took pleasure in volunteering the information that they would be anxious to have:

> Speaking of catalogues, I have just received one from a shop I visited when I was last in London, called "The Serendipity Shop." It is located in a little slum known as Shepherd's Market, right in the heart of Mayfair. It may be that my readers will be curious to know how it got its name. "Serendipity" was coined by Horace Walpole from an old name for Ceylon—Serendip. He made it, as he writes his friend Mann, out of an old fairy tale, wherein the heroes "were always making discoveries by accident and sagacity, of things they were not in quest of."

The last query stimulated by the Serendipity Shop appeared in *Notes and Queries* on March 22, 1930, and its author, Aneurin Williams (of Menai View, North Road, Carnavon) was himself a fairly frequent contributor of information to *Notes and Queries*. He asked: "Serendipity Shop in Mayfair: Can any light be thrown on the building thus named, time of erection, and the business carried on in that quarter of London?" There is no record of a reply to Williams's inquiry; perhaps he found it himself in the back numbers of *Notes and Queries*.

Once more, in 1932, Edward Heron-Allen wrote in to that periodical to ask if serendipity had ever been used to mean "luxurious delicacy and refinement of taste." The editors supplied him with the correct definition and added that the *New English Dictionary* gave no such meaning as the one suggested.

So much for the discoverers of serendipity who were sufficiently famil-

---

[19] A. Edward Newton, *A Magnificent Farce, and Other Diversions of a Book-Collector* (Boston: Atlantic Monthly Press, 1921).

iar with the resources of the literary world to satisfy their necessary curiosity about the "new" word by writing to *Notes and Queries*. Others who discovered the word, especially in the era before it found its way into dictionaries, had a more difficult time consolidating their claims to it. One man's frustrated attempt to recapture even the word itself was revealed by the anonymous "Spectator," a regular contributor of chatty columns to the American weekly magazine *The Outlook*. (All efforts to break the Spectator's pseudonym have been unsuccessful.) The Spectator made his search for the elusive word public:[20]

> Wanted—one lost word and a lost recreation! Everyone knows how elusive the right word is at times, and how we vainly attempt to find it by a dozen well worn methods of suggestion, often abandoning all to let the mind settle and, in the clearing process, come to its own. The Spectator's lost word, however, is not one of the kind which thus disappears from memory; it stays by him only too faithfully. His trouble is that he cannot 'place' it; nor can his friend to whom he is indebted for it.
>
> "I've found a new word, and a new amusement for you," said this friend the other day. "It's seradipity." The Spectator acknowledged that he didn't recognize the article and craved enlightenment. "I don't know exactly what it means," was the answer, "but it's an amusement. I know that because I found it among the *Recreations* attached to the brief biographies in 'Who's Who.' I'll show it to you." But for once the tenacious memory of the Spectator's friend failed him; he could not recall the famous man who had invented this remarkable recreation. And after glancing over a number of the 1,532 pages of "Who's Who," the Spectator also finds it not. His confidence in his friend's memory is not, however, shaken. That word, or something like it, will yet turn up.

Then the Spectator goes on to describe the incidental amusement he had in reading about the recreations of famous English people, who, unlike contemporary Americans, were willing to admit that they had spent their leisure quite "unproductively."

If The Spectator hoped that his public confession of failure to find *seradipity* might bring a reader of *The Outlook* to his rescue, his hopes were realized. To his aid came H. K. Armstrong, whose own possession of serendipity evidently gave him considerable pleasure, and who himself reported a search:[21] "I am glad to be able to assist the Spectator [see *Outlook*, for July 18]," he wrote,

> and to put him upon the track of what he has lost—and more. 'Serendipity' (not 'seradipity') is the reported recreation of Mr. Wilfrid Meynell. The word

<hr/>

[20] Anonymous, "Spectator," *The Outlook* 74 (18 July 1903): n. 12.
[21] *The Outlook* 74 (15 August 1903): n. 16.

is an attractive one, the more so, perhaps, because of the difficulty in finding
out what it means. Having searched dictionaries in vain, I was much pleased
to get the following information from Mr. George William Harris, librarian
of the Cornell University library: "Serendipity" is a word coined by Horace
Walpole, meaning the gift or luck of discovering things the finder is not in
search of. Walpole says he once read a silly fairy tale called "The Three
Princes of Serendip," who, as they travelled, were always making discoveries
by accident and sagacity of things they were not in search of, and from this
he made up the word serendipity. Serendip is said to be an Indian name of
Ceylon.

This was detailed information, indeed, and it is hard to imagine how
Armstrong came to know about Meynell's recreation.

Since the appearance of serendipity in unabridged dictionaries, rudimentary information on its meaning and derivation has been easy to find.
Of those many users who consider it necessary to tell their readers the
facts about Walpole's coinage, few or none describe their trip to the dictionary in search of these facts. Some do, however, mention the dictionary definition as the first step in a longer search for the fairy tale that
inspired Horace Walpole to his coinage. One such search was described
with much wit by Bronson Ray, a neurosurgeon at New York Hospital, in
an address to the Cornell Medical School freshman class in October
1941. Dr. Ray began by telling of how he had heard Dean Keller of the
Yale Art Department describe William Beaumont as having serendipity.
Keller had painted a portrait of Beaumont, the nineteenth-century doctor
who made important discoveries about the digestive process, for the Yale
Medical Library. "Said Keller, 'He was adventurous, industrious, intelligent, strongly liked and vigorously disliked by his contemporaries. Like
any mortal he had good and bad traits but above all he had serendipity.'
'Yes,' said I, 'He certainly had serendipity,' being somewhat uncertain
whether that was good or bad and wondering what it was that Beaumont
had that perhaps I didn't have." Dr. Ray then recalled that he had heard
Elliot Cutler, of the Harvard Medical School, use the word to designate a
desirable quality in medical students, but Cutler had not enlarged upon
it—"more or less threw it in as a high sounding word as he probably
thought became a professor from Harvard."

After his second encounter with the word, Dr. Ray decided to find out
what it meant:

The dictionary says the word was coined by Horace Walpole and I became
interested in running down its origin. I found the workers in the Public
Library courteous and passing pleasant in listening to my desire to learn
more of the origin of this word but either the funds for employing a lexicolo-
gist at our building have been diverted, or he inhabits one of the far corners

of that mysterious building. It was almost by accident, therefore, that while thumbing through voluminous writings (many of them letters) by Horace Walpole, I came on a letter to Horace Mann, dated January 28, 1754, in which he tells of a happy discovery, and says, 'This discovery indeed is almost of that kind which I call *Serendipity*. . . . [And Ray proceeds to quote a large part of the letter.]

Now I thought the meaning of the word clear enough and the word a good sounding one, but the illustration a bit unconvincing, for anyone acquainted with mules knows very well their actions are unpredictable. A mule with two good eyes may for all I know often as not eat only the bad grass on one side of the road, and to suggest that it is a good quality for a medical student to jump to similar ill-founded conclusions is not without some danger.

So thinking that the wandering princes must have had more experiences, some of which might better justify the good qualities meant to be implied by Walpole's word, I asked the librarian for the *Three Princes of Serendip*. With a suspicious look at me she left and returned after some time saying that there was no such story but that Serendib referred to the Island of Ceylon in the time of the Tales of Arabian Nights, and I could find those stories in the Children's Room. In the Children's Room, eyed curiously by children, I perused the Arabian Nights but did not find my story. When I told the librarian there that I had not found what I wanted and asked if there were any more Arabian Night stories, she answered that there were some unsuitable for children but these were kept in the Oriental Room. Somewhat abashed and losing strength rapidly, I scouted the Oriental Room but left without my story.

This experience will serve to illustrate to you one type of investigative effort. It was perhaps poorly conceived, and now sounds like a sterile kind of endeavor, but I have learned a new respect for serendipity.

Apart from the sheer fun of hunting in the library, Dr. Ray's motivation appears to have been that common to many of those newly in possession of a rare word: the desire to know as much as possible about it in order to use it most aptly and effectively, and to be able, if necessary, to defend that usage against those who are skeptical of "high-sounding words."

A rather similar search was conducted by Winifred Rugg, the lady whose "pet word" serendipity was. Like Dr. Ray, Mrs. Rugg was misled into searching for an Arabian Nights story. She "happened to read in Viola Meynell's book about her family and Francis Thompson that her brother Everard had a Serendipity Bookshop in London. In a footnote she explained that the word serendipity had been coined by Horace Walpole "out of an *Arabian Nights* story about the Princess [*sic*] of Serendip whose adventures led them into many unexpected discoveries. Surely it was a good name for a browsing kind of bookshop." Mrs. Rugg found the word defined in *Webster's Dictionary*, and the *Oxford English Dictionary* led her to Walpole's letter to Mann. But in no edition of the Arabian Nights was she, of course, able to find the fairy tale she sought. So she concluded, "When I do find the princes—as I think I shall—it will probably be by serendipity." But unsuccessful as her quest was, she came close enough to appreciate the precious, frilly, exotic atmosphere in which Walpole lived, and so she, too, came to have a firmer grasp on "her" word, serendipity.

In one instance there is only fragmentary evidence of the elaborate tracking down that must have been done, nor does the writer reveal the history of his interest in the usage of serendipity. Howard Becker communicates only his pleasure in the word, and lists the references he has found. In a footnote in his article "Vitalizing Sociological Theory," Becker says, apropos of an accidental discovery described in the body of the text:

> Hurrah for serendipity! (Merton's research-scientific usage we know; but for an amusing reference to the theme, more recent than the tale alluded to by Horace Walpole, 'The Three Princes of Serendip,' see Ellen Burns Sherman, *Words to the Wise—and Otherwise* [*sic*] New York: Holt, 1907, pp. 115–137; Michael [i.e., Arthur Michael Samuel], *The Mancroft Essays*, London: Cape, 1939 [*sic*], 113–117, essay entitled "Serendipity"; Leslie Hotson, "Literary Serendipity," *ELH, Journal of English Literary History*, IX (June 1942), pp. 79–94; and eleven items from 1875 to 1922, in *Notes and Queries: A Medium of Communication for Literary Men, General Readers . . .* , London: Francis. Use Index.)[22]

Professor Becker's interest in serendipity remains obscure. In general, however, looking up the word and tracking down its meaning seem to be stock responses among those who not only take pride in their vocabulary but also feel it to be an obligation to express their ideas as forcefully and precisely as possible. For them, the word serendipity, when fully understood, is a valued aid to such expression.

Assumption of Familiarity with the Word

The careful choice of words is an occupational necessity for the professional scholar and writer. His success will, in large part, depend on his sensitivity to the problems of expression and communication, among them, an awareness of the problems connected with the use of words of

---

[22] Howard Becker, "Vitalizing Sociological Theory," *American Sociological Review* 19 (August 1954): n. 4.

varying currency. A precise and well-informed use of words is de rigueur for the scholar, and is similarly felt to be essential by the amateur litterateur, who emulates the scholar in his zeal for "documentation." But for people in most other occupations, such perfectionism in the use of words is optional—desirable but not necessary. The scholar or amateur word lover should know fairly accurately how much familiarity with a given word may be assumed, and he would feel uncomfortable if he risked the pretense of familiarity with a word that he did not really know. Thus, though Bronson Ray was unwilling to reveal his ignorance of the word serendipity during his conversation with Dean Keller, he soon afterward sought full information about it.

In the period when serendipity was a word with highly restricted currency among a few literary people, there was, of course, no opportunity for the layman to encounter it at all, let alone to make mistakes either about its meaning and derivation or about its currency. But as the currency of the word increased, there began to appear among the customary explanations of it some more or less garbled ones.[23] Also, some of what we might call the lay users of serendipity make pardonable errors regarding the currency of knowledge about the word itself and about the fairy tale that played a part in its invention.

One source of confusion is Horace Walpole's relationship to *The Three Princes of Serendip*, and in several cases he comes to be identified as the actual author of Christoforo Armeno's fairy tale. Vivien Leigh, the British actress, is forthright and wonderfully creative in telling where she found the word, which, like Wilfrid Meynell long before, she uses to describe her "Recreation" in *Who's Who in the Theatre*. At least, a reporter of the *Minneapolis Sunday Tribune* (March 23, 1952) has her say: "I found it in Walpole's 'Princes of Serendip.'" In another newspaper, the *Detroit Free Press*, the columnist James S. Pooler devoted his column "SunnySide" to the theme "Civilization Owes Debts to 'Serendipity.'" Pooler's folksy column is obviously more concerned with conveying some bits of folk wisdom and some "human interest" stories than in maintaining strict scholarly accuracy about the material, and so, with columnist's license, Pooler says of serendipity: "It's a 199-year-old word borrowed from an eighteenth-century writer, Horace Walpole, who wrote a fairy tale about the princes of Serendip who 'as their highnesses traveled, always were

---

[23] Until very recently, serendipity has rarely appeared without some explanation; it is interesting to note that the sensitive editor of the semipopular *Science News*, no. 25, which reprinted N. W. Pirie's article, "Concepts Out of Context: The Pied Piper of Science", in which Mr. Pirie used serendipity, and which had originally appeared in the *British Journal for the Philosophy of Science* 2 (February 1952): 269–280, added an explanation of serendipity for the benefit of the readers of *Science News*, though Pirie had considered any explanation unnecessary for his scholarly audience.

making discoveries by accidents, and sagacity, of things they were not in quest of.'" Even Harvey Breit's correspondents who claimed to have known serendipity so much longer than David Guralnik appear to have had a long rather than an intimate acquaintance with the word, for, according to Mr. Breit. "Most correspondents . . . located it either as coined or popularized by Horace Walpole's 'The Three Princes of Serendip.'"

If Walpole has won some unexpected publicity as a result of his coinage of serendipity, and, on occasion, of his alleged authorship of *The Three Princes of Serendip*, this fairy tale itself has become much more widely known than its bibliographic history would warrant. Although the book may have been reasonably accessible during the boom of interest in the Orient in the eighteenth century, what Edward Solly and Col. Prideaux said of it in the late nineteenth and early twentieth century is still true today: Copies of the fairy tale are very scarce and very few people have actually seen or read a copy. (We know of only four libraries in this country that have copies: Harvard College, the University of Chicago, Newberry Library, and the University of Michigan.) But as serendipity has diffused beyond that small circle of literary scholars and bibliophiles, who, presumably know about the limited accessibility of the book, people in other social circles who have come to use serendipity have assumed that not only they themselves but many others too really "know" the tale of *The Three Princes of Serendip*. And so they attribute to the three princes adventures that princes endowed with serendipity should have had.[24]

In an interview with a feature writer from *Fortune* magazine in the fall of 1946, Frank Rieber, "one of the most successful inventors in the U.S." (as the article describes him), thus summarized the nature of his work: "I'm like those three mythical princes of Serendip, who went traveling abroad," he muses. "Everywhere along the way the princes encountered new and useful things for which they had not been searching at the time."

George Merck, head of the pharmaceutical company Merck and Co., reflected with similar cogency, and incidental inaccuracy, in an interview recorded in *Time* (August 12, 1952): "Remember the story of the Three Princes of Serendip who went out looking for treasure? They didn't find what they were looking for, but they kept finding other things just as

---

[24] We do not mean, merely, that these people take Walpole's word for it when he says the princes made happy accidental discoveries, though actually the princes were skilled detectives who reconstructed events from a few well-observed clues. Even so distinguished a scholar as Wilmarth Lewis takes Walpole at his word when he says, "looking for one thing and finding another, which is what the princes of Serendip were continually doing" (*Collector's Progress*). But those who assume general familiarity with the fairy tale are not relying on Walpole's illustration from the story for their accounts of the princes' activities.

valuable." (*Time* thought it best to add an explanatory footnote: "An old Arab tale which inspired Horace Walpole to add a word to the language in 1754.") Two years later, Merck and Co.'s advertising copy writers made use of the philosophical remarks of the president of the company: a full-page Merck advertisement in *Business Week* (Christmas 1954) shows three Arabs riding on camels, and part of the copy reads: "No field of human endeavor illustrates better than chemistry the story of *The Three Princes of Serendip*. These princes went out searching for treasure. They never found what they were looking for, but they found other things just as valuable." There follows a case of a discovery in chemical research in which the by-product of an unsuccessful research venture turned out to be of great value.

Another advertisement by a pharmaceutical house, this one by the Pfizer Company, took the form of a two-page "article" in the *Journal of the American Medical Association* (July 17, 1954) titled "Serendipity: The Happy Accident." The author was thoroughly familiar with the phenomenon of serendipity in science, but his introduction showed that he assumed a similar and general familiarity with the *Three Princes*: "It is told how the 3 princes of Ceylon (or Serendip) always met with good fortune, whatever they undertook and wherever they went. So impressed was Horace Walpole that in 1754 he invented *serendipity* as a name for incorrigible good luck such as theirs." Of course, this was neither what impressed Walpole nor what the princes of Serendip did—but for the purpose of the article it does not matter very much, just as it in no way affects the main purport of the musings of Frank Rieber and George Merck.

Less common than the assumption that the obscure tale *The Three Princes of Serendip* is well known is the assumption that serendipity is a well-known word. Only lately, as we have said before, has the word been used without an accompanying explanation. But as far back as 1918, Arthur Michael Samuel, later Lord Mancroft, began his essay "Serendipity" in the *Saturday Review* (May 25, 1918): "When Horace Walpole coined the excellent word 'Serendipity,' it was straightway adopted by a grateful country." In part, certainly, this was said for its emphatic effect— Samuel wanted to indicate how grateful *he* was for Walpole's word, and to suggest, perhaps, that Walpole deserved nationwide gratitude for his coinage. (Samuel's style, generally, is somewhat energetic.) In part, also, possibly, Samuel overrated the general currency of the word because of the familiarity of his literary friends with it. In any case, it made a good opening line for an article—at the price of a little accuracy, Samuel enticed his audience into reading on.

Yet other errors have been made by those more interested in the aptness of serendipity or in the eye-catching value of the word, rather than in the details of its history. Like Rieber and Merck, Joseph Rossman in his *Psychology of the Inventor* is interested in the fact of accidental discovery. He does not consider it to be an important factor in invention, and so he adds, as if incidentally, to his remarks to the subject: "This lucky find is termed 'serendipity' after the Persian god of chance."[25] The mysterious "Persian god of chance" is an incidental error in an already incidental sentence. C. Norman Stabler, who writes a column on the stock exchange for the *New York Herald Tribune,* also has only a secondary interest in serendipity. In his column on September 24, 1953, he uses the word as a subheadline and says casually: "If your dictionary is old enough, you will find it defines 'serendipity' as the art of finding things you weren't looking for." Once having caught his readers' attention, he tells them they are going to need some luck in the current bear market. Stabler knew the word had been coined two hundred years earlier, and blandly assumed that it must be hidden away in *old* dictionaries.

These various errors and false assumptions about the history and currency of serendipity have been made by people from the world of advertising, from the stage, from journalism, and from science and industry. George Merck knows his pharmaceutical industry, Vivien Leigh knows her parts to perfection, Norman Stabler cannot afford to make crude errors of fact about the stock market, advertising copy writers know how to catch the reader's attention, and chatty columnists, whether literary or folksy, know how to hold their readers' interest, and while each of them may find in serendipity a word of incidental usefulness, none of them needs make it his "business" to be accurately informed about it.

Pleasure in the Experience of Serendipity

It is redundant to say that any discovery of good things is a pleasant experience, but it can be noted that the pleasure of the experience is enhanced if the discovery is unexpected. Corresponding to the enhanced terrors of unexpected ill fortune are the joys of unexpected good luck. Except for the gambler by choice, uncertainty tends to be menacing, and when out of uncertainty and the absence of control there emerge good things, they are doubly welcome—they suggest that the gods are smiling. Thus serendipity is a particularly pleasant experience, the more so if the discovery made is a significant one, but the context of pleasure surrounding even small accidental discoveries endows them, perhaps, with a kind of significance that they do not intrinsically merit. Moreover, the pleasure of the experience rubs off, however intangibly, on the word itself and

---

[25] Joseph Rossman, *The Psychology of the Inventor* (Washington, D.C.: W. F. Roberts Co., 1931), p. 120.

undoubtedly accounts, in part, for the delight some people take in know-ing and using it. The pleasure is contagious—not infrequently one tale of serendipity sets off other such tales, as if to increase the general enjoy-ment of unexpected good luck.

The joys of serendipity have been described in a variety of ways, by people in very different walks of life. Andrew Lang long ago (1881) spoke of himself and his fellow bibliophiles when he said: "All collectors of out of the way books know the pleasure of the exercise of serendipity but they enjoy it in different ways. One man will go home hugging a volume of sermons, another a bulky collection of catalogues."[26] Ellice McDonald spoke for many scientists when he said: "Fortunate indeed is the laboratory which has serendipity." Gerald Johnson, a well-known journalist, has a chapter titled "Serendipity and the Commoner" in his book about the 1930s, *Incredible Tale*,[27] in which he describes some of the joys that the common man experienced along with his many woes. A source of joy, especially, were the unexpected boons that resulted from the social planning of the New Deal:

> But the New Deal brought into our national life another novelty extremely difficult to describe, because it was unplanned, unexpected and unauthorized by the New Dealers, and is by its nature intangible and elusive. It is possible, indeed, that the great and important never perceived it at all, but the little fellow, the common man, the undistinguished citizen saw it vividly, and it was one of the features of the regime that enchanted him. . . .
>
> It can be argued plausibly that one thing about the New Deal that fasci-nated the common man was its serendipity. Through it, he was always dis-covering interesting, or amusing or delightful things, while hunting for something quite different; and this tended to keep him keyed up and expec-tant. Sometimes what he discovered was a personality; sometimes it was an institution; sometimes a lifting of his mental horizon; but always it was something by which he was astonished or pleased.

One such case of serendipity reported by Johnson concerned the joys as well as the profits that many youths derived from their work in the Civil-ian Conservation Corps: "We were looking for something else, and the profit was picked up was simply a pleasant thing that we ran across acci-dentally. It was so much velvet; and picking up velvet is one delightful form of serendipity." David Guralnik expressed the emotional climate that surrounds serendipity when, for Harvey Breit's readers, he defined it as "the opposite of accident-prone"—serendipity does have the opposite of the unhappy connotations that "accident" commonly has. If a "recre-

---

[26] Andrew Lang, *The Library* (London: Macmillan, 1881), pp. 2–3.
[27] Gerald W. Johnson, *Incredible Tale: The Odyssey of the Average American in the Last Half Century* (New York: Harper, 1950).

ation" is taken literally as a revitalizing experience, Wilfrid Meynell and Vivien Leigh could, indeed, have hardly chosen a better recreation than serendipity, and Howard Becker expressed the sense of pleasure surrounding happy accidental discoveries when he exclaimed: "Hurrah for serendipity!"

Instances of serendipity are among the experiences that people like to remember and report, as they happened both to themselves and to others. Encountering the word serendipity may serve as stimulus for describing such happy accidental discoveries. As far back as 1881, "Vebna" responded to Edward Solly's explanation of serendipity in *Notes and Queries* by describing his accidental discovery of an elusive quotation from Cicero. Again, when Mancroft's article on "Serendipity" appeared in the *Saturday Review* in 1918, a Mr. Edward Hodge from Cape Town, South Africa, was moved to write a letter saying: "Among the lighter articles I have particularly enjoyed 'Serendipity' . . . [w]ith its many interesting instances from Archimedes to Elia." After correcting some of Mancroft's facts, Mr. Hodge proceeded to give some examples of serendipity from the field of mining engineering, which was his profession. The appearance of Winifred Rugg's article on serendipity in the *Christian Science Monitor*, moved P.J.H.-H., the editor of the Home Forum page on which Mrs. Rugg's article appeared, to describe a recent experience of his own: "I had an excellent example of serendipity myself just the other day. I was searching a magazine for a reproduction of a Gainsborough portrait, which I believed to be there; instead, I came upon a map showing the distribution of the Nine-Banded Armadillo. We are all too familiar with maps of population, food or rainfall distribution, but it is few among us who are privileged to look on a map showing the distribution of the Nine-Banded Armadillo. That I am one of these few, I owe to serendipity."[28]

Serendipity is obviously something pleasant to have and to tell about, and the pride that, as we noted earlier, people appear to take in the possession of the word may be aroused not only by its aptness and its rarity, but by the aura of good luck with which its possession associates them.

Familiar Usage of Serendipity (Absence of Any Response)

Whereas during the first seventy-five years or so of the usage of serendipity its occurrence was almost invariably accompanied by explanations of it or by curiosity about its meaning and history, in the last few years it

---

[28] *Christian Science Monitor* 45 (16 November 1953): 10.

has become so thoroughly domesticated in certain circles that explanations have of course disappeared.

Familiar usage of this kind is found most often among natural and social scientists, but occasionally also among literary people. (In part, this is a consequence not merely of the familiarity of the scientists with the word, but of their lesser interest in anything apart from the denotation of words.) As early as 1937, to be sure, Carolyn Wells, a writer of numerous detective stories, a bibliophile, and a collector of anthologies on humor, used serendipity without explanation in her autobiography, *The Rest of My Life*. In a chapter titled "A Little Learning," she deplores the general feeling inculcated by Alexander Pope that a "smattering" of knowledge is objectionable: "This is not always true. If one hears the word *serendipity* mentioned, is it not better to have an idea what it means even though one may not often have occasion to use it?"[29] But though Carolyn Wells knew what serendipity meant, she was not assuming any familiarity with the word on the part of her readers—on the contrary, she tossed the word out in this fashion because she assumed they probably had no idea what it meant. Another such usage of serendipity, in literature, as something of a toy word occurs in Rumer Godden's novel *A Breath of Air*,[30] in which, for no apparent reason, the cook is called Serendipity. A strange literary use is Mary McCarthy's in a "Letter from Portugal" in the *New Yorker* (February 5, 1955). Miss McCarthy does not define the word, and whatever meaning it has for her remains private. Describing an unusual story told to her by a Portuguese acquaintance, she says: "The plot of the story was familiar; I had heard it on occasion in my own country. No doubt these experiences do befall people who, as it were, act as lightning rods for them; it is a case of serendipity."

The only occurrence of serendipity without explanation (in literary circles) that is in keeping with the commonly accepted meaning of the word is in an anonymous review in the *Times Literary Supplement* (January 22, 1954) of Ralph Kirkpatrick's biography of Domenico Scarlatti. The reviewer writes: "Until the publication of this book all that was known of the younger Scarlatti amounted to a handful of snippets of eighteenth-century musical gossip by Burney and others. By an exercise of serendipity which amounts almost to genius, Mr. Kirkpatrick, temporarily checked in the hunt for Scarlattiana in Madrid, opened the telephone directory, perceived there the name of Scarlatti, made a call, found that the subscriber was a direct lineal descendant of Domenico's second son

---

[29] Carolyn Wells, *The Rest of My Life* (Philadelphia: J. B. Lippincott Company, 1937), p. 34.

[30] Rumer Godden, *A Breath of Air* (New York: Viking Press, 1951).

and was soon enabled to photograph the considerable collection of Scarlatti family documents."

Among scientists, natural and social, such usage as that of Kirkpatrick's reviewer, which takes for granted both the writer's and the reader's familiarity with the word, is somewhat more frequent. As we noted earlier, N. W. Pirie, in an article in the *British Journal for the Philosophy of Science* (February 1952), writes of the false assumption that "all scientists are fully equipped with serendipity." In an American scientific publication, too, serendipity occurs thus unexplained: In an article by L. D. Hamilton, S. H. Hutner and L. Provasoli,[31] the authors remark, "It may be mentioned that the immediate motive for isolating and studying these organisms [protozoa and algae] was generally interest in them for their own sake; their occasional applications to analysis represent instances of serendipity." Another instance of such usage is that by Ernest Jones in his biography of Freud.[32] Chapter 16 opens in this fashion: "By general consensus *The Interpretation of Dreams* was Freud's major work, the one by which his name will probably be longest remembered. Freud's own opinion would seem to have agreed with this judgment. As he wrote in his preface to the third English edition, 'Insight such as this falls to one's lot but once in a lifetime.' It was a perfect example of serendipity, for the discovery of what dreams mean was made quite incidentally—one might almost say accidentally—when Freud was engaged in exploring the meaning of the psychoneuroses." Jones obviously assumes that most readers will know what serendipity means; only for the few who do not does he insert an unobtrusive explanation.

Among social scientists, too, explanations of serendipity now seem superfluous. Some sociologists now use the word with only a footnote reference to the explanation by the sociologist who introduced serendipity to his professional colleagues. Irving Spaulding's article "Serendipity and the Rural-Urban Community,"[33] in which the word occurs only in the title, and Svend Riemer's "Empirical Training and Sociological Thought,"[34] in which it is said that "we rely increasingly upon serendipity for the advancement of our science," have such footnotes.[35] In Allen Wallis's re-

---

[31] L. D. Hamilton, S. H. Hutner, and L. Provasoli, "The Use of Protozoa in Analysis," *The Analyst: The Journal of the Society of Public Analysts and Other Analytical Chemists* 77 (November 1952): 618–628.

[32] Ernest Jones, *Sigmund Freud: Life and Work* (London: Hogarth Press, 1957), p. 305.

[33] Irving Spaulding, "Serendipity and the Rural-Urban Community," *Rural Sociology* 16 (March 1951): 29–36.

[34] Svend Riemer, "Empirical Training and Sociological Thought," *American Journal of Sociology* 59 (September 1953): 112.

[35] Spaulding refers to R. K. Merton, *Social Theory and Social Structure* (New York: Free Press, 1949), pp. 12 and 98; and Riemer to R. K. Merton, "The Bearing of Empirical

port on the 1953–1954 Program of University Surveys of the Behavioral Sciences sponsored by the Ford Foundation (February 1955), the explanation of serendipity is almost as inconspicuous as that given by Ernest Jones. Wallis says, "If, on the other hand, the progress of science is more continuous and mostly the product of serendipity—the knack of spotting and exploiting good things while searching for something else—then it is important that . . ." Walpole and *The Three Princes of Serendip* have dropped out of the picture entirely as far as these scientists are concerned.

It would be an exaggeration to say that the word serendipity has by now become commonplace. Its currency has increased considerably, however, from the time when information about its meaning and history was hard to come by and diligently sought. Now, this information, even in abridged form, is sometimes no longer considered necessary proof that the user really knows what he is talking about when he confronts his readers with so "odd" a word as serendipity. The pleasures of the "exercise" of serendipity have not diminished—on the contrary, familiarity with the word that describes this experience so well may have increased the awareness of these pleasures for some people—and except for the lovers of esoterica, the pleasure of owning so "expressive" a word is still there. But with increasing use, the exchange of detailed information about serendipity is surely dwindling.

Research upon the Development of Social Theory," *American Sociological Review* 13 (1948): 505–515.

Chapter 5
THE QUALITIES
OF SERENDIPITY

*T*he responses to serendipity we have considered so far might easily result from any encounter with this unfamiliar, etymologically puzzling, and seemingly expressive word. Responses such as that possessiveness that makes the word peculiarly one's own, the prompt urge to search out its origin, and the multiplying of ideas about its current usage probably occur quite often among people with a certain self-consciousness about their vocabulary. Should the reactions to the word seem strong, they might nevertheless be proportionate to the degree of the word's unfamiliarity, its puzzling aspect, and its usefulness.

Serendipity is, of course, a coined word. Now coined words need not be puzzling to those with even a modicum of etymological knowledge. But such knowledge can be put to use only for words that have been coined on sound etymological principles. Even when they are a bogus coinage, their components can often be easily identified—sometimes, perhaps, more easily by the college-educated man in the street than by the rigorous philologist. (The latter might understandably be perplexed when confronted, for example, with *heliport*.) Among other word coinages, those most likely to puzzle are the common words that have been derived from proper names or place names. Some of these, to be sure, are so common that their meaning is taken for granted and their derivation practically never comes up for discussion: A cardigan is a cardigan, and who cares how it got its name; Roquefort is a cheese that may come either from Wisconsin or be "imported"; and if anyone ever wondered about the name of the gardenia, he probably decided it had something to do with a garden, not with the eighteenth-century Dr. Alexander Garden, who developed it. Other names-into-words are far less familiar, and if the name on which they are based is obscure, it is almost impossible to puzzle out their meaning. How could one stress that a Samoyed is a dog, let alone that its name is short for Samoyed dog, "the Samoyeds being a race of Mongols living in Siberia"? Or that tupperism is "the principles and practice of *tupperian* or even *tupperish* philosophy, which tends to *tupperize* everything to the trite moral and mental mediocrity of *Proverbial Philosophy* (1st ser., 1838, 2nd, 1842, 3rd, 1867, 4th, 1876)" by the "complacent preacher in so-called verse, Martin Tupper (1810–89)"?[1] Words such as these have a kind of history that goes beyond the etymological transformations ordinarily of interest to philologists. However, Eric Partridge, a lexicographer who enjoys his eccentricities, has compiled a sizable dictionary titled *Name into Word*[2] (in which we found the histories of *Samoyed* and *tupperism*), and scholars such as Ernest Weekley and Bruno Migliorini have studied these works intensively.

However obscure the proper name or place name from which a common word derives, people who encounter such a mystifying word are apt to assume that in some way or other it does have a meaningful history. The word will sometimes be assumed to have some real and intimate connection with the characteristics of a person or place, or to conform to obscure linguistic principles, or, as a last resort, to make onomatopoeic sense.

Confronted with *tupperism* and even without access to Partridge's dictionary, some of us might infer the existence of Mr. Tupper, although we could never guess what particular characteristics of the man or town were incorporated in tupperism. After all, we might reason, there is Micawber and *micawberism*; perhaps Tupper, too, is a Dickensian character—it has a decidedly Dickensian sound—and *tupperism* will then refer to Mr. Tupper's most characteristic trait. Or we might try, in the same half-tutored way, to relate *Samoyed* to a samovar. Or, to return to *tupperism*, we might try to make something of the onomatopoeic significance of the syllable *tup*. Whatever the grounds for our guess might be, we would *not* assume that the strange word was an arbitrary collection of consonants and vowels. Implicitly, we would make an assumption that Professor Migliorini has made into an explicit proposition, namely that "to say that for a word to take hold, the accompanying factors must be several, is the same as saying that it must be as little arbitrary as possible. Consequently, rare

[1] Martin Farquhar Tupper, *Proverbial Philosophy: A Book of Thoughts and Arguments* (London: Joseph Rickerby, 1838 and 1842).
[2] Eric Partridge, *Name into Word: Proper Names That Have Become Common Property* (New York: Macmillan, 1950).

indeed are the cases of words formed really *ex nihilo*, without reckoning upon the traditions fixed in national languages."[3]

The proposition puts at least one category of responses to serendipity in a particular and revealing light, namely, the category of responses that are little more than exclamations at the word's etymological irregularity. The curiosity it evokes, the expressions of surprise and puzzlement, are certainly, in part, conventional responses to its unfamiliarity. But the sense of its being odd, strange, or even absurd may stem from the fact that serendipity is that rare thing—a word created almost ex nihilo. Not only have most people never heard of the archaic word Serendip, but the connection between serendipity and Serendip is a tenuous one indeed.

To be sure, Horace Walpole explained his "derivation" of serendipity from the fairy tale of *The Three Princes of Serendip*, but that the provenance of the princes happens to be Serendip is altogether incidental to the meaning of serendipity. With nothing else to go on, Dr. Cannon's friend could venture the facetious guess that the word sounded like a cross between *serenity* and *stupidity*, the most revealing feature of that guess being, perhaps, that the friend could not retain the sequence of letters in serendipity long enough to realize that *d* and *p* in *stupid* come in reverse order to those letters in serendipity. At first sight, serendipity seems to some people like a collection of nonsense syllables.

At any rate, many of those who encounter the word serendipity seem compelled to comment on its oddity. Some of them do so even when they are thoroughly familiar with its meaning, as if to cushion the shock for others about to meet it for the first time. Earlier we mentioned John Hebb's writing in 1903 to *Notes and Queries* for an explanation of the word: "A shop has recently been opened at No. 118 Westbourne Grove with the extraordinary name of 'Serendipity Shop.'" Hebb's attitude was a little like that of a lay observer reporting to a scientific authority his astonished discovery of what appeared to be an aberration of nature. He had obviously never seen a word of this kind before. But the writer who condensed for the *Literary Digest* (February 5, 1916) a piece written by W.M. in the *London Times* was at least familiar with W.M.'s maldefinition—"An English writer thinks [Horace Walpole] meant the 'mania for collecting' turned into a vice"—and nevertheless finds the word rather odd.

His article begins: "Serendipity is a curious word of which Horace Walpole is said to be the coiner." An explanation of its meaning evidently did not make the word any less curious. Still, it might be said that, after all, this writer was encountering serendipity for the first time, and it was

---

[3] Bruno Migliorini, "The Contribution of the Individual to Vocabulary," in *Saggi linguistici* (Florence: Le Monnier, 1957), p. 328.

his unfamiliarity with it that made it seem curious. But then what are we to make of Hendrik Van Loon, when he introduces the reader to "my old friend 'serendipity,'" and immediately tells the reader that he may find "that strange-looking word" in the dictionary.[4] Van Loon may, of course, be identifying himself with his reader's assumed reaction to the word, but his ability to do so indicates that even old friendship with serendipity has not allowed him to forget how strange the word must appear to others.

If Van Loon was aware that the word is strange looking, it is not inappropriate that to Mrs. Beatrice Hunter of Evanston, Illinois, the word should initially have appeared "rather silly." Mrs. Hunter wrote a letter to Mrs. Elizabeth Kingsley, who used to construct the Double-Crostic puzzles in the *Saturday Review of Literature*, after Mrs. Kinsley had mentioned the word serendipity in the column she ran along with the puzzles: "A few years ago a copy of the Marlborough School paper, Los Angeles, was sent to me. Then later I found it in Van Loon's 'History of Art' [*sic*]."[5] Van Loon's use of the word probably made the word a little more respectable, a little less absurd.

"Extraordinary," "curious," "strange-looking," "silly"—all these adjectives are used to describe the bizarre quality that serendipity has for many of those who use the word, often unaffected by the degree of their familiarity with it. Even habitual usage does not, evidently, help very much to diminish the self-consciousness that is felt about the unusual structure of the word—it is difficult to take completely for granted. The descriptions of serendipity by other users follow the same pattern. When Ellice McDonald found occasion to use the word in a report to his fellow biologists, he referred to it as "this strange word."[6] Gerald Johnson has a chapter in his *Incredible Tale* (1945) titled "Serendipity and the Commoner," and though he knows something of the history of serendipity, he speaks of "[borrowing] an expression, and a staggering one, from Horace Walpole." Again, in a news feature in the *New York Times* (August 21, 1955) the author expresses his sense of something alien about the word serendipity by calling serendipity "the phantom word."

The oddity of serendipity has, of course, a special impact on those who are particularly sensitive to language. In this connection Nora Archibald Smith describes a revealing episode: in contrast to Mrs. Hunter's first dismissal of the word as "rather silly," Mrs. Smith tells how "an eminent British author, when asked to set down his likes and dislikes in one of these biographical booklets too often presented to literary lights by their

[4] Hendrik Van Loon, *The Arts* (New York: Simon and Schuster, 1937), chapter 5.

[5] Beatrice Hunter, letter to Elizabeth Kingsley, *Saturday Review of Literature* (July 24, 1943).

[6] *Journal of the Franklin Institute* (January 1939).

admirers, alleged his favorite occupation to be 'Serendipity.' The novel and tantalizing term immediately caught the attention of those curious [about] words."[7]

For some, then, the word is odd-silly, and for others odd-interesting, while yet others might be impressed with it for oddity's sake. Christopher Morley may be one of the last group, for in some brief reminiscences about serendipity, Joseph Henry Jackson, the literary editor of the *San Francisco Chronicle*, writes (in a private letter to us): "I remember very well the last time it gained brief but wide currency; it was just before World War I, when I was in college, and the man who brought it into special fashion was Christopher Morley, who picked it up, as he always liked to pick up such oddities."

The most literal reference to the alien character of serendipity comes from one of us, who footnoted his usage of serendipity in the text of an article with remarks about the currency of this "outlandish term." And when this article was reprinted in a book,[8] he added further remarks on the growing currency of the word, "for all its etymological oddity," and wondered about the sources of the apparent cultural resonance of this "contrived, odd-sounding, and useful word." It was no longer a case of serendipity being odd-silly or odd-interesting; rather, it was the beginning of the sociologist's interest in the appeal of such a word as serendipity, and one aspect of that appeal is, perhaps, its oddness, its seeming descent upon the language ex nihilo.

Although serendipity was not, in fact, constructed altogether out of nothing, although there was a measure of methodical wordcraft as well as of whimsy in Walpole's creation, the element of whimsy has for some people been its most striking feature. In some cases, we surmise, such whimsicality was received with the same pained distaste that is frequently encountered by those who take pleasure in punning. We have seen earlier in this study that the Victorians frowned on Walpole and all his works for their frivolity, and such a word as serendipity may very well have epitomized all that they disapproved of in him. More lately, his frivolity has been tolerated and even enjoyed, yet even among the Walpolians who enjoy him there have been some who did not take him seriously. Odell and Willard Shepard's book, *Jenkins' Ear*, for example, cleverly satirized Walpole in all his petty preoccupations. So too, serendipity, the creature of frivolity and whimsy, has on several occasions become the victim of playful indignities. It is not so much the "light" essays about serendipity

---

[7] *The Bookman* (May 1925).

[8] Robert K. Merton, "The Bearing of Empirical Research on Sociological Theory," in *Social Theory and Social Structure* (Glencoe, Ill.: Free Press, 1949), p. 256, n. 5.

that we have in mind here (such as Mancroft's essay or the *Cleveland Plain Dealer*'s editorial "All Good Weathermen Should Serendipidize," August 22, 1955), nor even the literary fillip, half-irrelevant to the subject in hand, that the word is occasionally permitted to give to dry, scholarly papers; rather, we are dealing with the out-and-out playful usage, the wordplay, that has been inspired by serendipity.

We have already seen that for people sensitive to etymology, serendipity must look odd, and that its oddity may then be identified with silliness. It may also be the stimulus to flights of fancy. As Leslie Hotson put it, serendipity is a "fanciful derivation," and it has begotten further departures from strict and purposeful "sense." Vivien Leigh was simply having some fun when she listed her recreations in *Who's Who in the Theatre* as "Flower arranging and serendipity."[9] According to the syndicated interview with her by Cynthia Lowry, "Actually, it wasn't quite fair," she confessed, "but I did it because I was tired of the usual things. I do like arranging flowers, but serendipity didn't exactly fit." Tongue in cheek, Miss Leigh just threw in serendipity for good measure; and it seems likely that "R.", who wrote to *Outlook* in order to "correct" H. K. Armstrong's explanation of serendipity, was following the same facetious track. "R." indulged in the following bit of nonsense: "Your correspondent [Armstrong] is all wrong about *serendipity*. Why, 'seren' in Welsh means star, and 'dip' in its root means little, like 'tip' of a tail; and 'ty' means *house*. Therefore, *serendipity* means 'the star of the little house.' See! If serendip has an Indian derivation, that only goes to prove a family relation, that is all."[10]

In several instances, also, serendipity has inspired the writing of light verse, of amusing little pieces that attempt to make the most of the fanciful, exotic, or simply odd qualities of the word. Betty Bridgman recaptures something of the romantic and fantastic background of the word in her humorous verse:

> Serendips stroll through dictionaries,
> Cookbooks and itineraries.
> They trade the cow for the magic bean
> And come on treasures unforeseen.[11]

Such a far-fetched, fanciful word seems to have led Mark A. DeWolfe Howe, also, to humorous fantasies about odd intellectual adventures, as witness the following jingle:

---

[9] *Who's Who in the Theatre* (London: Pitman, 1952).
[10] *Outlook* (12 September 1903): 140.
[11] *Christian Science Monitor* (16 November 1953).

A classicist in the Antipodes
Devoted his life to Euripides.
　His bottle and jigger
　Played hob with his figure,
But gave him some sweet serendipities.[12]

The contagion of the whimsy in serendipity spread also to some schoolboys at Phillips Exeter Academy in New Hampshire: William S. Bayer II and William G. Lambrecht concocted an advertisement for a clothing store that is a combination of the utilitarian nonsense used by Madison Avenue to catch the eye of readers and of sheer nonsense, deprived of all utility. As a preparatory step, they inserted a small announcement in the *Exonian* (February 10, 1955), saying: "Join the Serendipidous [*sic*] Society." When this produced the desired response (e.g., a letter of inquiry to the *Exonian* signed by four other students), they ran the following poem:

Serendipity

By Prince Fashionplates

Hey! Where you going?

Where d'you think?

To the Serendipidous Society's first meeting.

Huh? The WHAT society? Serendipidous?

Come on, I'll show you.

What are you? A member?

Not yet. This is only the first meeting.

Where's the meeting? I'll bet it's all a boff.

(Indignantly) It is not. Come on. You can hear every-body tell about their serendipities.

What's a sippity?

That's what happens when nothing's happening and yet it's happening anyway, but not really.

What? Where are you REALLY going?

To Jovial's. There's something happening there that did happen and is now happening backwards.

Huh?

The Reversible. Prices go down to 40% off, and now they're coming back again. Only 30% off today, and you'd better hurry. Tomorrow will be too late.

Say, that's a REAL Serendipity! (adv.)

It is fitting, finally, that one of the best-known living masters of word-play, Ogden Nash, should have been inspired by serendipity to write a bit

---

[12] *Time* (29 August 1955).

of nonsense verse titled "Don't Look for the Silver Lining, Just Wait for It."[13] The verse begins:

> The rabbit loves his hoppity and the wallaby loves his hippity.
> I love my serendipity.
> Let none look askance;
> Serendipity is merely the knack of making happy and unexpected
>   discoveries by chance.

Then Nash tells of discoveries of this kind that he had made recently—finding a funny paper with something funny in it, or noting that a bore he was talking to has providentially stepped on some chewing gum—and he concludes:

> If your coat catches on a branch just as you are about to slip over a
>   precipice precipitous,
> That's serendipitous.
> But when you happily and unexpectedly discover that you don't have to go
>   to the dentist or the chiropodist,
> That's serendopitist.

Nash's pleasant abuse of serendipity is perhaps the best proof possible of the ingredient of whimsy in the makeup of Walpole's coinage.

Serendipity: An Engaging Word

It is rather difficult to conceive of a word stripped of its meaning and of all the associations to that meaning—a word-in-itself, as it were, a collocation of sounds making a sensory impact on the hearer, reader, or speaker. But it is as just such an entity that we shall now try to consider serendipity: not as a word whose aptness of meaning is pleasing, not as a word that is rare and therefore precious, not as a word that is pleasant because the pleasure of the experience of happy accidental discovery has rubbed off onto it. Rather, we shall consider what distinctive characteristics it may have, what we might call "psycho-aesthetic" characteristics, that render it attractive apart from any denotative or connotative qualities it may have. Just as it is not unfamiliarity alone that makes serendipity an "odd" word, so is it also not merely its aptness or its associations that make it, apparently, an "attractive" one.

There is testimony to the fact that people *like* the word serendipity. H. K. Armstrong said of it that "the word is an attractive one, the more so perhaps, because of the difficulty in finding out what it means." Inci-

[13] Ogden Nash, "Don't Look for the Silver Lining, Just Wait for It," *New Yorker* (6 October 1951): 42.

dentally, Armstrong reveals himself as something of an esotericist, but more important, he experiences the word simply as "attractive." Mancroft refers to it as an "excellent word," and it is, perhaps, not reading in too much to say that this superlative, "excellent," carries his appraisal of the word beyond mere approval of its usefulness. For Winifred Rugg it is a "charming word," and its charm, as we shall see, lies in large part in the sound qualities of the word. N. W. Pirie, or perhaps the editor of *Science News*, gratuitously throws in the word *pleasant* when he explains serendipity in a footnote: "A pleasant word coined by Horace Walpole." Again, after reading a manuscript in which Allen Wallis used serendipity (a report on the Program of University Surveys of the Behavioral Sciences), Milton Friedman, of the University of Chicago economics department, jotted down the comment "wonderful word—thanks" in the margin.[14] And finally, as if won almost against his will, Leslie Hotson, in his article "Literary Serendipity," labels serendipity a "sweet and insinuating word." From these expressions of pleasure alone we would conclude that the word seems to be an attractive one.

What, then, are the sources of this attractiveness? They appear to lie in certain sensuous qualities the word has, qualities pleasing primarily to the ear, but, perhaps, also to the tongue. It is not easy to pin down the psychological and aesthetic correlates of different sounds and of various arrangements of sounds. In the case of serendipity, several factors have been suggested by a linguist as contributing to a pleasing effect. The first, not in any order of importance, is the neutralization of the last three consonants of the word, so that the *d* sound lies between *d* and *t*, the *p* lies between *b* and *p*, and the *t* again lies between *d* and *t*. With this neutralization, every provision has been made for saying *dipity* quickly and easily, and it seems to make *dipity* fun to say. The second factor making for the pleasurable sound qualities of the word is the repetition of the vowels, first of the two *e*'s, and then of the three *i*'s. Third, the symmetrical rhythmic pattern of the word, with its two relatively unstressed syllables, its peak of stress on the middle syllable *dip*, and its last two unstressed syllables might also make it appeal to the ear. And last, our linguist noted that in its sounds, at least, serendipity is thoroughly English: It draws, that is, only from the reservoir of English sounds, the so-called English paradigm, and syntagmatically it is also English—that is, it has no combinations of sounds that are prohibitive in English. At the least, then, the word would be unlikely to arouse any hostility in an English audience, and for some people this familiarity of sounds coupled with etymological strangeness might have a kind of fascination.

As far as the relationship between sound and sense is concerned, this

---

[14] Personal correspondence, Professor Wallis and the author.

linguist rejected the notion that the sounds in serendipity had any particular meaningful associations. In writing, perhaps, *seren* might be reminiscent of *serenity*, but when the word is spoken, the differently placed accent would block even that association. As for the possible significance of
the *dip* in serendipity, this too was rejected as a likely common association. We wonder, nevertheless, whether Horace Walpole did not, consciously or half-consciously, choose the tale of the princes of Serendip as
the basis for his coinage because the *dip* syllable was associated with the
process of finding things wherever he "dipped" for them. Be that as it
may, there appears to be little or no sound linguistic basis for giving the
word onomatopoeic significance.

The several users of serendipity who remark on the attractive sensuous
qualities of the word are only in a general way aware of its pleasing impact on the senses; they do not enlighten us as to the specific sources of
that impact. Their comments are mostly rather general allusions to the
pleasure of hearing, or, in one case, of saying the word. Rudolf Flesch
refers to it as "that beautiful word," and even though he is presumably an
expert in the art of clear expression, he does not explain wherein lies, for
him at least, the beauty of the word. Winifred Rugg felt that the word
"chimed melodiously in the part of me that hears silent sounds," and
although she failed to look up its meaning, "once in a while it would ring
its sweet little bell." She did not trouble to analyze the bell-like quality of
the word, but perhaps it was the repetition of vowels that made it appear
to chime. Mrs. Elizabeth Kingsley was "allured by its onomatopoeic
quality," but since she does not explain why it appeared to be onomatopoeic, we may assume that this was only a vague allusion to a collocation
of sounds that pleased her. And Gladys M. Wrigley refers to serendipity
merely as "this useful word with the pleasant sound."[15] Laymen have
little experience in describing sound qualities; while they might attempt
to analyze in elementary terms a pleasing visual impact, they do not appear to have even rudimentary tools for dealing with the qualities of
something that is heard.

A very few users did attempt to describe serendipity. An anonymous
article in the Standard Oil publication *The Lamp* refers to the "poetic
cadence" of the word, and the writer may have in mind the rhythmic
qualities we mentioned earlier. Ellen Burns Sherman describes it as a
"spacious word," and again we have an inkling that she did not think of
the word merely as multisyllabic, but as having several syllables arranged
in a graceful way. And in the column "Your Health," which originates
with the Medical Society of the State of Pennsylvania and the Centre
County Medical Society, the writer, again anonymous, makes the most

[15] *Geogr. Review* 42: 511.

direct allusion yet to the ease of saying the word: "The word serendipity serenely *drips off* the tongue."[16] This is almost an echolalic image of the word.

Even for a poet, far better equipped than the layman to deal with the sound qualities of words, the appeal of Serendip has proved elusive: Serendip rather than serendipity, for this is part of the title of a poem by Anne Atwood Dodge.[17] While the sound qualities of Serendip, alone, are, of course, different from those of serendipity, the poem is revealing both for the impact on the poet's ear of the root, Serendip, and for the difficulty her "lazy wit," as she calls it, had in explaining her pleasure. The full title of the poem is "Serendip and Taprobane," Taprobane being another obsolete name for Ceylon.

> Serendip and Taprobane—
> Words as argent-chimed as rain,
> Words like little golden beads,
> Apple and pomegranate seeds,
> Strung upon a silver thread,
> Little drops of lacquer-red,
> Tintinnabular and sweet,
> Little words with crystal feet
> Running lightly through my mind.
> If my lazy wit could find
> Gilded phrases to express
> Their perfected loveliness,
> I would make a cage of words
> Where, like bright heraldic birds,
> They should strut and flaunt and preen,
> Scarlet, silver, gold and green,
> Elegantly strange and vain—
> Serendip and Taprobane.

Anne Dodge communicates her sense of the exotic, the colorful, the graceful, about these words, and also her impression that they command a kind of attention that she is not fully able to accord them—she cannot, or will not, quite do them justice. Her linking of these two names for Ceylon brings out vividly the question of the onomatopoeic significance of serendipity raised before: Could Horace Walpole as easily have coined the word *taprobanity* to express his peculiar gift, and would that word have "caught on" as serendipity did?

[16] "Your Health," *The Centre Daily Times* 53, no. 56 (7 June 1950): 4.
[17] Anne Atwood Dodge, "Serendip and Taprobane," *Harper's* (August 1927); reprinted in *Literary Digest* (13 August 1927).

Many questions remain unanswered about the sources of the attractiveness of serendipity. It is difficult even to suggest reasons why people might have had the reactions of pleasure that they did. They themselves hardly know, and the science of linguistics appears to be able to give us only rather vague suggestions about the sound qualities of serendipity (its rhythm, its Englishness, its neutralized consonants and repeated vowels) that might arouse approval and pleasure.

Serendipity: Word Formations

We shall probably never know the more intimate details about the coinage of serendipity: when and on what occasion Horace Walpole became aware of the gap in the English language that needed to be filled; why the camel episode in *The Three Princes of Serendip*, however inaccurately remembered, should have stuck in his mind; whether he was familiar with other oriental tales with the same "theme," or whether this was his only encounter with it. We do not know either in what measure it was chance and in what measure design that Walpole chose for the basis of his coinage the name of the place the princes came from, Serendip, with the syllable *dip*, which has a meaningful relationship to the notion of unplanned discovery. But since in the sentence immediately preceding the unveiling of his coinage he speaks of his "talisman, which Mr. Chute calls *Sortes Walpolianae*, by which I find everything I want, *à pointe nommée*, wherever I dip for it," it seems too much of a coincidence that serendipity should have that syllable *dip* in it. As we suggested in an earlier section, it seems likely that consciously or unconsciously Walpole was influenced by the *dip* in *Serendip* in the coinage of his word. Ultimately, however, such a coinage as this is based on an unfathomable mental process, and it is only possible to look in a somewhat different direction for an explanation of its occurrence. It appears that the coinage of words through hybridizing was at a low ebb in the eighteenth century, "the overcorrect eighteenth century" as George Nicholson calls it,[18] and Walpole's motive for coining serendipity may well have been his desire to attract attention by eccentricity. People would surely comment on his "stooping" to coin a hybrid word—and a totally irregular one, at that.

It is likewise difficult to make even an informed guess as to why Walpole should have chosen the suffix *-ty* for his neologism, once he had settled on *serendip* for its root. It may be that he was aware of the contribution that suffix made to the euphony of the word—the syllables *dipity*

---

[18] George A. Nicholson, *English Words with Native Roots and with Greek and Latin or Romance Suffixes* (Chicago: University of Chicago Press, 1916), p. 2.

are pleasant to hear and to say (see earlier discussion). But since Walpole according to Logan Pearsall Smith also coined *"greenth,"*[19] which was doomed from the start, it may not have been his ear that guided him to the use of the suffix *-ity*. It is possible that he was using more or less consciously the principle of analogy, by which serendipity could be added to such other of his endowments as urbanity, sagacity, and temerity (or perhaps it should be timidity). It is even possible that Walpole saw the almost unique suitability of *-ty* for the concept he had in mind, for according to Samuel Haldeman,[20] only *-ty* and *-ness* mean *both* a quality *and* a power; how better could one convey the ambiguity of the role of the actor in a happy accidental discovery than by a suffix that makes the actor either passively endowed or actively accomplished? Or, finally, it may be that it was without any design and by mere chance that he chose that particular suffix, rather than constructing *serendippery*, or *serendiption*, or *serendipness*.

So much, then, for the possible factors contributing to Horace Walpole's original coinage of this particular word, serendipity, to describe the complex phenomenon of happy accidental discovery. Walpole coined only this noun, and his derivation implied that it was a quality that some people like himself or Lord Shaftesbury had, which enabled them to make happy accidental discoveries. Walpole does not suggest that the word may be used to designate the discovery itself (he himself speaks nowhere of *a* serendipity), he provides no noun to describe such an agent of happy accidental discoveries as himself, nor does he construct verbal, adverbial, or adjectival forms for his neologism.

Since serendipity has acquired a certain currency, quite a few of the users of the word have felt the need for additional word forms, and they have, accordingly, invented constructions to fill this need. The fact that serendipity itself is a neologism has, perhaps, given them the courage for further excursions into word coinage, where in the case of a familiar word that lacked certain forms they might simply have resorted to circumlocution. These neo-neologists, Walpole's epigoni, as it were, have acted, by and large, in accordance with accepted principles of word formation. But this does not mean that they have invariably arrived at the same word formations. The variation that we find is, in part, due to the fact that many different English suffixes may be used to convey the same meaning and that only ad hoc theories seem to be available to explain the preferential use of one or another in any particular case. (Logan Pearsall Smith

---

[19] Logan Pearsall Smith, *The English Language* (New York: H. Hold and Company, 1912), p. 92.

[20] Samuel Stehman Haldeman, *Affixes in Their Origin and Application, Exhibiting the Etymologic Structure of English Words* (Philadelphia: E. H. Butler, 1865).

uses the "Genius of the Language" to explain the prevalence of "gaiety" over "gayness" or "cruelty" over "cruelness.") In part, also, the diversity of words formed on serendipity is due to the etymological puzzle that is presented by that word, and which is "solved" in different ways by those deriving new word forms. The surprising thing is, perhaps, not that there should be so many different formations, but, in the case of the adjectives and adverbs at least, that there should be so many identical ones.

The adverbial and adjectival formation that we encounter most frequently is "serendipitous." (Frequently, to be sure, in this case means three or four times.) An anonymous article in *Chemical and Engineering News* reports Walter B. Cannon's revival of the term serendipity, and goes on to say, "In Dr. Cannon's own field of physiology many serendipitous discoveries have been and are still being made."[21] Ogden Nash thinks that "If your coat catches on a branch just as you are about to slip over a precipice precipitous, / That's serendipitous," and it seems likely that he first decided on the formation *serendipitous* and then found *precipitous* to rhyme with it. The advertisement in the *Exonian* features the "Serendipitous Society," and a previous advertisement in the same paper for the "Serendipidous Society" only serves to illustrate our linguist's analysis of the neutralization of the *t* in serendipity. The headline for Vivien Leigh's interview with the reporter Cynthia Lowrey reads: "Vivien Leigh Thinks Oscar Quite 'Serendiptous,'" and the elision of the *i* may be either a typographical error or a genuine variation that assumes that the *i* is not an integral part of the root word and may conveniently be eliminated.

Besides these formations in *-ous*, we have only three others, two of them serious, and one humorous. The Reverend Samuel McChord Crothers writes (in his collection of essays, *Humanly Speaking*)[22] of having a "serendipitaceous" mind, and one Jasper Abbot has a note titled "A Serendipital Portrait," which describes how a certain portrait of one of his ancestors is now "serendipitally in the hands of his five great grandsons."[23] Finally, Ogden Nash invented the humorous superlative *serendopitist* to describe the happiest possible accident—not having to go to the dentist or chiropodist.

All these adverbial and adjectival suffixes (except for the last) appear to be entirely legitimate. Following again the definitions of suffixes given in Haldeman's book, *-ous* means "having," *-aceous* may mean "of, like, having," and *-al* means "relating to, like, capable of being." And the authors

[21] *Chemical and Engineering News* (25 April 1943), reprinting an article from *Industrial Bulletin of Arthur D. Little, Inc.* (July 1940).

[22] Samuel McChord Crothers, *Humanly Speaking* (Boston: Houghton Mifflin, 1912), p. xiii.

[23] Jasper Abbot, "A Serendipital Portrait," *Notes and Queries* new series, vol. 1 (June 1954): 247.

of these word formations might also have constructed *serendipitious* for *-itious* is "the quality of being," or *serendipitive*, with *-ive* meaning "having the quality of." It seems impossible even to guess why the majority settled on *-ous*. What we may surmise, however, is that to those who felt the need of adjectival or adverbial forms, serendipity looked like a word that must be taken *as a whole*, and not broken up into such components as *serendip* and *ity*, for if the latter had occurred we might well have found the formation *serendiply*.

For evidence that such an interpretation of the etymology of serendipity is not entirely fantastic, we need only look at a present participle and a noun that have been formed on just this basis. To begin with the verbal forms, we find besides the *Cleveland Plain Dealer*'s editorial headline (August 22, 1955): "All Good Weathermen Should Serendipidize" (again, incidentally, evidence of the neutralization of the *t*), Odell Shepard's reference to the activity of "serendipping." Or, when we come to nouns describing the agent of a happy accidental discovery, we have besides Leslie Hotson's *serendipitist*, Betty Bridgman's *serendip* and Emma Carleton's *serendipper*. The *Cleveland Plain Dealer* and Hotson appear to see serendipity as a foreign word that is best combined with a Romance or Latin suffix; but Shepard and Emma Carleton take seriously the *dip* root in serendipity, and add to this English "root" an English suffix. They tend to confirm, too, our hypothesis that Horace Walpole's coinage of serendipity with a *dip* in it was not entirely fortuitous. Betty Bridgman's *serendip* can be explained only in terms of a poet's license.

Other noun formations have been required to express serendipity in the plural and in the negative. In neither case has there been very much scope for inventive variation. The plural formation has invariably been *serendipities*—we find it thus in Mark Howe's limerick, in Mancroft's essay, and in Odell Shepard's *Jenkins' Ear*. There is one instance of *serendipitiana*, in an advertisement of the shop called Serendipity Three, and presumably it is intended to suggest that a large number of happy accidental discoveries may be made at that shop. For the negative of serendipity, John Masters, in his book *Bhowani Junction*,[24] coined *unserendipity*, and Leslie Hotson constructed the perhaps slightly more euphonious *inserendipity*.

One final word about noun formations: If serendipity was originally coined to mean a quality of the actor in a happy accidental discovery, it has with use become coterminous with the whole event of accidental discovery, and even with the object of such a discovery. Such usages, then, as "an instance of serendipity," "the serendipity pattern," or "a happy Serendipity, indeed" occur more and less frequently. They are not

---

[24] John Masters, *Bhowani Junction: A Novel* (New York: Viking Press, 1954).

properly, however, word formations, but, rather, they represent changes in the meaning of the word, and as such we shall discuss them in the appropriate section of this study.

The foregoing section on the words formed on serendipity is intended not so much to shed any light on the processes of word formation in general as to show what befalls such an odd neologism as serendipity when it is molded to the immediate needs of its users.

Chapter 6

DICTIONARIES AND "SERENDIPITY"

*A*mong people with a concern for propriety and precision in the use of language, there are few occasions when idiosyncratic erudition or sophistication in these matters takes precedence over the authority of "the dictionary." This authority is multifold: The dictionary (for the moment, let us assume that any modern dictionary may be so designated) legitimizes by sheer inclusion the use of a word as part of standard literary language; it rules on legitimate meanings that may be attributed to such a word; most dictionaries may be expected to have at least some accurate information about the history and etymological derivation of a word that is included; and last (least important for our purpose), the dictionary rules on permissible spelling and pronunciation.

Some Historical Sidelights on Lexicography

For those who thus feel it to be an intellectual responsibility to consult the dictionary on all the problems related to the legitimate or even the preferred usage of words, the rulings of the dictionary have an authority tantamount to that of the rulings of a court of law in another sphere. But unlike the authority of the law court, the legitimizing function of the dictionary is of comparatively recent origin. Early dictionaries were not concerned with the proper use of *English*; they were much more akin to that other kind of dictionary we use when we are concerned with learning or translating a *foreign* language. Down through the sixteenth century, indeed, dictionaries (or, rather, glossaries and vocabularies, as they were called) were aids to the translating of Latin into the vernacular and vice versa, or simply aids to the learning of Latin. Also, in the sixteenth century, contemporary foreign languages began to be treated in the same way as Latin. But though later scholars might learn much about the use of English words from these dictionaries, contemporaries had no such interest in them. As James Murray puts it, "no one appears before the end of the sixteenth century to have felt that Englishmen could want a dictionary to help them to the knowledge and correct use of their own language. That language was either an in-born faculty, or it was inhaled with their native air, or imbibed with their mother's milk; how could they need a book to teach them to speak their mother-tongue?"[1]

The first dictionary that did concern itself entirely with the English language appeared early in the seventeenth century: It was Robert Cawdrey's *The Table Alphabeticall of Hard Words* (1604). In 1623 the "work which first assumed the title of 'The English Dictionarie,'"[2] prepared by Henry Cockeram, was published. Part 1 of Cockeram's *Dictionarie* contains difficult words and their explanation in ordinary language; part 2 does the reverse—it gives the hard equivalents of ordinary words; and part 3 is described by Murray as "a key to the allusions to classical, historical, mythological, and other marvellous persons, animals, and things to be met with in polite literature."[3] This dictionary had in embryo the vast authority of modern dictionaries—it began, at least, to have the authority to define words and to establish legitimate synonyms.

Yet even by the beginning of the eighteenth century "the notion that an English Dictionary ought to contain *all* English words had apparently . . . occurred to no one."[4] It was, however, well in keeping with the spirit of the eighteenth century that a new need should be felt for standardization of the language.

> The language had now attained a high degree of literary perfection; a perfect prose style, always a characteristic of maturity, had been created. . . . The age of Queen Anne was compared to the Ciceronian age of Latin, or of Aristotle and Plato in Greek. But in both these cases, as indeed in that of every known ancient people, the language, after reaching its acme of perfection, had begun to decay and become debased. . . . The fear was that a like fate should overtake English also; to avert which calamity the only remedy appeared to be to *fix the language* by means of a "Standard Dictionary," which should

---

[1] All these remarks on the history of lexicography draw upon James A. H. Murray, *The Evolution of English Lexicography* (Oxford: Clarendon Press, 1900); the quotation is on page 26.

[2] Ibid., p. 28.

[3] Ibid., p. 30.

[4] Ibid., p. 34.

register the proper sense and use of every word and phrase, from which no polite writer henceforth would be expected to deviate.[5]

The first English dictionary that attempted to be inclusive was Nathaniel Bailey's *Universal Etymological English Dictionary*, published in 1721. The most important such dictionary was, of course, Samuel Johnson's *Dictionary of the English Language* (1755): A masterful man constructed a dictionary that was to become, with its successors, the writer's master as well as his servant. Although it may now seem a "childlike and pathetic" notion that any dictionary should fix the language for ever after, the authority of the dictionary has not waned with the recognition of the inevitability of change in the language. Rather, with the abandonment of the notion of fixity, the influence of the dictionary has been enhanced, for it has become the arbiter of legitimate flexibility in the boundaries of the literary language.

Only one significant lexicographical innovation remained after the Johnsonian dictionary (apart from the introduction of pronunciation by William Kenrick in 1773): The history of the words, whether told in the form of etymological derivations or by a series of quotes, had to be made factually correct. What Dr. Johnson and, later, Noah Webster did by inspired guess and memory became a matter for research. Charles Richardson's dictionary (1836–1837) adhered to this new conception in principle, but fell short in practice. And R. C. Trench put in motion the scholarly machinery that was to accomplish a *New English Dictionary* in fact, a dictionary that could rule authoritatively on the proper use of words, on their currency or obsoleteness, on their history, their spelling, and their pronunciation.

The methodical labor that went into the making of the *New*, or *Oxford English Dictionary* (henceforth referred to as *OED*) was tremendous. Thousands of readers sent in millions of slips recording word usages. They were instructed as follows: "Make a quote for *every* word that strikes you as rare, obsolete, old-fashioned, new, peculiar, or used in a peculiar way. Take special note of passages which show or imply that a word is either new and tentative, or needing explanation as obsolete and archaic, and which thus help to fix the date of its introduction or disuse. Make as *many* quotations *as you can* for ordinary words, especially when they are used significantly, and tend by the context to explain or suggest their own meaning."[6] From the deluge of slips that they received, the makers of the *OED* had to choose those that they decided merited inclusion, and eliminate those that they thought did not, and they state ex-

[5] Ibid., pp. 36–37.
[6] See introduction to *OED*.

plicitly their rationale for what must ultimately be, in part, arbitrary decisions: for the lexicographer must "draw the line" somewhere; he must include all "Common Words" of literature and conversation,

> and such of the scientific, technical, slang, dialectical and foreign words as are passing into common use and approach the position or standing of "common words," well knowing that the line that he draws will not satisfy all his critics. For to everyman the domain of "common words" widens out in the direction of his own reading, research, business, provincial or foreign residence, and contracts in the direction in which he has no particular connection: no one man's English is *all* English. The lexicographer must be satisfied to exhibit the greater part of the vocabulary of each one, which will be immensely more than the whole vocabulary of *any* one.[7]

Even in the most scientific of all dictionaries, then, the intuitive familiarity of the lexicographer with the boundaries of the "common language" must play a large part in the highly important decision as to what is included and on what terms (i.e., as a colloquialism, an archaism, a current literary word, etc.).

No other major dictionary has a statement of comparable detail and explicitness as to this aspect of lexicographical methodology as does the *OED*. Henry C. Wyld, the editor of the *Universal Dictionary of the English Language* (1932), speaks vaguely of the necessity of including "purely scientific and technical words and terms" as well as the more important words "current in literary and colloquial use at the present time." In the introduction to the 1934 edition of *Webster's New International Dictionary*, William Allen Neilson says concisely: "Space has not been wasted on words and phrases which are too technical, too rare, too ephemeral or too local, or which are self-explanatory." Only one-third of the definitions in *Webster's New International* are literary and nontechnical. For decisions as to the suitability for inclusion and proper definition of technical words, Webster's calls in experts in different fields, and this is certainly a refinement in procedure over the intuitive quasi-omniscience of Murray's lexicographers at the turn of the century.

After this brief overview of the manner in which the lexicographer exercises his considerable authority, it will be interesting to see how a very specific item fares as it passes through lexicographical channels, to wit, our word serendipity. We shall trace the story of its "official" recognition by various dictionaries, of its definitions and its "biographies," and so obtain a word's-eye view of the workings of the science of lexicography.

[7] Ibid.

Another Look at Horace Walpole's
"Derivation" of Serendipity

In the hope of retaining a more or less stable point of reference in the welter of definitions and derivations of serendipity that will soon follow below, another look at Horace Walpole's description of his coinage of the word, and of the meaning with which he intended to endow his word-child seems warranted. In his letter to Horace Mann, Walpole said: "This discovery [of the Capello arms] is almost of that kind which I call *Serendipity*, a very expressive word, which, as I have nothing better to tell you, I shall endeavour to explain to you: you will understand it better by the derivation than by the definition." Walpole seems to be suggesting here that it may not be entirely easy to explain serendipity to his friend Mann; and he is putting Mann on notice that an understanding of his derivation of the word is an important part of understanding the meaning of it, for on mere inspection the word will make no sense. Then follows the story of the derivation: "I once read a silly fairy tale called *The Three Princes of Serendip*: as their Highnesses travelled, they were always making discoveries, by accidents and sagacity, of things which they were not in quest of: for instance, one of them discovered that a mule blind of the right eye had travelled the same road lately, because the grass was eaten only on the left side, where it was worse than on the right—now do you understand *Serendipity?*" If Horace Mann had a certain perspicacity, he could begin to see several of the important ingredients in serendipity: Walpole revealed the root of his whimsical made-up word by telling of its connection with the story of princes who happened to come from Serendip, and he either takes it for granted that Mann knows that Serendip is the old name for Ceylon, or considers it unimportant that he do so; next comes the nub of a real definition—"they were always making discoveries, by accidents and sagacity, of things which they were not in quest of." For the initiated, this alone might suffice as a pithy summary of the concept, but Walpole adds an example of the princely discoveries. And then, it is perhaps not too much to surmise that Walpole realized what a poor example it really was, that it was, in fact, an example of simple inductive ability; perhaps it dawned on him that "deriving" serendipity from the dimly remembered story of the three princes was not a very good idea, and that he had better help out poor Horace Mann with a better example. He made a quick recovery and proceeded: "One of the most remarkable instances of this *accidental sagacity* (for you must observe that *no* discovery of a thing you *are* looking for comes under this description), was of my Lord Shaftsbury who, happening to dine at Lord Chancellor

Clarendon's, found out the marriage of the Duke of York and Mrs. Hyde, by the respect with which her mother treated her at table."

While this example still falls far short of being unambiguous, it does help to clarify Walpole's conceptions both of what is accidental and of what constitutes sagacity. A discovery is accidental if the object discovered is not being sought—but whether it is merely not being sought *at the time* of its discovery, or whether for true serendipity its existence should be unknown to, and even unsuspected by the discoverer, is not clear. As for sagacity, the example of Lord Shaftesbury's discovery seems to permit the inference that for Walpole a sagacious man was one who knew his way around, as it were, and who had a kind of intuitive familiarity with the field in which he might ultimately make an accidental discovery. The frivolous and whimsical spirit in which he presents his "derivation" does not permit him to consider the relationship between sagacity and accident. But Walpole's very misuse of the story of *The Three Princes of Serendip* suggests that it was a quality of mind that interested him, a quality that enabled the sagacious individual to take advantage of unanticipated occurrences.

It is not made clear by Walpole in this one short passage in which he defines serendipity, whether "making discoveries, by accidents and sagacity" is in all cases a recurrent experience. It is in his own case, as he describes it: He says he has a "talisman," "which Mr. Chute calls the *Sortes Walpolianae*, by which I find everything I want, *à pointe nommée* wherever I dip for it." (In his own case, be it noted, modesty does not impel Walpole to refrain from letting sagacity, or "fate," almost entirely overshadow the component of accident.) The three princes, too, were "always making discoveries"; but for all Walpole informs us, Lord Shaftesbury's instance of serendipity may have been unique.

In view of the variety of definitions and interpretations to be found in dictionaries, to say nothing of the less "official" interpretations we shall deal with in another chapter, it may be worth pointing out here what Walpole did *not* say about serendipity. Two characteristics, especially, that were later attributed to serendipity were completely ignored by Walpole: (1) nowhere does he say anything about what kinds of things are found through accidental sagacity, nor does he say anything about the value or interest of the items that are discovered; and (2) he says nothing at all about making accidental discoveries while looking for something else: as far as he was concerned, it was important that the item discovered had not been the object of search, but there is no implication that the discovery is the *by-product* of some *other* search.

The History of Dictionary Definitions

The dictionary that included a definition of serendipity at the earliest date was *The Century Dictionary and Cyclopedia*. It was first published in 1889–1891 in six volumes, and it has been partially revised many times since, but never entirely revised and reset. In 1901 it was expanded to ten volumes, and in 1909 two more volumes (ca. 100,000 new words) were added, bringing the total to twelve. As the title page has it, the dictionary was prepared under the superintendence of William Dwight Whitney, late professor of comparative philology and Sanskrit at Yale University, and revised and enlarged under the superintendence of Benjamin E. Smith. Serendipity made its first appearance in the two supplementary volumes of 1909. The reference is a lengthy one:

> serendipity (ser-en-dip'-i-ti) *n*. [A humorous formulation with an allusion to *dip*, from *Serendip*, a form of *Serendib*, a former name of Ceylon, + *ity*. The name of Serendib figures in Eastern romance. The name is from Ar. *Serendib, Sarandib*, also *Sarandip* (LL: Serendivi, pl., as the name of the people), MGr. Skt. *Simhala-dvipa*, the island of Ceylon, *Simhala*, Ceylon, (*simha*, lion), + *dvipa*, island. The Skt. *Simhala* is in Pali *Sihalan*, whence *Silan*, Old Tamil *Ilam*, whence Malay *Sailan*, European *Seilan, Zeilon, Ceylon*.] The happy faculty or luck, of finding by "accidental sagacity" interesting items of information or unexpected proofs of one's theories; discovery of thing unsought: a factitious word humorously invented by Horace Walpole. [The Quotation follows, from "This discovery . . . treated her at table."]

In spite of, or perhaps because of, its emphasis on the humorous character of the word serendipity, the *Century Dictionary* gives a most elaborate etymology of Serendip. It is hard to believe that anyone looking for an explanation of serendipity would be much concerned with the Pali or Malay versions of the old name for Ceylon; but this scholarly apparatus makes Walpole's invention all the more "humorous," not to say a little ludicrous. The implication that whoever was responsible for this definition was not altogether in sympathy with words such as this one is reinforced by the use of the term *factitious* to describe it: *Factitious* is well on the way to becoming a pejorative word, growing out of its meaning of "artificial" and "unnatural." The *Century Dictionary* says nothing about the fairy tale that Walpole had read which moved him to this coinage, and the absence of this historical datum, of course, makes Walpole's invention seem even odder than it is. So much for the derivation of the word in this particular dictionary.

The definition, also, is an extensive one; in fact, it is comprehensive to the point of confusion: Serendipity is a "happy faculty" or "luck," that is to say, it is a personal quality or an array of external circumstances,—and these two possibilities are together equated with the making of discoveries by "accidental sagacity." Is then "accidental sagacity" the equivalent of one or the other, or of both? In distinguishing the types of matters that are found by serendipity, the *Century Dictionary* is not quite so catholic, for the nature of the discoveries seems to be chiefly intellectual rather than material. To be sure, the "interesting items of information" tend to be the discrete interesting or amusing finds without further systematic implications that are characteristic of the realm of literature and collecting, while the "unexpected proofs of one's theories" are more nearly related to the strand to be developed in the scientific realm. But treasures of a more tangible kind are not considered. In one respect the *Century Dictionary*'s explanation of serendipity is unique, and that is in its highlighting the allusion to *dip* in the word. Walpole himself makes the same allusion, almost unwittingly, for before he ever "derives" serendipity for Mann's benefit, he tells of his own propensity for finding things "wherever I dip for them." The *dip* in serendipity may have quite a lot to do with its cultural resonance.

The *Century Dictionary* may have been sensitized to this allusion to *dip* because one of its scouts turned up the nonce word *serendipper*, which is thus defined below serendipity:

serendipper (ser-en-dip'er) *n.*, [serendip (ity) t-er (after *dipper*).] One who has the gift of serendipity, or who finds things unsought by merely "dipping." See *serendipity*, N.Y. *Times Sat. Rev.*, April 29, 1905, p. 282. [Nonce word.]

The coiner of *serendipper* (as we saw, one Emma Carleton, who wrote a letter to the *New York Times Saturday Review* pointing out that Ellen Burns Sherman's use of serendipity actually had the meaning of "coincidental sagacity"; more on this later) was evidently trying to keep the word English rather than choosing the latinized word form, *serendipitist*; *serendipper* also seems to strike a more light-hearted tone than the Latin form. The dictionary further stresses the humorous and casual connotations of the word by defining a serendipper as one who finds things by "*merely 'dipping.*'" In sum, one might say that the *Century Dictionary* treats serendipity and its word forms with more than a touch of mock-seriousness. By including the word at all, it grants it a certain legitimacy, but what the right hand giveth, the left taketh, at least in part, laughingly away.

Three years after its inclusion in the *Century Dictionary*, serendipity made its next entry into a dictionary: In 1912, it appeared in one of those fascicles that the *OED* issued prior to the publishing of a whole volume. The fascicle in which serendipity appeared was published as part of a volume the following year, in 1913. The entry in this great multivolume dictionary "on historical principles" ran as follows:

serendipity (serendi.pĭti). [fr. Serendip, a former name for Ceylon, + ity. A word coined by Horace Walpole, who says, (Let. to Mann, 28 Jan. 1754) that he had formed it upon the title of the fairy tale '*The Three Princes of Serendip*' the heroes of which 'were always making discoveries by accidents and sagacity, of things they were not in quest of.']
The faculty of making happy and unexpected discoveries by accident.
    1754 *Let. to Mann*, 28 Jan., This discovery, indeed, is almost of the kind which I call Serendipity
    1880 E. Solly, *Index Titles of Honour*, Pref. 5
The inquirer was at fault and it was not until some weeks later, when by the aid of *serendipity*, as Horace Walpole called it—that is, looking for one thing and finding another—that the explanation was accidentally found.

The treatment that the *OED* gives to serendipity differs quite markedly from that of the *Century Dictionary*. Where the *Century Dictionary* gave much attention to the history of Serendip, the *OED* concentrates more pertinently on serendipity, and it accurately presents the circumstances of Walpole's coinage of the word. The *OED* is reticent about the etymological oddity of the word and treats it with perfect seriousness. As far as defining serendipity is concerned, the two dictionaries differ again, whether as a matter of lexicographical principle or not we do not know. The definition in the *Century Dictionary* is comprehensive and permits itself considerable latitude in the interpretation of serendipity; that in the *OED* is much more concise, but it is sufficiently general to allow the reader scope for a variety of interpretations.

To the rather bare bones of the *OED*'s definition of serendipity, however, the quotation of Edward Solly's use of the word adds something of importance. Solly's use of serendipity, familiar to careful readers of *Note and Queries* since the 1870s, reinforces strongly an implication in the meaning of the word that Horace Walpole at most merely hinted at in his letter to Mann, namely, that serendipity comes into play while one is *looking for something else*. It might be best to note here, at the expense of chronology, that in the *Shorter OED on Historical Principles*, which appeared in 1933, the derivation and definition from the *OED* are reproduced almost exactly, with very minor omissions, but the quotations are omitted entirely, so that the implication of activity in the meaning of serendipity is concealed as it were from the readers of the abridged dictionary.

The late nineteenth and early twentieth centuries were a golden age of lexicography, and besides the *Century Dictionary* and the *OED*, another great dictionary was created at that time, Funk and Wagnall's *Standard*

Dictionary of the English Language (1893). Twenty years later it was thoroughly revised and then contained 450,000 words, but since 1913 there has been no substantial revision of this dictionary. Funk and Wagnall's *New Standard Dictionary of the English Language*, prepared by more than 380 specialists and other scholars under the supervision of Isaac K. Funk, editor-in-chief; Calvin Thomas, consulting editor; and Frank Vizetelly, managing editor (all this according to the title page), includes a reference to serendipity:

> ser″endip′ity, 1. ser″-en-dip′-i-ti 2. sĕr″-ĕn-dĭp-i-ty. *n*. The ability of finding valuable things unexpectedly: from a fairy tale, *The Three Princes of Serendip*, the heroes of which were continually finding valuable articles by chance; a word coined by Horace Walpole [*Serendip* = *Serendib*, former name of Ceylon].

The *New Standard Dictionary* has converted the discoveries that the *OED* termed merely "happy" into things of value; and to make everything as neat as possible, the discoveries that the good princes of Serendip made (such as the fact that the mule was blind in the right eye!) consisted of valuable articles, too. Apart from the historical inaccuracy of this construction, it makes the discoveries involved in serendipity very concrete, quite different from the "information" and the "proofs" of the *Century Dictionary* and the undefined "discoveries" of the *OED*. For those who consult this particular dictionary, this concreteness may well limit the applicability of serendipity to concrete objects that are either invented or discovered. The reference of the *New Standard Dictionary* seems of limited adequacy in its derivational aspect as well as its definitional one: It could easily lead to considerable puzzlement as to the relationship between Horace Walpole and *The Three Princes of Serendip*. In only one regard does this dictionary leave nothing to be desired: The pronunciations of serendipity are dealt with as extensively as seems possible.

So far, the dictionaries in which serendipity has appeared have all been "big" dictionaries, and almost twenty years elapsed before its appearance in dictionaries that aimed at something less than maximum inclusiveness. Of course, just as there are many different kinds of people who use dictionaries, so there are different kinds of dictionaries to meet their needs. Users of dictionaries vary in their level of education and in their interests, from scholars and writers with a professional stake in the optimal use of words, to readers of popular literature, who occasionally come across unfamiliar words they want to look up, to students who need to refer frequently to a convenient guide to proper usage. For scholars, the *OED*, or Funk and Wagnall's, or *Webster's New International* serve well; the student and the casual reader might best use a portable dictionary. But there is a group of widely read people for whom the one kind of dictionary is too comprehensive, the other too meager, and it is for this group that Henry Cecil Wyld put together the *Universal Dictionary of the English Language*, published in New York by E. P. Dutton in 1932. The introduction to this dictionary says explicitly that it is intended to be smaller than Webster's and larger than the *Concise OED* (not to be confused with the *Shorter OED*—the *Concise OED* is a portable dictionary). Wyld's remarks about the kinds of words included show that he has in mind that large minority of well-educated and well-read people, rather than either the tiny minority of scholars or the so-called masses: "The greatest importance is naturally attached to words current in literary and colloquial use at the present day. At the same time, a large number of purely technical and scientific words and terms have been included, since many of these necessarily play a considerable part in the lives of important sections of the community." Here is the *Universal Dictionary*'s reference to serendipity:

> serendipity, *n.* [1. sèrendípity 2. sérendipiti] Coined by Horace Walpole, Letters to Mann, January, 1754; fr. *Serendib, -dip*, Arab. name of Ceylon, in tale of three princes of Ceylon. The faculty of finding valuable or interesting things by chance or where one least expects them.

Anyone whose curiosity is limited to the definition of serendipity will most likely be satisfied with what he finds in the *Universal Dictionary*. The definition at least makes sense. To the purist who has Walpole's own explanation in mind, the dictionary may be taking certain liberties with the meaning of the word, insofar as it specifies that the things found by chance are valuable (though even Walpole himself would not deny that they are interesting), and in that it would have the unexpectedness of the discovery relate to the *place* in which it is found. Previous dictionary definitions had not related the factor of unexpectedness either to time or to place, and if there are any suggestions in Walpole in this regard, they would seem to imply that he had unexpectedness with respect to time rather than place in mind.

Those users of the *Universal Dictionary*, however, who are interested in the derivation of the word are likely to be more than a little puzzled. The derivation is lacking in vital information both as to the relationship between Walpole and the tale of the three princes, and as to that between the princes and the making of happy accidental discoveries. To be sure, the resourceful reader can go to Walpole's letter to Mann if his curiosity is not satisfied. Unhappily, even this possibility is no longer open to anyone consulting the 1952 impression of the *Universal Dictionary*. The reference there to serendipity is identical with the one in the earlier printing, with the one exception that Walpole's letter to Mann is said to have been written in 1752.

The *Shorter Oxford English Dictionary*, which appeared in 1933, is intended for "the student" and "the general reader," according to the introduction by C. T. Onions—students and general readers, we might suspect, with literary and historical rather than scientific interests. Whereas the original *OED* contains a half-million words and a million and a half quotations, this abridgment, according to Onions, "presents . . . a quintessence of those vast materials. The method reflects exactly that of the principal work: It is historical in its representation of the chronological sequence of the development of meaning. It gives the etymologies of words in such a form as to exhibit every significant stage of their history from their place of origin. The meanings are illustrated by quotations either exactly dated or assigned to their authors." The preparation of the *Shorter OED* was begun in 1902 by William Little, and completed after his death in 1922 by H. W. Fowler and others, and in consequence of this long period of preparation it was possible to supplement the older *OED* references with materials collected since the completion of the original *OED*.

While the *Shorter OED* carries a reference to serendipity, in this case, at least, any sense of the historical development of the word has been lost. Not only does the *Shorter OED* make no reference to any new usages of serendipity that had turned up since the publication of the parent *OED*, it does not even consider the word of sufficient importance to warrant the use of space for the quotation from Edward Solly used by the *OED*.

> serendipity (serendi.pĭti) 1754. [f. *Serendip* (-*b*) former name of Ceylon + ITY, coined by Horace Walpole upon the title of the fairy tale *The Three Princes of Serendip*, the heroes of which "were always making discoveries, by accidents and sagacity, of things they were not in quest of."] The faculty of making happy and unexpected discoveries by accident.

The definition is identical to that in the *OED*, and the derivation is substantially the same, although the reference to Walpole's letter to Mann is omitted. The dictionary reader who encounters the word for the first time in this "dictionary on historical principles" might well suppose that serendipity is a nonce word.

Webster's New International Dictionary comes relatively late into the history of dictionary definitions of serendipity. Although the first edition of this dictionary was published in that golden age of dictionaries, the early twentieth century (specifically, in 1909), that edition did not include our word. In 1934 a new edition was published, with William Alan Neilson as editor-in-chief, and Neilson, a forceful man, describes neatly the rationale for the inclusion or omission of words: "Space has not been wasted on words and phrases which are too technical, too rare, too ephemeral or too local, or which are self-explanatory." No one has ever accused serendipity of being self-explanatory, but evidently it did not fall into any of the other categories either, for it is included in the 1934 edition of Webster:

> ser′-en-dip′-i-ty (-dĭp′-ĭ-tĭ) *n.* [see Serendib] The gift of finding valuable or agreeable things not sought for; a word coined by Walpole, in allusion to a tale, *The Three Princess of Serendip*, who in their travels, were always discovering by chance or sagacity, things they did not seek.

The *New International Dictionary* is accurate but very brief in its derivation of serendipity: No words are wasted on *Serendip*, but the reader is given the necessary clue to looking it up; and Walpole is stripped of his first name—how many readers will assume that Robert or Hugh is involved, rather than Horace? Where the *OED* and the *Century Dictionary* speak of a "faculty" and Funk and Wagnall's of an "ability," Webster's calls serendipity a "gift"; it seems unlikely, however, that most people consulting the dictionary would draw any very different conclusions from the use of one or another of these terms. As for the objects of discovery, here Webster's dictionary seems to support the implication that they tend to be concrete: "valuable and agreeable" suggests, at least to these readers, gratification of the senses. Finally, the *New International* plays hob with one very important ingredient of Walpole's definition of serendipity: where Walpole speaks of "accidents *and* sagacity," the *New International* says "chance *or* sagacity," and the vital suggestion that the two factors operate in conjunction is, thus, lost.

With its inclusion in *Webster's New International* in 1934, serendipity had found a place in all the most reputable "big" and medium-sized dictionaries. It appeared next in several rather specialized dictionaries. In 1945, Joseph T. Shipley published a *Dictionary of Word Origins*, and in it he says of serendipity:

> serendipity. This happy faculty of finding what one did not seek was named by Horace Walpole from the fairy tale, *The Three Princes of Serendip* (Serendip is the former name of Ceylon). Thus, Saul set forth to find his father's asses and found a kingdom. *Serendipity* is the treasure of every artist.

This is a confusing definition indeed—initially it seems deceptively simple, but the explanatory illustrations complicate the matter badly for the reader who has never encountered serendipity before. It might be argued, to be sure, that such a reader would be unlikely to seek enlightenment from Shipley's dictionary, and that those who go to Shipley merely for a possible further opinion on the word (having first successfully consulted a more conventional dictionary), may find his illustrations "suggestive." "Suggestive" they are, though of quite different things: After serendipity is defined as "finding what one did not seek," apparently, then, being simply lucky, the example of Saul and the kingdom serves to recast the picture quite radically, implying as it does an unexpectedly great reward for dutiful activity. Then, to confuse the matter further, Shipley suggests an intimate connection between serendipity and art, but here, instead of being merited, serendipity appears to be vaguely synonymous with inspiration. Salvation may be considered to be deserved by good works, but those inspired are surely mysteriously chosen. There is no evidence that Shipley is aware of the different meanings suggested by his two examples, different meanings that, as we shall see in the next section, have both been attributed to serendipity, but not blithely, as in Shipley's case, by one and the same person.

Another specialized dictionary that includes serendipity has been put together by Eric Partridge, the well-known lexicographer with something of a penchant for the byroads of lexicography. It is called *Name into Word*, and subtitled *Proper Names That Have Become Common Property*, one we mentioned earlier in another context. It was published in 1949. In the foreword, Partridge tells something about his principles for including words: "Most of the relevant terms current in everyday speech of those who speak English (whether British or American), not jargon, are here." He also elaborates somewhat his criteria for establishing currency: There are two such criteria, widespread use and potential usefulness. (The later would seem to be a criterion of potential everyday currency rather than actual currency.)

Here is Partridge's reference to serendipity:

Serendipity means the faculty or the habit of making felicitous discoveries by accident. A very literary word, it was coined by Horace Walpole in a letter to Mann on Jan. 28, 1754 from precisely this faculty as possessed by the titular heroes of the fairy tale, *The Three Princes of Serendip*, an old name for Ceylon, also written *Serendib*, from Aryan *Sarandib*, which appears to denote "the Singhalese island." In an interesting book, *An Index of Titles of Honour*, 1880, E. Solly defines the term as "looking for one thing and finding another" (cited by the OED).

If we compare Partridge's definition with previous ones, it differs in only one respect: He suggests that serendipity may be a *habit* as well as a faculty, and since habits may be acquired, this is the first time it is implied in a dictionary that serendipity is defined as cultivable. Curiously, this is a meaning of the word that had attained some currency among scientists by 1949, though at that very time Partridge was labeling it a "very literary word." Perhaps we may infer from this label that in the case of the inclusion of serendipity in this dictionary, the criterion of potential usefulness far outweighed that of widespread usage in Partridge's estimation.

Since currency plays a part in the formulation of criteria for the inclusion of words in all dictionaries, it is a measure of the increasing currency of serendipity that while it is not to be found in the 1921 edition of Ernest Weekley's *Etymological Dictionary of the English Language*, the 1952 edition (which has added the word *Concise* to the title) does include the word. Weekley states in his introduction that defining is not the principal interest of this dictionary; only a few out-of-the-way words are defined. He does define serendipity, so, like Partridge, he must think of it as a word of quite limited currency.

> serendipity. Gift for finding one thing when looking for another. Coined by Horace Walpole from fairy story of *Three Princes of Serendip*, i.e., Ceylon.

About all that is interesting about this reference is that it provides, by sheer inclusion, one more instance of validation of the use of the word.

"Freshness" and "originality" are generally considered admirable qualities in intellectual endeavor, but from our word's-eye view of dictionaries, the definition of serendipity in *Swan's Anglo-American Dictionary* suggests that in lexicography "freshness" and "originality" have their drawbacks. (These are the qualities explicitly claimed for the definitions in this dictionary in the introduction.) Further, it is claimed that "every effort has been made to give definitions which are adequate without being flamboyant or unnecessarily profuse." This is a dictionary that is trying, self-consciously, to be "different." Serendipity is defined as follows:

> serendipity *n.* The sheer luck or accident of making a discovery by mere good fortune or when searching for something else.

This definition seems not only somewhat inadequate, but also a bit profuse, for its repeated emphasis on the aspect of luck, "sheer" luck and "mere" good fortune, considerably distorts the meaning of serendipity. After all that "sheer luck" and "mere good fortune," the possibility of making discoveries "when searching for something else" does little to redress the balance in favor of some minimal amount of activity in the pursuit of serendipity. *Swan's Dictionary* makes no effort to explain the origin of the word.

An even less adequate treatment of serendipity than that in *Swan's* is to be found in *Webster's New World Dictionary of the American Language*, edited by Joseph Friend and David Guralnik, with Harold E. Whitehead as etymological editor. The 1951 edition did not contain serendipity, but it was included in 1954. (In an interview with Harvey Breit of the *New York Times*, March 22, 1953, Mr. Guralnik said of serendipity: "It's in the air now, I've come across it quite a few times recently and maybe we'll have to include it next year." And he defined it as "the opposite of accident-prone.") The dictionary is a two-volume work with at least 130,000 entries. The introduction emphasizes the inclusion of idioms

and colloquialisms, and states that the dictionary gives "fuller etymologies" than other comparable works.

> serendipity. (seren-dip-ti) n. coined by Horace Walpole (c. 1754) after his
> tale *The Three Princes of Serendip* (i.e., Ceylon) who made such discoveries,
> an apparent aptitude for making fortunate discoveries accidentally.

It may be possible to have differences of opinion about the adequacy of a
definition, but about certain facts relating to derivation of serendipity
there can be no dispute, and the New World Dictionary is simply uninformed about them: Its caution about the date of Walpole's coinage is
amusing, since it happens to have got hold of the correct date, and more
especially since it is so blatantly in error in attributing *The Three Princes of
Serendip* to Walpole. As for the definition, we can only wonder on what
ground the *New World Dictionary* feels justified in implying disbelief in
the aptitude for making fortunate discoveries by calling it an "apparent"
aptitude. If it *is* only apparently an aptitude, what would the editors suggest it is *really?*

Ten years after publishing the *Dictionary of Word Origins*, Joseph T.
Shipley brought out a *Dictionary of Early English* (1955). He says that
"gathered in this Dictionary are, in the main, words that have dropped
from general use," though "a few still current words" are included "because of their old associations, or because of older meanings lapsed from
use." And in describing his basis of selection, Shipley lists in one category: "Words not in the general vocabulary today, but that might be
pleasantly and usefully revived." It is evident from these remarks that if
serendipity is to be found in this dictionary, as it is, Shipley must both
exaggerate its obsolescence and consider it a potential revivee. The reference substantiates these predictions:

> Serendipity—The faculty of making happy finds. This is too good a word to
> have been wholly forgotten, for from Saul (who went to look for his father's
> asses and found a kingdom) to the most recent work of art, serendipity
> reigns. Horace Walpole, who coined the word, said in a letter of Jan. 28,
> 1754, that he took it from the title of a fairy tale, *The Three Princes of Seren-
> dip*; the princes "were always making discoveries by accident and sagacity, of
> things they were not in quest of." *Serendip* was an early name for *Ceylon.*
> Any resemblance between *serendipity* and *heredipety* (q.v.) is purely coinci-
> dental. What Ogden Nash did with the word in *The Private Dining Room*
> cannot be called serendipitous.

There is clearly a family resemblance between this reference and the one
in Shipley's earlier dictionary: Again, we find allusions to Saul and the
kingdom, and to art, but in this case the allusions are even more cryptic
than in the *Dictionary of Word Origins*. The derivation, however, is far

more elaborate in the present dictionary. As for the implication that serendipity has been almost wholly forgotten, we suspect that the word has had at least a small boom in the last ten years, that it is too good a word to have been even almost wholly forgotten. Shipley seems to treat serendipity rather facetiously, and whether or not he felt compelled to do so because he smuggled into this dictionary a "good" word whose state of obsolescence does not entirely warrant admission is a matter of conjecture—in any case, the reference to Ogden Nash and the jocose comparison with *heredipety* suggest that serendipity should not be taken too seriously as a going word. (*Heredipety*, incidentally, means legacy hunting, from *herediu petere*; adj. heredipetous; accent on *dip*.)

All the dictionaries we have described so far have been either large, comprehensive works or smaller but specialized ones, but so far we have not encountered a reference to serendipity in a small, portable dictionary. It must be confessed that chronology has been slightly violated in order to set off the one such dictionary that does include the word: It is to be found in the 1951 edition of the *Concise Oxford English Dictionary*. This dictionary was adapted by H. W. Fowler and F. G. Fowler from the *OED*, and the first edition was published in 1911. Subsequent revisions were made in 1929, 1934, and 1944, but we believe (without absolute certainty, since the 1944 edition could not be examined) that the 1951 edition was the first to include serendipity. The purpose of this dictionary, as stated by H. W. Fowler, is to stress the full treatment of common words, illustrating at length their idiomatic usages, and to give relatively short shrift to uncommon words. The treatment of serendipity is, accordingly, quite brief:

> sĕrĕndĭp′ĭty, n. The faculty of making happy and unexpected discoveries by accident. [coined by Horace Walpole after *The Three Princes of Serendip* (Ceylon) a fairy tale].

It is the sheer inclusion of serendipity in the dictionary that is most interesting here, for this seems to be a dictionary designed to help "ordinary" people use the language properly, and the inclusion of the word suggests that in the editors' opinion the likelihood that these ordinary people would encounter the word was increasing.

In America, abridged dictionaries similar to the *Concise OED* in size, and, roughly speaking, in purpose, have yet to include references to serendipity. Thus, the word is not to be found either in *Webster's New Collegiate Dictionary* nor is it to be found in Funk and Wagnall's *New College Standard Dictionary*, though in both cases the parent dictionaries have recognized the word.

Conclusions Drawn from the Study of the Dictionary
Existence of Serendipity

By tracing the dictionary references to just one word over the past fifty years or so, one gets a keen sense of the important role played by dictionaries in the life of the language. The dictionary acts upon the language: It has the authority to determine, at any given time, which words may be used by people without erring out of bounds into the colloquial, the obsolete, or the esoteric, and it standardizes the usage of words of sufficient intellectual dignity and necessary currency to be included in it, assigning them one or more legitimate meanings, pronunciations, and spellings, and furnishing them with etymological and historical credentials. But this process of validating word usage is not a unilateral one: Powerful as it is, the dictionary is in turn acted upon from various quarters by people who coin new words or resurrect old ones, in order to express themselves with more precision or force. The dictionary may hesitate, for a variety of reasons, to accept such coinages and resurrections, but if they acquire sufficient currency, and especially currency among educated people, the dictionary must grant them recognition by inclusion. Once granted standardized acceptance, the word must conform to the meaning given it in the dictionary, unless there is strong outside pressure to change it. And so the process continues.

So far we have spoken of *the* dictionary, as if there were in fact only one such arbiter for a language. Happily for the flexibility of the language, there are several dictionaries, and although their agreement in the treatment of words would seem to be substantial, it is not complete. Also, the several dictionaries are constructed to meet the needs of different publics, publics that vary most significantly in their level of education and in their intellectual interests. Different words have differential currency among these groups, and by noting the inclusion or absence of a given word in dictionaries designated for the use of one group or another, one may get a fairly good sense of the distribution of its use in the population. This distribution may, of course, change. To coin an expression that all dictionaries would disapprove of, slang expressions and colloquialisms very likely percolate *up* the educational scale, while literary and scientific terms more likely percolate *down*.

In the case of serendipity, the various dictionaries overlap to a large extent in the meaning and the history they attribute to the word. No matter which of them the reader happened to consult, he would find a meaning for serendipity that had something to do with the making of happy accidental discoveries; and almost all of them will inform him that the word was coined by Horace Walpole upon the title of *The Three Princes of Serendip*. Beyond this common core of meaning and history there are, to be sure, almost as many minor variations as there are dictionaries, and they undoubtedly exert some influence on the definition and derivation the reader carries away, but to get the full import of these variations would require closer inspection and exegesis of each treatment than most users of the dictionary would be likely to think necessary. In all these dictionary references we have examined there is only one outright error: the attribution of *The Three Princes of Serendip* to Horace Walpole by *Webster's New World Dictionary*.

As for the differential currency of serendipity in different segments of the English-speaking population, only very crude inferences are possible. The most striking bit of evidence is its virtual exclusion from all abridged dictionaries until 1951, and its inclusion in only one since that date. This would suggest that serendipity is still a pretty esoteric word for most people, even for most college-educated people. Still more impossible is it to draw any conclusions about its selective adoption among different groups within that small public that uses unabridged dictionaries. Most of the "big" dictionaries began to include it at about the same time. We do not know the patterns of preferential use in these groups of one of these "big" dictionaries as against another, nor is it possible to tell whether each of these dictionaries has a distinct audience in mind within the small, highly educated public that it serves. It would have been somewhat surprising if the *OED* had not included the word, oriented as it is to the English literary group; correspondingly, it is not so very surprising that *Webster's New International* should be the last of the unabridged dictionaries to make room for serendipity. But what influence this late inclusion in Webster's might have on the diffusion of the word is almost beyond conjecture. It is very hard to say whether or not most of those few who used the word before, say, 1945, might not have done so regardless of its inclusion in any dictionary. It might be safe to say, perhaps, that as its currency has increased, legitimation by inclusion in one or another dictionary has assumed increasing importance. In the course of its downward percolation, serendipity has probably been encountered by more and more rather timid people, who, when they use an unfamiliar word, need that moral support which a dictionary reference provides. Esotericists, on the other hand, may be more likely to ignore, if not actually to disdain, the dictionary.

Great as is the authority of the dictionary in determining the usage of words, the subtler aspects of its influence on the diffusion or the changes of meaning of a particular word are an unknown quantity.

Chapter 7

THE SOCIAL HISTORY OF SERENDIPITY

*O*ver the past eighty-odd years, since the contributors to *Notes and Queries* rescued Horace Walpole's word-child from oblivion, about 135 different people have used serendipity in print, several of them on more than one occasion. What records we have of the responses and attitudes of these people toward the word have already been described and analyzed in an earlier part of this study (see Chapter 4), but little has been said as yet about the social and intellectual backgrounds of these users or of the routes they traveled, literally and figuratively, which brought them in contact with serendipity. And though we did suggest earlier a certain relationship between sheer currency and acceptability, we have yet to investigate the factors in the background of the user that tend to make him receptive to a word that means the making of happy accidental discoveries.

The 135 people who used the word serendipity did not appear at random in the English-speaking population: While undoubtedly many different kinds of people are represented in this group of 135, many more are inconspicuously absent. What many of these users of serendipity seem to have in common, what sets them apart from other men, is their participation in some activity in which the making of unanticipated discoveries is a frequent occurrence. They come from social milieux where happy unanticipated discoveries constitute a salient feature in their common patterned experience, and consequently they tend to be receptive to such a label as serendipity with which to tag this aspect of their experience.

When we speak of people belonging to a social milieu, we mean simply that they participate in similar activities, activities that may be vocational or avocational, public or private. People who belong to the same milieu do not necessarily know one another. Within such a social milieu, however, it may be possible to discover one or more social circles, that is, groups of people who do know one another or at least have read one another's works. Although the members of a milieu may individually pick up a "useful" word such as serendipity and adopt it for its sheer aptness, members of a social circle have additional motives for accepting (or, in some instances, rejecting) the same word. Within the circle, serendipity may become more than an apt word; it may take on something of the character of a password. The use of the word, then, becomes a symbol of identification with the group, and the word itself becomes part of a relatively specialized vocabulary that sets apart those who are "in," intellectually and socially. To say that the desire for identification with a certain social circle may heighten the individual's receptivity to some of the more esoteric ingredients of the circle's vocabulary does not necessarily mean that unusual words are adopted for this reason alone. In some very rare instances in the history of serendipity, however, such conformism appears as the only detectable reason for the use of the word.

The social history of serendipity, then, tells of the different social milieux and circles in which serendipity has been adopted, and of some of the reasons for the receptivity of the members of these groups to such a word as this. It is a history that is only roughly chronological: It is chronological in that the social milieux and circles are taken up in the order in which they first came in contact with serendipity. But several of the milieux existed side by side at the same time, and in some cases we find an overlap of personnel between groups.

Collectors

Since Horace Walpole coined serendipity to describe a kind of experience he had as a collector, it is etymologically just that the word should have been resurrected by another ecumenical collector, Edward Solly. Although there is, of course, a great variety of collectors, only two kinds, to our knowledge, have used serendipity, namely bibliophiles and quasi-antiquarians such as Solly. (The quasi-antiquarians connected with serendipity have not been interested in actual antiquity, but in more recent history, and especially the less familiar aspects of recent history. To call them historians would seem to attribute to their work a coherence it lacks—they tend to value the discrete historical item for its own sake. For convenience, we shall call them simply antiquarians.) Solly was a collector of highly diverse items of historical information, items that frequently must have proved elusive, and serendipity for Solly described that unknown quantity that intervened between his assiduous efforts and certain success.

All collectors are familiar with different aspects of this experience, the recurrent discrepancy between efforts and results. The experience may take the form of a windfall, of a valuable item that "just drops into one's lap," as it were; or, conversely, it may take the form of an endless, devoted, and ever-unsuccessful pursuit of a valued item; or, finally, and most commonly, the discrepancy between efforts and results may be less extreme than the totally unexpected windfall or the ever-elusive Grail, yet "just a little bit of luck," good or bad, will make a great deal of difference to the satisfactory outcome of a collector's enterprise. In a collector's life the conditions of success are to an unusual extent unknowable, and the notion of serendipity (whatever its particular interpretation may be, of which more in a later chapter) serves to make some sense out of this uncertainty.

Among collectors, antiquarians may be somewhat more exposed to the occurrence of happy accidents than bibliophiles. Bibliophily is, after all, relatively well organized: The criteria of value in books are elaborately specified; items of potential interest to the bibliophile are periodically catalogued; and literary scholarship works hand in hand with bibliophily— in fact, there is considerable overlap of personnel between the two. The amount of common standardized knowledge about books and their authors reduces the likelihood of unanticipated discoveries and also makes it possible to establish quite clearly when an unexpected discovery of a valuable item has been made.

For the antiquarian, life is both more and less rigorous. Antiquarians have far wider interests than most bibliophiles and much less codified information available on the subjects that interest them. Compared to the vast literature available to most bibliophiles, antiquarians have recourse to only a few such works of reference as *Notes and Queries* or *The Antiquarian* in their search for information. Since these periodicals consist themselves of just such disconnected items of information as the antiquarian is in search of, his chances of finding material on any particular subject are small. A successful antiquarian, then, must build up on his own a huge stock of knowledge, a stock whose adequacy or deficiency he has no way of checking. Edward Solly's stock was such that he could contribute authoritative information on such diverse subjects as the following (this is a selection only from the notes he contributed to *Notes and Queries* in 1878): Friar Sebastian Michaelis's *History of Magic* (1612); Sir Nathaniel Bacon; a letter concerning the death of Canning; a sarcastic dictionary of the early nineteenth century; Dean Swift's attitude toward Communion; serendipity; the grasshopper crest of the Gresham family; and Montague, an eighteenth-century bookseller. Solly evidently had the resources with which to dig up information on almost any literary-historical subject.

Another such well-informed antiquarian, and one who was also familiar with the word serendipity, was Arthur Michael Samuel, Lord Mancroft. In 1923, Mancroft published a collection of essays that had previously appeared in the *Saturday Review*, essays on oddly assorted subjects such as: the trade in canaries; an eighteenth-century antique dealer by name of Gavin Hamilton; serendipity; and the vagaries of Count Rumford. Mancroft wrote two full-length books: One is on the herring fisheries, the other on Piranesi. As one reviewer of the *Mancroft Essays* said of the author: "He is devoted to the history of posy rings, of fans, of weathercocks, of wigs."[1] Such a historian must certainly have unusual independent intellectual resources if he is to be successful in his researches. And since little or no cumulative work exists on many of the subjects, it seems unlikely that the author did not benefit by happy accidents in collecting his data. This supposition is confirmed by the same reviewer mentioned above: "Mr. Samuel is evidently devoted to serendipity, and it accompanies him in all his desultory excursions. Readers of 'The Mancroft Essays' will be mildly thrilled by many a felicitous coincidence and fortunate chance in their adventure." Simply his pursuit of scholarly byways rather than highways, then, tends to expose the antiquarian to the excitement of unexpected discoveries.

These explorations of unfamiliar intellectual territory, then, make considerable demands on the antiquarian's individual skill and ingenuity as a searcher; but this same departure from the scholarly beaten track also makes it somewhat easier for the antiquarian to find *something* that will please him than it is for the bibliophile. Antiquarians tend to hunt for "odd" items and to value them, in large measure, for their very oddness. Scarcity is, of course, a factor in the value of any collector's item, but most collectors do not value the uncommonness, the peculiarity, the strangeness of their items above all else. Rarity enhances the value of a bibliophilic treasure, but it must have other "points." The oddness and quaintness that delight the antiquarian defy categorization by their very nature: Oddness cannot be systematically evaluated; odd items, whether physical or intellectual, have a value that is often largely private. The antiquarian's treasure is the ordinary man's junk or, at best, trivia; it is the undifferentiated residue that remains after what is conventionally defined as valuable has been skimmed off. Frequently, therefore, the antiquarian's happy accidental discovery is happy and accidental only to him; on occasion, he may turn up something that was discarded by mistake, as it were.

---

[1] *Saturday Review* (16 June 1923): 809.

In any case, the antiquarian seems to be serendipity-prone, even for a collector.

The pleasant surprises of the antiquarian are illustrated by Alan Walbank from his own experience as a "junk hunter":

> I count among my serendipitous experiences the search for material illustrative of family life in the early Victorian era and the findings there of a set of jig-saw puzzles. They were maps of England, Scotland and Ireland, dissected on mahogany, each in its wooden box with a pictorial sliding lid. . . . Inside the lids, after rubbing away a little dust, I brought to light the names and birthdays inscribed in ink of the nephews and nieces of that Sewell family whose best known member wrote the children's classic, *Black Beauty.* It was exactly what I wanted and had *not* been looking for.[2]

Antiquarians are exempt from explaining too precisely why they want a certain item; hence, many odd and pleasant things come their way unexpectedly.

In the milieu of the bibliophiles it is somewhat different. Bibliophiles tend to be somewhat more precise in the definition of what they want than do antiquarians, and depending on the nature of that definition, their methods of collecting and the likelihood that they will turn up unexpected treasures will vary. Bibliophiles invariably do make some happy unanticipated discoveries, and as a group they have been markedly receptive to serendipity as an apt term for these experiences. But for some bibliophiles the factor of accident is probably more prominent than for others. The more the collector's interests are narrowly focused on well-known and frequently collected authors, the less likely it seems that he will turn up the unexpected, while those who collect the works of less popular authors or who have relatively unspecialized collections seem more likely to run into unexpected good luck. Perhaps we are not reading in too much if we find confirmation of this hypothesis in the following remark by the earliest bibliophilic user of the word serendipity, Andrew Lang, to which we referred earlier: "There is a faculty which Horace Walpole named 'serendipity'—the luck of falling on just the literary document which one wants at the moment. All collectors of out of the way books know the pleasure of the exercise of serendipity but they enjoy it in different ways. One man will go home hugging a volume of sermons, another a bulky collection of catalogues."[3]

In the early years of the twentieth century, a shop was opened in London to cater to these very bibliophiles who wanted "out-of-the-way books," books by not-so-well-known authors at rather moderate prices. The first

[2] Alan Walbank, "Joys of the Junkshop," *The Saturday Book* 14 (1954): 261–267.
[3] Andrew Lang, *The Library* (London: Macmillan, 1881), pp. 2–3.

mention we have found of it comes, not surprisingly perhaps, in the form of a query to *Notes and Queries*. As we saw earlier, in 1903 one John Hebb writes: "A shop has recently been opened at No. 118 Westbourne Grove, with the extraordinary name of 'Serendipity Shop.' What is the meaning of 'Serendipity'? I may add that the shop appears to be intended for the sale of rare books, pictures, and what Mrs. Malaprop (was it Mrs. Malaprop?) calls 'articles of bigotry and virtue.' "[4]

The Serendipity Shop was the property of Everard Meynell, one of the sons of Wilfrid and Alice Meynell. We noted in an earlier chapter that serendipity was a favorite word of Wilfrid Meynell's, and since Wilfrid is known to have been a rather dominating man, it is not surprising to find his son following his taste in words. As for Everard Meynell's taste in books, it was a slightly offbeat taste at that time. He was particularly interested in the poets of the sixteenth and seventeenth centuries; as his sister Viola says, he was "an expert on the inside and outside of his copies of Donne and any seventeenth-century poets."[5] It is improbable that Meynell was a very keen businessman, for in a short-lived periodical, *The Imprint* (about seven issues appeared in 1913), one of the articles he contributed was on the subject of "Starting a Shop."[6] In this article he described his daily experiences in the shop, his few customers, his inability to discipline the boy who assisted him, and the kind of little incident that made it all worthwhile to him, though it might not have satisfied a bookseller who wanted to make money: "I put autographs of Mr. Lawrence Binyon where they may catch the eye of his admiring juniors in the B.M. Print Room, and Mr. Binyon himself stops to learn from my window that he is more valuable than Sir Herbert Tree and Mr. Maurice Hewlett, than Mr. Walter Crane and Sir Arthur Pinero. None of these, nor the manuscripts of the prospective poet laureate, tempt him; but the Serendipity Shop has his blessing and is content."

Meynell and his shop were appreciated by others besides Mr. Binyon. In one of his chatty essays, the great American bibliophile A. Edward Newton tells his readers:

> Speaking of catalogues, I have just received one from a shop I visited when I was last in London, called "The Serendipity Shop." It is located in a little slum known as Shepherd's Market, right in the heart of Mayfair. [This must be 7 Chapel St.] It may be that my readers will be curious to know how it got its name. "Serendipity" was coined by Horace Walpole from an old

---

[4] *Notes and* Queries series 9, no. 12 (31 October 1903): 349.

[5] Viola Meynell, *Alice Meynell: A Memoir* (New York: Charles Scribner's Sons, 1929), p. 272.

[6] Everard Meynell, "The Plain Dealer: VII: Starting a Shop," *The Imprint* (17 July 1913): 28.

name for Ceylon—Serendip. He made it, as he writes to his friend Mann, out of an old fairy tale, wherein the heroes "were always making discoveries by accident and sagacity, of things they were not in quest of." Its name, therefore, suggests that, although you may not find in the Serendipity Shop what you came for, you will find something that you want, although you did not know it when you came in. Its proprietor, Mr. Everard Meynell, is the son of Alice Meynell, who with her husband, did so much to relieve the sufferings of that fine poet, Francis Thompson, and who is himself a poet and essayist of distinction. Is not every bookshop in fact, if not in name, a Serendipity Shop?[7]

And Percy Mule, a great English authority on book collecting, writes:

> One of the nicest bookshops I was ever in was Everard Meynell's Serendipity Bookshop, just off Curzon Street. Alas, it exists no longer, and its owner is dead; but, before ill-health drove him to California, his shop offered one the wide choice, the occasional surprise (a slight shock at finding some oddity offered for sale, and the final response to its attraction, "Well, after all, why not?") and the comparative inexpensiveness that could only be the effect of the good taste of a shopkeeper who began with few if any preconceived notions of what it was proper for a book-collector to buy.[8]

And again, in a recent article, one George Sims, himself a bookseller, reminisces fondly about the "perfect" catalogues put out by Everard Meynell of the Serendipity Shop.[9]

Whether these few patrons of the Serendipity Shop whom we happen to know about shared their knowledge either of the shop or of the word with their friends, we do not know. Edward Newton was an expansive, friendly man, who knew a great many literary people and bibliophiles, and among his good bibliophilic friends were several who knew the word serendipity; all of them were, certainly, independently receptive to it. One of Newton's friends was Christopher Morley, and he uses the word serendipity in his preface to Boswell's *London Journal* to describe Professor Claude Abbott's discovery of the Boswell papers at Fettercairn House.[10] Wilmarth Lewis knew Newton well, and Lewis cannot help but know serendipity, but, again, we do not know whether it was a word that they enjoyed sharing. In the social circle to which all these American biblio-

---

[7] A. Edward Newton, *A Magnificent Farce, and Other Diversions of a Book-Collector* (Boston: Atlantic Monthly Press, 1921), p. 87.

[8] Percy Muir, *Book-Collecting: More Letters to Everyman* (London: Cassell and Co., 1949), p. 80.

[9] George Sims, "Three Booksellers and Their Catalogues," *The Book Collector* (1955): 291–298.

[10] Christopher Morley, "Preface," in James Boswell, *London Journal, 1762–1763* (New York: McGraw-Hill, 1950), p. xxi.

philes belonged, common acquaintance with the word serendipity does not appear to have created any special bond; there is no evidence that any one of these bibliophiles (except, possibly, Christopher Morley) attached any particular significance either to the word itself or to the experience to which it refers, nor is there anything to suggest that any member of the group used it *because* someone else had done so.

Returning to the English social circle to which the Meynell family belonged, we again have not found any evidence that serendipity had any special significance for anyone but the Meynells themselves. Wilfrid Meynell knew "everyone of importance in England for half a century, and the majority have been his friend," according to Cameron Rogers.[11] He was particularly intimate with Catholic writers such as Meredith, Thompson, Blunt, Henley, and Patmore. Indeed, the sale of Patmore's library was one of the last events that took place in the Serendipity Shop, in 1921. But except for Francis Thompson's one reference in a letter to Everard Meynell to "the shop whereof the name is mystery which all men seek to look into," we have found no references either to serendipity or to the Serendipity Shop among the Meynells' literary friends.

We suggested earlier that the Serendipity Shop would be most likely to appeal to collectors of out-of-the-way books, and, as P. H. Muir remarked, its charms would be more likely to be appreciated by collectors with eclectic tastes than to those who set out systematically to specialize. Collectors such as Wilmarth Lewis or Harvey Cushing tend to have rather definite ideas of what it is they want, and though they certainly make happy unanticipated discoveries, these can hardly play as large a part in their lives as they do for collectors who have few preconceived ideas of what it is that will make them happy. It is relevant in this connection that Mr. Lewis should remark on the frequency of accidental discoveries in the collection of association items: "Coincidence is so frequent in this branch of collecting that the collector of association items is led to believe that he has occult powers and that the person he is collecting is seeing to it that books, manuscripts, prints, snuff boxes, and so on, formerly in his possession come to the attention of the collector."[12] Such items are surely less well catalogued and cannot be tracked down as efficiently as can the author's works. Another area in which the specialized collector seems more likely to turn up unexpected treasures is that of unpublished material by his author, and so, since Horace Walpole was so voluminous a letter writer, Mr. Lewis is exposed to an unusual degree to the making of happy unexpected discoveries. Indeed, the dramatic dis-

---

[11] See "Punning in Paradise," in *Oh Splendid Appetite!* (New York: John Day Company, 1932), pp. 112–125; the quotation is on page 115.

[12] Wilmarth Lewis, *Collector's Progress* (New York: Alfred A. Knopf, 1951), p. 88.

coveries that Mr. Lewis describes so well in his *Collector's Progress* fall almost entirely into these two categories: unpublished materials and association items.

So much, then, for the collectors. In their social milieu unexpected good luck is a relatively frequent occurrence, and in retrospect, at least, it is fairly easy to see why we should find concentrated among collectors people who are familiar with the word serendipity.

Writers of Fiction and Nonfiction

In a very narrow sense, perhaps, writers can be considered to be collectors. Writers are collectors of words, but they do not collect them for their own sake, but rather as means to the optimal expression of ideas and feelings, to the capturing of the reader's interest and imagination. If we find, then, that the word serendipity occurs with more than random frequency among writers, it must be deemed to have special qualities of aptness and evocativeness by those who try to be stern judges of words. We discussed earlier the different kinds of responses that people have had to serendipity—curiosity about it, pride of possession, a sense of its oddness and whimsicality, a liking for its sound qualities—and it is responses such as these that writers seem to be projecting onto their readers when they elect to use the word. Sometimes the word is used loosely or whimsically, and it seems as if the writer used it to startle his readers into attention or curiosity. Or it may be used to flavor a piece of writing with preciosity, to give it an air of subtle refinement, and, perhaps, a touch of bluestocking pride in erudition. And, again, euphony may combine with aptness to recommend the word to a writer.

Among writers of fiction, especially, the effect obtained by using an odd word appears to be a significant consideration, and its aptness may be a somewhat secondary one. The first writer of fiction to use serendipity was Grant Allen, a very versatile man, at once a philosopher, a popularizer of science, and a novelist. His novels were really potboilers. In 1898, the year before Allen died, one such novel called *Miss Cayley's Adventures* was serialized in the *Strand* magazine. The narrator of the story is a lady journalist, and she says of herself: "I believe I must have been born with serendipity in my mouth instead of the proverbial silver spoon, for wherever I go, all things seem to come out exactly right for me."[13] Allen is at least minimally concerned that his readers understand what the word means. Not so Rumer Godden, as we saw earlier: In her novel *A Breath of Air* (1951), the cook is called Serendipity, without

---

[13] Grant Allen, "Miss Cayley's Adventures," installment 9, *Strand* (December 1898).

reason or explanation. John Masters, again, in *Bhowani Junction* (1954) uses a word form of serendipity very aptly, but with only enough explanation to whet the reader's curiosity, not to satisfy it. The author is describing the fine animal heads that one of his characters, Patrick Taylor, had collected, and he remarks: "It might have been incredible good luck—but the words 'Patrick Taylor' and 'good luck' were not in the same dictionary. If he got those heads with his clumsiness and his unserendipity, he had got them by will power, determination, and nothing else."[14]

The writers of nonfiction—that is, the essayists—tend to use serendipity somewhat less casually than do the novelists. In other words, they seem to be somewhat more concerned that the meaning they attribute to serendipity be understood by their readers, and, further, that this meaning be taken seriously. However, essays for the lay reader should not be "heavy going," and so the oddness and whimsicality associated with the word come usefully into play. Serendipity works as a kind of leaven to an essay: subtle and apt, yet slightly frivolous, an ideal combination for the middle-brow reader.

The very first essay in which serendipity played a part was, indeed, only about a rather dimly remembered version of the word. "The Spectator," a regular columnist for the American periodical *The Outlook* in the early twentieth century, and a man with an unbreakable pseudonym, wrote in the issue of July 18, 1903:

> " 'I've found a new word and a new amusement for you', said [a] friend the other day. 'It's serendipity.' The Spectator acknowledged that he didn't recognize the article and craved enlightenment. 'I don't know exactly what it means,' was the answer, 'but it's an amusement. I know that because I found it among the *Recreations* attached to the brief biographies of 'Who's Who.' I'll show it to you.' But for once the tenacious memory of The Spectator's friend failed him; he could not recall the famous man who had invented this remarkable recreation. And the Spectator also finds it not."

Characteristically for this kind of essay, the Spectator's vain search for the famous man in *Who's Who* presents an opportunity for some general remarks about English and American attitudes toward recreation: In England, recreations are permitted and are, therefore, listed in *Who's Who*; in America, recreations are deemed too frivolous to be listed. (The famous man who gave serendipity as his recreation was, of course, Wilfrid Meynell—as one of The Spectator's readers, H. K. Armstrong, hastened to inform him in the issue of August 15.)

Again in 1905, serendipity became the subject of an essay by Ellen Burns Sherman in *The Criterion* (which in 1907 was reprinted in her

---

[14] John Masters, *Bhowani Junction: A Novel* (New York: Viking Press, 1954), p. 344.

collection of essays titled *Words to the Wise and Others*). Miss Sherman, who died in 1956 at the age of eighty-nine, was an intellectual lady from New England, an early graduate of Smith College (class of 1891), who devoted her life to writing essays and poetry. Judging by a review of *Words to the Wise and Others* that appeared in *The Outlook* (November 30, 1907), her essays were appreciatively received, and her style, which today would be perhaps rather generally considered flowery and precious, was especially commended by the reviewer. The essay on serendipity toys in a quasi-scientific manner with ideas about subconscious intelligence and subliminal memories, and describes rather trivial episodes in which by coincidence odd little discoveries were made or problems solved. It is the sort of essay that might be relished at tea time by serious ladies who scorn idle chatter about fashions and servants; and it might strike a responsive chord in certain gentlemen who enjoy taking tea with these same ladies. We know of at least one interested and well-informed reader of Miss Sherman's essay in *The Criterion*: On April 29, 1905, there appeared a long letter in the *Saturday Review of Books, Supplement of the New York Times*, by one Emma Carleton of New Albany, Indiana. Miss (or could it be Mrs.?) Carleton enjoyed the Sherman essay very much, and she proceeds to give a very accurate account of the invention of serendipity by Horace Walpole and to take issue with Ellen Sherman with regard to her interpretation of the meaning of the word.

Serendipity was well suited not only to the thought but also to the style of yet another essayist, this time a preacher-essayist, Samuel Mc-Chord Crothers. Crothers was for a third of a century (1894–1927) the minister of the First Unitarian Church in Cambridge, Massachusetts, and there he preached a gospel of common sense in his sermon-essays. The essays had both wit and seriousness, subtlety and playfulness, and, according to Crothers's admiring biographer in the *Dictionary of American Biography*, Ephraim Emerton, they "[made] their appeal to a choice circle of discriminating readers."[15] Crothers's use of serendipity serves as a nice illustration of the way he combined seriousness of thought with novelty in language. In the introduction to his volume of essays called *Humanly Speaking* (1912), he recommends serendipity to his readers as an "auxiliary virtue,"[16] in addition to the ordinary Christian virtues, and he enlarges upon the rewards this auxiliary virtue will bring to the individual—all with a mixture of seriousness, grace, and good humor.

Crothers's sermons impressed not only his parishioners and the de-

---

[15] *Dictionary of American Biography* (New York: Charles Scribner's Sons, 1929), pp. 572–573.

[16] Samuel McChord Crothers, *Humanly Speaking* (Boston: Houghton Mifflin, 1912), p. xii.

votees of magazines such as the *Atlantic Monthly*. Evidence can be found elsewhere of the impact of his "half-apologetic seriousness."[17] In *Collector's Progress*, Wilmarth Lewis recalls:

> I never met the man who first helped direct (unconsciously) the course of my collecting. He was the Reverend Samuel McChord Crothers of Boston. He talked to the Thacher School one evening nearly forty years ago in the interval between supper and the evening study hour. Speakers on those occasions usually tried to save us from ourselves, but Dr. Crothers, even though he was a clergyman and an essayist, merely talked to us pleasantly about the delights of knowing another age as well as our own. A dozen years later Professor Tinker said something similar to me one day and Dr. Crother's little talk returned with the force of revelation.[18]

Ephraim Emerton was not the only Harvard professor who admired Crothers. We have it on the authority of Walter B. Cannon's son-in-law, John K. Fairbank, that it was Crothers who introduced Cannon to the word serendipity, and who was, therefore, the all-important link between the literary and the scientific users of the word.

Earlier, we mentioned that Lord Mancroft's proclivities for collecting made him receptive to serendipity. But Mancroft, like The Spectator, also contributed chatty essays to the *Saturday Review*, and serendipity, both as a word and as a notion, gave him a useful jumping-off place for a rambling essay about all manner of happy accidental discoveries.[19] In this case, the appreciative reader who wrote a letter to express his pleasure in the article was a civil engineer, Edwin Hodge, cut off from civilization in Cape Town.[20]

In the literary world there are still others who have used serendipity, more or less aptly, in order to enliven their writings with an unusual and interesting word: Leonard Bacon, a poet and critic with a somewhat aggressive bent, used it in an essay titled "The Long Arm of Coincidence" in the (American) *Saturday Review of Literature* (December 14, 1946); Winifred Rugg wrote a rather bubbly essay for the *Christian Science Monitor* (November 16, 1953), apostrophizing the sound and sense of serendipity; and Mary McCarthy used it to describe an odd incident that occurred while she was traveling in Portugal, an incident which, curiously enough, bears no resemblance whatsoever to a happy accidental discovery (*New Yorker* [February 5, 1955]: 87). Serendipity seems to serve all these writers as a device to stimulate their readers to a kind of double-take, a double-take

---

[17] See Emerton's entry on Crothers in *Dictionary of American Biography*, pp. 572–573.

[18] Wilmarth S. Lewis, *Collector's Progress* (New York: Alfred A. Knopf, 1951), pp. 68–69.

[19] See Mancroft's essay in *Saturday Review* (26 May 1918).

[20] See *Saturday Review* (25 September 1918).

that will make them more likely to pay willing attention to the writer's words. The attractiveness of the word lies, for the writers, in its arresting quality, and by its use the writers seem to hope to produce silent exclamation of pleasure, of curiosity, of amusement, or simply of satisfaction at getting to know such a rare bird in the English vocabulary.

Literary Scholars and Lexicographers

It is the professional concern of literary scholars and lexicographers to be as widely informed as possible in their chosen fields, and so if one of these scholars has come upon the word serendipity, it is most likely to be a discovery made in the course of professional researches. The word turns up as one more fact about eighteenth-century literary history, perhaps, or about word coinages, and the scholar takes note of it along with other facts. Its history may legitimately arouse his interest as a scholar; but if he is struck by its whimsicality or its pleasant sound, he is responding as a private citizen, as it were. Correspondingly, the use of serendipity is rigorously limited by its relevance to the scholar's subject matter; whimsy and frivolity are extraneous to scholarship, the reader's attention (or lack of it) is taken for granted, and the word may be introduced only if the fact of its existence is significant.

Thus, as a matter of relevant information, serendipity is brought to the reader's attention in Martha Pike Conant's *The Oriental Tale in England in the Eighteenth Century* (1908). In her chapter on "The Imaginative Tales," Miss Conant writes: "After the *Arabian Nights*, the *Persian Tales*, and the *Turkish Tales*, the best imaginative oriental tales are the English versions of the so-called pseudo-translations. The first to appear in English was *The Travels and Adventures of the Three Princes of Serendip*." Here the following footnote is inserted: "Cf. Horace Walpole's coinage of the word 'serendipity,' meaning 'accidental sagacity,'" and a precise reference to Walpole's letter to Mann in the Toynbee edition of Walpole's letters.[21]

Another well-informed scholar who can supply information about serendipity when it is needed is Marjorie H. Nicolson. In reply to a query by one G.M.B. of Niagara Falls, New York, addressed to May Lamberton Becker's "Reader's Guide" column in the *New York Herald Tribune Book Review* (July 7, 1946), Miss Nicolson provides the desired information about that elusive fairy tale, *The Travels and Adventures of the Three Princes of Serendip*. But we have no evidence that Miss Nicolson ever

---

[21] Martha P. Conant, *The Oriental Tale in England in the Eighteenth Century* (New York: Columbia University Press, 1908), pp. 29–31.

used serendipity in her books about the relationship of science and literature: Although it might be apt, it could not, apparently, be considered
relevant. Not until serendipity enters the milieu of scientists and social
scientists do we find it in use as a reputable technical word.

In the world of literary scholarship, even such a devotee of serendipity
as Leslie Hotson seems to shy away from using the word where it does
not in itself constitute a bit of factual material. In his pleasant article
"Literary Serendipity," Hotson describes well that kind of scholarship of
which all his own books are fine examples: literary detective work. But in
these same books, for example, *The Death of Christopher Marlowe*, or *The
First Night of Twelfth Night*,[22] he never once uses serendipity to describe
what are, without doubt, happy finds of the literary prospector.

Even more directly than literary scholars, lexicographers make it their
business to know the facts about the history of the use of words, and we
know about quite a few of these scholars who are familiar with the word
serendipity. Eric Partridge includes the word in his dictionary *Name into
Word* (1949), Joseph T. Shipley has it both in his *Dictionary of Word
Origins* (1945) and in his more recent *Dictionary of Early English*
(1955), and David Guralnik of *Webster's New World Dictionary* has a
passing acquaintance with it. Ernest Weekley, a great English etymolo-
gist, used serendipity in a book titled *Adjectives and Other Words* (1930)
to describe the addition of words to the language by happy misunder-
standings; in 1921, Weekley had not included it in his *Etymological Dictionary of the English Language*, but in the revised edition of that dictionary, in 1952, it is included. This would suggest that there are many
more lexicographers who know the word serendipity than can be inferred
simply by examining dictionaries to see if they do or do not include the
word. One of the criteria that determines the inclusion of a word in a
dictionary is currency, and by that criterion serendipity may well have
been excluded although the editor of the dictionary was quite familiar
with it. (See also Chapter 6.) Among lexicographers as among literary
scholars, sheer knowledge of a word is insufficient reason for exhibiting it
when it does not constitute relevant information.

The Medical Humanists

The first man who transferred serendipity from the milieu of humanistic
endeavors to that of scientific research had to be a man with both imagination and courage. Imagination was required not so much to perceive

---

[22] Leslie Hotson, *The Death of Christopher Marlowe* (London: Nonesuch Press, and
Cambridge, Mass.: Harvard University Press, 1925); Leslie Hotson, *The First Night of
Twelfth Night* (London: Rupert Hart-Davis, 1954).

the role of accidental discoveries in the advance of science—that had been done by eminent men before—but to see the advantage that might be gained by calling that phenomenon by a new and eye-catching name. Courage was required to use with great seriousness a word that had heretofore been, at best, a bit of literary bric-a-brac, and, at worst, a bit of a joke. Walter Cannon, who performed the successful operation of grafting the word from the language of literature onto the language of science, was, not surprisingly, a scholar who was at home in both fields. Like several of his distinguished contemporaries, he was a scientist with a wide knowledge of English literature. He had much in common with other great medical humanists such as Sir William Osler, Harvey Cushing, and Edward Clark Streeter, all of whom loved and collected books. One has only to read the biographies of Osler and Cushing to see with how many of their medical colleagues they shared their detailed knowledge of books. But until Cannon learned of serendipity from the Rev. Samuel McChord Crothers and proceeded to use it to describe certain experiences in the career of an investigator, no other medical humanist had, to our knowledge, used it in any context.

It is hard to know just when Canon began to use serendipity and to tell his colleagues and students at the Harvard Medical School about it. His own first use of it in print came quite late: In December 1939, in an address before the American Science Teachers Association, his topic was "The Role of Chance in Discovery," and he told his audience all about serendipity. This address was printed in the *Scientific Monthly* in 1940, and Cannon used it almost unchanged as his chapter "Gains from Serendipity" in *The Way of an Investigator* (1945).[23] (The chapter has been twice reprinted: In 1952, Samuel Rapport and Helen Wright included it in *Great Adventures in Medicine*,[24] and in 1956, Leonard Engel put it into his *New Worlds of Modern Science*[25] and described it as on the way to becoming a classic.)

However, five years before Cannon's address to the American Science Teachers, Milton J. Rosenau, another doctor with humanist inclinations, and dean of the School of Public Health at Harvard, made "Serendipity" the subject of his presidential address before the Society of American Bacteriologists.[26] Rosenau said in his talk: "I first learned of serendipity from my colleague, Dr. Walter B. Cannon," and this sounds as if Rose-

[23] Walter B. Cannon, "The Role of Chance in Discovery," *Scientific Monthly* (19 March 1940): 204–209; and Walter B. Cannon, *The Way of an Investigator* (New York: W. W. Norton, 1945).

[24] Samuel Rapport and Helen Wright, *Great Adventures in Medicine* (New York: Dial Press, 1952).

[25] Leonard Engel, *New Worlds of Modern Science* (New York: Dell, 1956).

[26] *Journal of Bacteriology* 29 (February 1935): n. 2.

nau had learned it a long time ago, so long that he himself felt quite sufficiently at home with the word and the idea to make it the subject of an important address. Rosenau was, as we already observed, an academic elder statesman in 1934 (he retired from Harvard in 1935), and it seems possible that, though both he and Cannon had used serendipity for a long time, they felt that only the dignity of an aging professor could launch it as a word to be taken seriously in public. (There is a confirmatory straw in the wind for this hypothesis: In a talk to the graduating class of the Yale Medical School in 1911 titled "The Career of an Investigator," Cannon exhorted the new-fledged doctors to consider the rewards of medical research as well as of practice, and he described the role that chance plays in research—but he did *not* use the word serendipity.)

It was surely the influence of Cannon and of Rosenau that made serendipity a byword in the social circle of the Harvard Medical School. Miss Holt, the chief librarian at the medical school, who has been working there a long time, thinks that, although Cannon and Rosenau were the only men who wrote about serendipity, many other teachers at the medical school would very likely have used the word in informal talks. Her supposition is borne out by some evidence we found independently. In a talk to the Cornell Medical School freshman class in 1941, Bronson Ray described how he had recently rediscovered the word serendipity in a conversation about William Beaumont with the artist Deane Keller of Yale University. Dr. Ray did not know what the word meant when Keller used it. "And then," he said, "I recalled that Dr. Elliott Cutler, Professor of Surgery at Harvard, had given a talk here several years ago to the students and had mentioned serendipity as a desirable quality for a student to have." Dr. Cutler had been educated at the Harvard Medical School, and in 1933 he had become Dr. Cushing's successor as surgeon in chief at the Peter Bent Brigham Hospital.

Another medical educator who made use of serendipity in an address before a group of students was Harold Brown, director of the School of Public Health at Columbia. Dr. Brown's acquaintance with the word seems likely to have been the result of his association with Dr. Rosenau, for he was at Harvard when Dr. Rosenau was there, and he also taught at the School of Public Health at the University of North Carolina, where Rosenau went as director after his retirement from Harvard. Dr. Brown chose "Medical Serendipity" as the subject of his address at the opening exercises of the College of Physicians and Surgeons in September 1953.

The very fact that serendipity was something of a byword at the Harvard Medical School, where its use was, perhaps, a small symbol of membership in the "club," may rather directly have led to the word's not having been used in an affiliated institution, namely, the Brigham Hospital. For twenty years Harvey Cushing reigned supreme at the Brigham, and Dr. Cushing was a proud and independent man. He was not likely to take up a "new" word that had been introduced by someone else. More than that, Cushing never felt that he belonged at Harvard; in all the years he was there, he never got over the sense of being a stranger at odds with the community. And this feeling of alienation may have enhanced his nonreceptivity to what was, at that time, a Harvard word. It is significant that Dr. Ray, who was a friend and associate of Cushing's on the staff of the Brigham between 1929 and 1931, first came into contact with serendipity when he heard Dr. Cutler use it—Dr. Ray received his medical training at Northwestern University.

Quite independent of the Harvard Medical School circle, another doctor-humanist, Ellice McDonald, discovered serendipity and saw how aptly it described some of the events that occurred in the research laboratory. Like Cannon, Dr. McDonald devoted himself to research (in his case, cancer research), although unlike Cannon, he had practiced medicine when he first got out of the medical school at McGill University. For the last twenty years of his career, until his death in 1955, McDonald was director of the Biochemical Research Foundation of the Franklin Institute in Philadelphia, and it was in his annual report on the work of the Foundation for 1938 that he first used serendipity.[27] Not until ten years later, in an article titled "Choice of Research Projects,"[28] did Dr. McDonald reveal where he had found the word: "While on holiday in Cuba in 1938, I found this word in a detective story by S. S. Van Dine." The story in question has, unhappily, never been located by us.

The medical humanists were writing and speaking to professional audiences, to fellow scientists, or, at least, scientists in embryo, who identified with them in their work, and who were ready to benefit from the insights into the process of scientific discovery of such distinguished men as Cannon, McDonald, and Cutler. The members of these sympathetic audiences were, no doubt, generally receptive to new insights into the role of chance in the making of discoveries, and, in some cases, happy to adopt the new word that was being offered to designate the role of chance. But it is only by inference that we can connect several users of serendipity in the biological sciences with the writings on that subject by Cannon or McDonald or one of the other medical humanists we mentioned. Almost never are the medical humanists credited with introducing the word, and their contribution was evidently not judged to be of the kind that requires acknowledgement, that is, it was not intrinsic to the substance of science. (We shall see presently that in social science the insights provided

[27] See *Journal of the Franklin Institute* 227 (January 1939).
[28] *Journal of the Franklin Institute* 247 (April 1949).

under the heading of serendipity, apparently, are of a different order of significance.)

Although it is not possible, then, to trace the process of diffusion of serendipity from its initial use by the medical humanists, the biologists and biochemists who used it were certainly part of the wider milieu in which the influence of Cannon especially, and of the others to a lesser extent, was felt. Thus we find a report in *Chemical and Engineering News*[29] of a talk by Roger Adams, of the University of Illinois, given at the Westinghouse Science Writing Award luncheon, in which Adams had told his audience about the important part serendipity played in research. Again, in a technical paper on "The Use of Protozoa in Analysis,"[30] the authors remark that "the immediate motive for isolating and studying these organisms was generally interest in them for their own sake; their occasional applications to analysis represent instances of serendipity." And a bacteriologist, Salvador E. Luria of the University of Indiana, describes the role serendipity played in his exploration of "The T2 Mystery," the mystery of a virus that dissolves bacteria.[31] Only Esther Everett Lape, in her discussion of the pros and cons of planned research,[32] refers directly to Cannon's conception of the role of serendipity in research. (She gives no specific reference to Cannon's writings, however.) As with so many other verbal innovations that come to be taken for granted, the role of the medical humanists who first transferred serendipity to the language of science may soon be forgotten.

Social Scientists

The rediscovery of serendipity by Robert Merton, who subsequently introduced it into the vocabulary of the social sciences, was itself a combination of accident and sagacity. Merton, as we have cause to know, belongs to that breed of dictionary readers who find the dictionary a repository, not merely of definitions, but of synopses of cultural history; who not only "use" the dictionary in the sense of "looking up" words that have come to their attention but which they do not understand, but also, systematically or sporadically, *read* dictionaries. Looking up a particular word becomes only the starting point for an excursion into the sur-

---

[29] *Chemical and Engineering News* 29 (15 January 1951): 194.

[30] L. D. Hamilton, S. H. Hutner, and L. Provasoli, "The Use of Protozoa in Analysis," *Analyst: The Journal of the Society of Public Analysts and Other Analytic Chemists* 77 (November 1952): 618–628.

[31] Salvador E. Luria, "The T2 Mystery" [with biographical sketch], *Scientific American* 192 (April 1955): 22, 94–94.

[32] *Medical Research* 1 (1961): 101.

rounding territory, and it was in the course of such an excursion that Merton came upon serendipity.

So much for the accidental component in this discovery. As for sagacity, that was involved too, for it was a receptive eye that fell, in this case, on serendipity. For more than ten years, since about 1934, Merton had been much interested in the "unanticipated consequence of purposive social action," and had written a paper on this subject in 1936. Serendipity, which refers to the search for one thing that turns up another, is obviously related to the problem of unanticipated consequences, and so was intimately connected with the discovery's absorbing interest. (Before the development of this intellectual concern, Merton was quite unreceptive to the word: He had encountered it, without noticing it, in or about 1931, while reading Joseph Roseman's *The Psychology of the Inventor*, which was published in that year.)

The resonance of serendipity for Merton lay not only in its relation to his theoretical interests, but as a neologism per se. He has had an abiding interest in the case that can be made for, and against, word coinages, especially in science, and he has defended the use of technical terminology that is frequently damned as "mere jargon." He himself has tried his hand at the coinage of neologisms in sociology, some of which have been ratified by subsequent usage by his sociological colleagues—and others have died a quick death. Examples of successful coinages are: *manifest* and *latent functions, homophily, pseudo-Gemeinschaft,* the *focused interview;* serendipity was to be another successful quasi-neologism.

The result of Merton's discovery in the dictionary was, at first, only a bare mention of serendipity: In a paper on "Sociological Theory," there is a qualifying footnote that reads, in part: "Fruitful empirical research not only tests theoretically derived hypotheses; it also originates new hypotheses. This might be called the 'serendipity' component in research, i.e., the discovery, by chance or sagacity, of valid results which were not sought for."[33] The following year, however, in a paper read before the American Sociological Society (March 1, 1946), this casual allusion to serendipity of a year earlier becomes elaborated into the "serendipity pattern" of research in sociology, a pattern that is described for several pages.[34] Also, Merton's footnote to serendipity now included references to Walpole's coinage of the word and to Cannon's usage of it in *The Way of an Investigator.* By 1949, the present lengthy study of serendipity was

---

[33] Robert K. Merton, "Sociological Theory," *American Journal of Sociology* 50 (1945): 462–473; reprinted in Robert K. Merton, *Social Theory and Social Structure* (New York: Free Press, 1949). Quotation on p. 255.

[34] Robert K. Merton, "The Bearing of Empirical Research on Sociological Theory," *American Sociological Review* 13 (October 1948).

foreshadowed: When "The Bearing of Empirical Research on Sociological Theory" was reprinted in *Social Theory and Social Structure*, a new, long footnote was inserted, in which the diffusion of the word had become a matter of explicit interest.[35] Thus, serendipity was launched by Merton as a word that deserved notice, both as referring to an important pattern in research and as the object of a small boom in the market of words.

Because the journals in which Merton's usage of serendipity first appeared, the *American Journal of Sociology* and the *American Sociological Review*, are read regularly by the great majority of his sociological colleagues, and because, moreover, the articles in which it was used both dealt with subjects of very general interest, it followed that the word *was* noticed. Not only was the word noticed, but frequently the theoretical contribution that its use by Merton represented was acknowledged by a reference to one or the other of his articles. (More often, it is the second [1948] article that is referred to, because that is the one that enlarges the concept of serendipity into a pattern.)

Alvin Gouldner was the first to pick up the word and to use it, in his "Discussion" of a paper by Wilbert Moore on "Industrial Sociology."[36] Gouldner was a student of Merton's at Columbia University, and so he was likely to be motivated to use serendipity not only by the merits of the insights conveyed by the word, but by his identification with Merton's larger conception of sociology.

Some of the other users of serendipity in sociology who refer to Merton's initial usage of the word share, to a lesser extent, Gouldner's motivation: They are appreciative of the theoretical advance represented by the concept of the "serendipity pattern," and they identify in a general way with Merton's "kind" of sociology. They constitute, in effect, a rather loosely cohesive social circle. Among the members of this social circle are: Irving Spaulding, who wrote an article with the title "Serendipity and the Rural-Urban Continuum"; Bernard Barber, who, in his book *Science and the Social Order*, describes the "serendipity pattern" quite extensively (Barber refers to Cannon's use of serendipity as well as to Merton's); Svend Riemer, in his article "Empirical Training and Sociological Thought"; and, most recently, Daniel Glaser, in "Criminality Theories and Behavioral Images."[37]

---

[35] Merton, *Social Theory and Social Structure*, pp. 255–257.

[36] *American Sociological Review* 13 (August 1948): p. 397, n. 2.

[37] Irving Spaulding, "Serendipity and the Rural-Urban Continuum," *Rural Sociology* 16 (March 1951): 29–36; Bernard Barber, *Science and the Social Order* (New York: Free Press, 1952), pp. 203–205; Svend Riemer, "Empirical Training and Sociological Thought,"

One sociologist, Howard Becker, while cognizant of Merton's usage of serendipity, appears to be an independent discoverer and admirer of the word. Earlier we quoted the footnote in his article "Vitalizing Sociological Theory,"[38] in which he writes: "Hurrah for serendipity! (Merton's research-scientific usage we know, but for an amusing reference to the theme, more recent than the tale alluded to by Horace Walpole, 'The Three Princes of Serendip,' see Ellen Burns Sherman, *Words to the Wise and Otherwise*, [*sic*]," and there follow references to the *Mancroft Essays*, to Leslie Hotson, and to *Notes and Queries*.

Some few social scientists have assimilated the word without any remarks about its provenience. Samuel A. Stouffer of Harvard University mentions it in an article titled "Measurement in Sociology"; Allen Wallis, an economic statistician, uses it repeatedly, without any references, in his recent textbook on statistics, *Statistics: A New Approach*, written in collaboration with Harry V. Roberts; and, finally, in his review of Herbert Hyman's *Survey Design and Analysis*, Edward A. Suchman of Cornell University uses serendipity as simply as he would any other technical term.[39] In just over ten years, serendipity has become a standard ingredient in the vocabulary of sociological theory.

Even among non-English-speaking sociologists the word is now being diffused, as a result of the translation into French of part of Merton's book *Social Theory and Social Structure* in 1952. Serendipity was transposed unchanged into the French translation as "la serendipity," and French reviewers have, in turn, used the word, however strange it must seem to them. Robert Pagès, in *Critique* (Mars 1954) evidently found it so alien that he inadvertently added an extra syllable to it: it became *sérendipidité*. The other reviewers reproduced it correctly: Georges Condamines in *Sciences Humaines* (Août 1954) mentioned "la notion de 'serendipity', autrement dit, la découverte par chance ou sagacité," and M. Dufrenne even provided a bit of etymological explanation: "la 'serendipity'—ainsi nommée en souvenir d'un conte sur les trois princes de Serendip."[40] If linguistic purity means as much to the French as is often

---

American Journal of Sociology 59 (September 1953): p. 112, n. 2; and Daniel Glaser, "Criminality Theories and Behavioral Images," *American Journal of Sociology* 61 (March 1956): p. 437, n. 4.

[38] Howard Becker, "Vitalizing Sociological Theory," *American Sociological Review* 19 (August 1954): 384.

[39] Samuel A. Stouffer, "Measurement in Sociology," *American Sociological Review* 18 (December 1953): p. 595, n. 6; W. Allen Wallis and Harry V. Roberts, *Statistics: A New Approach* (New York: Free Press, 1956); and Edward A. Suchman in *American Sociological Review* 21 (April 1956): p. 233, n. 1.

[40] *L'Année sociologique* (1955): 210.

alleged, their acceptance of serendipity, even into the technical vocabulary of sociology, represents something of an accolade.

Applied Research

The members of the scientific milieux we have considered so far, both natural and social, have been concerned with the development of science as an end in itself, and, in some cases, with the education of future scientists. They have been interested in the phenomenon of unanticipated discoveries because an understanding of such discoveries might facilitate the advance of science in general; and they have been moved by the intellectual imperative that demands the understanding of all areas of human activity as an end in itself. In the milieu of applied research, however, such an understanding of the process of scientific discovery serves as a means to a further end: the turning up of discoveries that will be useful. Their immediate usefulness may consist in their bringing profits to a particular company or in their providing protection for men in the armed forces. In either case, it is the obligation of applied scientists to discover useful items, and the question of *how* discoveries are made is asked not for its own sake but as a matter of practical strategy. If useful discoveries may be made in unexpected ways, it will pay off to take this possibility into consideration.

The scientists who are engaged in research that is being sponsored by industry or by the government are considered directly accountable for the applicability of the results of their work in ways that scientists in the universities are not held accountable. They must answer to stockholders or to government officials for the usefulness of their activities, and they must expect the company or the government rigorously, and sometimes unsympathetically, to scrutinize those activities. In consequence, applied scientists, and especially the spokesman for applied research, explain how they are trying to maximize the returns for the work that goes on in the laboratories. More particularly, they are found to explain how the absence of rigid control over the work that scientists do facilitates the discovery of important things by accident.

Soon after Cannon's article on "The Role of Chance in Discovery" appeared, the *Industrial Bulletin of Arthur D. Little, Inc.* (July 1940) published an article on the theme of serendipity. Arthur D. Little is a company in Cambridge, Massachusetts, that does pilot studies in applied research for companies that do not have the resources to support research laboratories of their own. The anonymous writer of the article evidently was not encountering serendipity for the first time in Cannon's article, for his concluding paragraph begins: "This article was submitted for comment to Dr. Willis Whitney, as one of the users and appreciators of the term serendipity." And Dr. Whitney's comment follows.[41]

Willis Whitney was for many years director of research of the General Electric laboratories, as well as a professor at M.I.T. We do not know of any use that he himself made of serendipity in print, apart from his comment in the Arthur D. Little *Bulletin*, but another report of his usage of the word comes from Irving Langmuir, a distinguished scientist and a longtime associate of Willis's at General Electric. At a colloquium held at the G.E. Research Laboratory on December 12, 1951, Langmuir said: "Dr. Willis Whitney discovered a word which is in the dictionary—serendipity." Part of this talk by Langmuir was printed as a full-page advertisement by G.E. inside the back cover of the *Scientific Monthly*;[42] we shall refer to it again when we discuss the role of serendipity in the explanation of science to the public. It was printed again in the July 1952 issue of the General Electric *Share Owners Quarterly*—but this time the phrasing of the article does not include an acknowledgement of Langmuir's debt to Whitney for serendipity.

From General Electric the word serendipity diffused to Standard Oil of New Jersey. In the Standard Oil trade publication *The Lamp* (September 1953), there is a two-page article titled "Serendipity," which makes specific reference to Langmuir's discussion of the word. Again, in January 1956, Standard Oil found the word useful in a letter distributed among university scientists along with a booklet explaining the uses of butyl rubber.

In the pharmaceutical industry, too, serendipity was found to be a useful word to explain the vagaries of research. It appears to be a word appreciated by George W. Merck, president of Merck and Co., who first used it in an article titled "Peacetime Implications of Biological Warfare."[43] Some years later, he used the word again in an interview,[44] and at Christmastime in 1954, there appeared a Merck advertisement in *Business Week* that prominently featured the word serendipity. The Pfizer Company, a competitor of Merck's in the pharmaceutical field, ran a two-page advertisement-article in the *Journal of the American Medical Association* (July 17, 1954), titled "Serendipity: The Happy Accident," calculated to inform *JAMA* readers about the importance of letting accidents happen. And quite a few investors must have learned about the valuable role of serendipity in drug research as they read the *Weekly Review* (April 16,

[41] Three years later, this article was reprinted in full in *Chemical and Engineering News* 21 (25 April 1943).

[42] *Scientific Monthly* 74, no. 3 (March 1952).

[43] George W. Merck, "Peacetime Implications of Biological Warfare," *Chemical and Engineering News* 24 (25 May 1946): 1348.

[44] *Time* (18 August 1952): 44.

1956, p. 4), published by the brokerage firm of Baker, Weeks and Co., at One Wall Street.

From the area of government-sponsored research we have only one instance of the use of serendipity. In the *Monthly Newsletter* issued by the Bureau of Supplies and Accounts of the Navy Department,[45] there is a report by T. J. Seery, head of the Clothing Section of Research and Development, with the title: "Serendipity—A New Name for an Old Art." In an explanatory footnote, Seery tells of Horace Walpole's coinage of serendipity and goes on to say that the word "was given widespread attention after use by Dr. Cannon at Harvard in describing unexpected scientific achievement." Here is one of the few references to Cannon's role in the diffusion of serendipity, and it comes from one apparently quite far removed from the academic world.

Science Writers

The "general public," unlike the investor, is not so much interested in reassurance that the most effective strategy is being pursued for the achievement of useful discoveries as in some enlightenment about "what's going on" in the world of science. The layman knows that the work of the scientists has, ultimately, a more or less direct effect on his own life, and so he seeks both some explanation of what scientists' activities mean in concrete, everyday terms, and some vicarious participation in these momentous activities. The popularization of science, which is the response to the layman's quest for enlightenment, is not a recent development—we have only to remember the work of Samuel Smiles— but perhaps the task of science writers has become increasingly difficult as scientific theory has moved ever farther away from commonsense understanding. If we find, then, several science writers using the word serendipity, they may have been moved to do so not only because it so aptly describes what can happen in the course of research, but also because the particular occurrences it refers to are closer to the layman's level of experience than many others that take place in the laboratory. The factor of accident perhaps lends a kind of "human interest" to science, and the word serendipity describes that factor aptly and strikingly.

The group that we have chosen to call science writers is not a homogeneous one: It consists chiefly of professional journalists and historians of science, but also of scientists, insofar as they are trying to communicate with the lay public rather than with their professional colleagues. Among the journalists, we find Robert W. Dorman writing an article titled

[45] *Monthly Newsletter* 16, no. 8 (15 August 1952).

"Watching for the Unexpected," in which he remarks that the " 'lucky' side of scientific discovery is called serendipity."[46] Robert K. Plumb, one of the science correspondents of the *New York Times*, sent in a dispatch from Oklahoma City (July 4, 1949), reporting an address by Sir Alexander Fleming: "Sir Alexander's experience," says the dispatch, "is frequently cited as an outstanding example of the importance of serendipity in science." In the syndicated column "Your Health," disseminated by the Pennsylvania State Medical Society, the subject of the article for June 7, 1950, was the role of serendipity in medical discoveries.[47] The piece about serendipity in the Standard Oil publication *The Lamp*, which we mentioned earlier, was used by two journalists to enliven their columns with some "inside dope" about science and as an opening for making some remarks about luck: C. Norman Stabler, the financial columnist of the *New York Herald Tribune* (September 24, 1953) was one of these, and James Pooler, who writes a folksy column called "Sunny Side" for the *Detroit Free Press* (January 7, 1954), was the other. A condensation of the article in *The Lamp* appeared in the *Democrat Digest* (March 1954), thus spreading to an avowedly political audience the fact that, as the title has it, "science owes a lot to serendipity"; but no political implications are explicitly drawn from this fact. Finally, among the journalists, we have Earl Ubell of the *Herald Tribune* reporting on two occasions the role of serendipity in cancer research: One such dispatch appeared on April 4, 1954, the other, by coincidence, on April 4, 1955.

More conventional scholars in the history of science do not appear to have used serendipity in their writings, but two outsiders, one from the field of fiction and one from poetry, have done so. The poet is Muriel Rukeyser, who in her biography of *Willard Gibbs* (1942) writes: "This quality, serendipity, is the luck of the discoverer, and those who find new applications have it. The knotted myth of the voyages of Gibbs' ideas is a myth of this quality."[48] Mitchell Wilson, who has written novels, put together a pictorial history of *American Science and Invention* (1954) and in it the section that describes the work of Irving Langmuir is called "A Case of Serendipity."[49]

Last of all, we shall mention some scientists who have tried to tell the lay public something about scientific developments and the role of acci-

[46] Robert W. Dorman, "Watching for the Unexpected," *Science Illustrated* (October 1946): 66.

[47] See "Your Health," *The Centre Daily Times* 53 (7 June 1950): p. 4, n. 56.

[48] Muriel Rukeyser, *Willard Gibbs* (Garden City, N.Y.: Doubleday, Doran, 1942), p. 421.

[49] Mitchell Wilson, *American Science and Invention, a Pictorial History: The Fabulous Story of How American Dreamers, Wizards, and Inspired Tinkerers Converted a Wilderness into the Wonder of the World* (New York: Bonanza Books, 1954), p. 372.

dent in these developments. In an essay in a 1946 collection, the Harvard astronomer Harlow Shapley tells of the event leading to the discovery of stellar atomic energy, and he mentions the element of serendipity.[50] The British biologist N. W. Pirie wrote an article for the semipopular *Science News*, in which he assesses the role of serendipity in the equipment of the scientist.[51] As we saw earlier, Ernest Jones, the famous psychoanalyst, in his biography of Freud speaks of Freud's major work, *The Interpretation of Dreams*, as a "perfect example of serendipity, for the discovery of what dreams mean was made quite incidentally—one might almost say accidentally—when Freud was engaged in exploring the meaning of the psychoneuroses."[52] And though Hendrik Van Loon was writing a history of the arts and not of the sciences, his usage seems relevant here too; his chapter in *The Arts* on Heinrich Schliemann, the great archeologist, is subtitled: "A Short Chapter Which for the Greater Part Is Devoted to an Explanation of the Word 'Serendipity.'"[53]

We have seen, then, that serendipity has been adopted by people in a variety of milieux: among collectors, writers, literary scholars, medical humanists and their orbit, among social scientists, people engaged in applied research, and finally, among science writers. In each group the reasons for interest in the word have been somewhat different. Also, the patterns of diffusion we have been able to discover have varied as the word moved from one milieu to another, and as it diffused among the members of a particular group. In the next chapter we shall see that as serendipity was adopted by different kinds of people, they endowed it with different meanings and interpretations.

---

[50] Harlow Shapley, "It's an Old Story with the Stars," in *One World or None*, ed. Dexter Masters and Katherine Way (New York: Whittlesey House, McGraw-Hill, 1946), pp. 7–10.

[51] *Science News* 25 (1952), published by Penguin Books.

[52] Ernest Jones, *Sigmund Freud: Life and Work* (New York: Basic Books, 1953), vol. 1, p. 350.

[53] Hendrik Van Loon, *The Arts* (New York: Simon and Schuster, 1937), chapter 5.

Chapter 8
MORAL IMPLICATIONS
OF SERENDIPITY

*I*n tracing the social history of the word serendipity, we have told, in effect, the story of the diffusion of serendipity among people in certain social milieux or "statuses," statuses that heightened the receptivity of their occupants to this particular word. The social history of *serendipity* shows that receptivity to a given word does not occur at random in the population: Rather, the run of experience in certain social statuses is so structured that a word may have more or less resonance for people in them, and, consequently, more or less chance for adoption. The sources of such resonance also vary: In some of the social milieux in which *serendipity* was adopted it was received more for the sake of its esoteric character, its bizarre derivation, or its attractive sound qualities; in others, the word was perceived to have a special aptness for describing certain experiences of the members of the milieu. Among those interested in serendipity for its novelty are writers and lexicographical scholars, people who are professionally concerned with words and for whom acquaintance with new, old, or pleasant words is part of their stock in trade. For collectors, humanists concerned with ethics and morality, or those connected in one way or another with the development of science, the word serendipity has a significance beyond the acquisition of a new piece of verbal equipment; it refers to important chance components in the success of their enterprises and it touches, therefore, on matters that require the making of value judgments about themselves and others. In this chapter we shall deal with the interpretations of serendipity by those individuals whose status makes them not only receptive to the word but sensitive to the implications of happy accidental discoveries. We shall see that among those who are alert to the part that happy accidents play in their enterprises, this experience of luck may have quite different significance.

The Problem of Unexpected Evil—and the Problem of Luck

The intellectual and moral problems that arise from the making of happy accidental discoveries are similar in important respects to those connected with the explanation and justification of good and ill fortune in general, in that such happy accidental discoveries raise the problem of legitimate expectations and deserts. Unexpected ill fortune may weigh on man more painfully than good, but we shall see presently in our discussion of the interpretations of serendipity that unanticipated good fortune may be hard to bear also. It is man's fate to recognize good and evil when it comes his way, and it is his lot as a valuing creature to try to establish the extent of his responsibility for it. The problem of *unexpected* good and evil is a part of the more general problem—the factor of unexpectedness has especial poignancy for men insofar as they place value on a predictable universe inhabited by rational individuals. Premature death and catastrophic illness, natural disasters, the frustration of efforts by "bad luck," all of these are to some extent incompatible with man's values, and to be borne they have to be explained as far as possible and justified. Good fortune, like ill fortune, is in some measure uncontrollable, and both create the problem of squaring things as they are with things as they should be, or, to put it another way, of answering the question of why they are as they are. It is the problem of finding meaning in the general irrational aspects of the world, and in the way particular people become the fortunate or unfortunate victims of those irrational events. The answers that are given to the general problem of man's responsibility for good and evil are related to the more specific problem of the extent of his responsibility for unexpected good and evil. These answers may, in turn, be related to the yet more specific problem of his responsibility for happy accidental discoveries.

It is especially when man suffers acutely and unexpectedly, or when he is overwhelmed by unexpected good fortune, that he cannot help but ask "why?" It is a two-fold "why" that seeks both some explanation of the sources of good and ill fortune and an answer to the question of deserts: How does it happen that good and ill fortune come into men's lives, and what has any particular individual done to deserve his lot?

The explanations and justifications of good and evil have varied widely, but this variation has been in emphasis rather than in the essential character of the answers given: The greater emphasis in explaining the sources of good and ill may be placed on the freedom and responsibility of the individual to prevent evil and to promote good, or on the predestined, necessary subjection of the individual to damnation or salvation, to misery or happiness, to good or bad luck; the greater weight may be given to the choices the individual can make that will bring him deserved rewards or punishments or to the innate condition of a mind, psyche, or soul that will determine what happiness or suffering will be his lot; greater emphasis may be placed on his capacity for rational mastery over a potentially threatening natural and social environment or on his dependence on natural and social forces beyond any possibility of ken or control. It is important to stress that the answers that are given to the problem of the sources of good and evil vary in emphasis only—no viable answer has ever been given that made salvation or damnation contingent either on faith alone or on good works alone; no such answer can make the individual totally responsible or totally irresponsible for success or failure; no such answer can depend entirely on the rational and moral activity of the individual, or on his devout or indifferent passivity. Total responsibility for good and evil is too great a burden for man, limited as he is, yet total irresponsibility is deeply disturbing psychologically and entirely disruptive from the point of view of society. (In Calvinist Scotland, the painful social and psychological consequences of radical predestinarianism were mitigated by the development of the concept of a covenant between God and man, a covenant that, in effect, limited God's omnipotence and bound Him to save righteous Calvinists; and the Soviet Union has come a long way, in theory as well as in practice, from those Marxist doctrines that held society entirely responsible for the good and bad actions of individual men.)

If simple explanations locating the sources of good and evil are impossible, so also are simple answers to the problem of individual deserts. The two questions are, of course, related—the amount of responsibility that is allocated to the individual for his share of the good and the bad in life is proportional to the responsibility that mankind is thought to bear for its fate. But beyond the general explanation of man's responsibility, it is necessary to take into account and justify the different fortunes of individuals. Does the individual deserve just what he gets? And is, therefore, even unexpected good or ill luck mysteriously deserved? To what extent do human fallibility and frailty account for misfortunes, and to what extent is misfortune the fruit of incompetence? To what extent is good fortune the result of "forces" beyond the individual's control, and to what extent do his rational efforts produce rewards? Again the answers to these questions vary in emphasis: To some extent, at least, the individual is considered to get what he deserves, but some would have him get less than he deserves, qua individual, because of the sins of his forebears, while others would have him get more than he deserves because of some inherent principle of progress in the world. If he gets more or less than it appears to his fellow men that he deserves, the discrepancy may be considered to be somehow justified *sub specie Dei* or *sub specie aeternitatis,* or it may be partly attributed to the irrational nature of things. Though all good and evil that befall individuals cannot easily be justified, neither can human efforts appear, in any large measure, futile nor can rewards appear random. It is a constant human problem to overcome apparent futility and to find order and justice in rewards.

In modern western society, man's responsibility for good and evil is generally defined as quite large. Success and failure in a man's life are to a considerable extent the products of his efforts, though they are also attributable to social and psychological forces beyond his control. There is a general expectation that people will "do their best," that results will be roughly commensurate with efforts, and that what discrepancies there are "cannot be helped." When the discrepancies involve premature death or unexpected suffering, such "tragedies" are deplored, and the hope is expressed that in the future they will be prevented. When the discrepancies involve striking oil or the sudden acquisition of "peace of mind," there is a tendency to ridicule the spurious or "phony" qualities of such good fortune. Too much luck, good or bad, is suspect; though effort and reward may not be congruent, the individual must be able to show that he has striven to overcome difficulties and has earned success. Too much bad luck, if not an outward sign of God's wrath and damnation, is likely to be taken as a sign of incompetence and inadequacy unless the individual can exonerate himself by demonstrating the invincibility of the "forces" against him. Too much good luck is equally likely to be detrimental to the evaluation of the individual, unless he can show either that his success was really earned or, if unearned, that it was not the result of Micawberish sloth but of accidental circumstances that fructified his efforts beyond expectation. In any case, a conspicuous discrepancy between effort and reward must be explained and justified if it is not to be judged discreditable.

Since ours is a predominantly secular society, the most significant judgments of men are those made by their fellow men: It is not God who compensates man for misfortune or who attaches strings to the good that befalls him, but men who judge the merits of each other's cases. In judging these merits, the factor of good and bad luck plays an important part. Compassion for the unfortunate hinges, implicitly or explicitly, on the demonstrability that their ill luck could not have been prevented by "reasonable" efforts and precautions. Admiration for the fortunate, in turn, depends on the demonstration that they have not been "merely lucky." The moral discomfort connected with *mere* good luck is evident when the lucky individual reminds his "judges" that, after all, he has had his share of bad luck in the past—over the long run, as it were, justice is being done. But this leaves the judges with a difficult, multiple problem of evaluation, a problem more easily solved *sub specie Dei* than by mere men: If it is claimed that good and bad luck cancel each other out, what of the responsibility of the individual who has encountered both? Did he deserve either? Both? Neither? To what extent is he to be judged for living up to the norms of hard work and competence, and to what extent is he the victim of fate?

These problems do not, fortunately, have to be solved by every member of the society ad hoc in each instance. The necessary explanations and justifications of luck are provided for most people by the moral policy makers of the society, whether these be theologians, philosophers, or social and political ideologues. People in all statuses—with the probable exception of avowed gamblers—need to have defenses against the denigration and self-doubt that come with too much good and bad luck, and their fellow men need guidance as to the evaluation of those who are conspicuously unfortunate or lucky. The moral policy makers take it as their concern to work out general answers to the problem of good and bad fortune; they attempt to reinterpret apparently undeserved luck as actually deserved. The theologians may try to interpret "the will of God" for this purpose. The philosophers, according to their stripe, may either give support to the individual by explaining that the burden of chance that weighs on him is the necessary price for the dignity of rationality, or, conversely, that the helplessness from which he suffers is the inescapable fate of humanity, ridden as it is by unreason. The social ideologues may talk of the means justifying the ends, and of the inevitable inequities attendant on the achievement of those ends; or, again, they may exalt the combination of "pluck and luck" and imply that you cannot succeed without having good personal qualities *and* some "breaks." According to such religious or secular doctrines, the onus of having too much good or bad luck is, in part at least, displaced from the individual onto supernatural powers or human nature or social processes, and, correspondingly, the stigma of failure or undeserved success is made somewhat less painful. This exemption from responsibility is conditional on the individual's maximizing his available resources to please his God, or to take optimal advantage of his rationality, or to control the historical processes affecting his destiny.

General prescriptions for dealing with the crises of good and bad fortune are, perhaps, sufficient for the majority of people, who are only occasionally exposed to the stress such crises produce. But in addition to this majority there are two kinds of minority for whom general moral and ideological analysis of man's fate are inadequate. One kind of minority consists of individuals who have an extraordinary run of good or bad luck, luck that is unconnected with their occupying any particular social status; the other consists of individuals whose status exposes them to recurrent brushes with unexpected, painful failure or with bittersweet (because "undeserved") success. Often, the members of these minorities do not find their needs met by the available religious or secular formulae. Instead, members of the first group may feel themselves to be singled out for some special mission, which they can fulfill only by intense dedication to a religious or social gospel. The second group, which is of greater interest from the point of view of this study, tends to work out a set of answers that takes into consideration the particular conditions of its various "exposed" statuses.

What makes for the extraordinary exposure of certain statuses to the *problem* of chance success and chance failure? The factors involved seem to lie along two dimensions: The first is the dimension of control over the conditions of success; the second is the dimension of responsibility. If the occupants of different statuses can be said to have more and less control over the conditions of success of their enterprises, those with little control are, by definition, more exposed to chance. The occupants of some statuses, therefore, must deal with situations that contain a considerable amount of structured uncertainty, while for other statuses the element of structured uncertainty is minimal. For example, for the railway switch operator the factor of uncertainty in his work has been rationally cut down to a minimum, while for the door-to-door salesman the conditions of success are far from being either formulated or insured. But structured uncertainty is not by itself enough to make unanticipated success or failure problematic—it becomes a problem only insofar as the individual holds himself, and is held by others, responsible for the effective use of available skill and knowledge. If he is so held responsible, and if the optimal use of his skills and efforts may bring either the expected result or a result that is apparently incommensurate with those skills and efforts (whether it be greater or less), then the individual's reputation must be guarded against unwarranted reassessment. Unless there is a special set of explanations and justifications of chance success and chance failure, the intervention of chance will impugn the basis of skill and responsibility on which estimates of such an individual's competence rest.

Among the statuses that involve both a high degree of structured uncertainty and a high degree of responsibility, medicine and natural and social science are prominent. Although medicine is an applied science, it is frequently involved with "frontier" problems, where the known, the knowable, and the unknown shade into one another. At the same time, the medical practitioner is held highly responsible by his professional peers and by his patients for achieving the best possible results that knowledge and skill will permit. Nevertheless, doctors are often incapable of specifying the conditions under which certain kinds of therapy will succeed, and in such situations they may obtain positive or negative results without knowing why. In addition, therefore, to the defenses that doctors must acquire to meet predictable failures—the inevitable loss of patients for whose ills medical science simply has no remedies—doctors must learn to cope with situations in which they think they know what can or cannot be done, and what, consequently, will happen, but which turn out, instead, better or worse than their expectations. It is only if doctors become familiar with the problem of structured uncertainty in the course of their medical education, if they learn to work as well as possible within the limits set both by the development of medical science and by human fallibility, that they can avoid becoming demoralized by accusations and self-accusations of incompetence and quackery.

The research scientist, whether he be concerned with natural or social science, also faces a situation of structured uncertainty. Indeed, since his primary concern is to advance knowledge, to innovate, his activities are constantly at that frontier of knowledge which the medical man faces only occasionally. His ability to predict results is, therefore, even more limited than the doctor's. His responsibility for results is, however, also less extensive: Although he is held responsible by his peers for the optimal use of present knowledge for the extension of that knowledge, he is not so directly responsible for serving the needs of the public. Nor is he, in consequence, so often accountable to the lay public for explanations of the unexpected success and failure of his efforts as is the doctor, who must give satisfaction to his peers and his patients. It is generally only in the case of significant unanticipated discoveries of immediate consequence to the lay public that the research scientist has to justify his luck to those who do not even understand the structure of the situation in which his work takes place.

Medicine and science differ in another respect when the problem of the intervention of chance between effort and result is concerned. In the practice of medicine it is surely the unexpected failure that is the harder to bear, since the questioning that comes with unexpected success is overshadowed by delight in the welfare of the patient. But both unexpected good and ill is possible and visible in the medical situation: The doctor is called upon to remedy ills, and the condition of his patient may, accordingly, improve or deteriorate, in accordance with expectation or otherwise. In the work of the research scientist it is the crises of good fortune, the unexpected successes, that have perhaps greater saliency than the unexpected failures. Scientists are called upon less directly to remedy ills than to produce innovations: The situation they face contains not so much potential deterioration as potential improvement, and it is this improvement that may come according to expectation—or otherwise. Among scientists the happy accidental discoveries are relatively conspicuous, whereas the happily missed discoveries may not come to their attention for a long time, if ever. (Missed discoveries may, of course, occasion considerable retrospective regret in science. Because the scientist's professional reputation hinges on the making of discoveries, the revelation of a missed discovery may be so hard to bear that the perpetrator of the unhappy miss may actually put in a claim to the discovery.)

By contrast with medical men and scientists, most collectors experience exposure to chance success and failure without the concomitant responsibility. To the extent, therefore, that they are thoughtful, they may try to explain the sources of their success, but they feel no need to justify it. Collectors, too, miss discoveries and experience regret at doing so, but claims to adumbration, let alone priority of discovery, are neither possible for them nor are they necessary. It is by contrasting, as we shall presently, the kinds of explanations of unexpected good fortune that are given by collectors and scientists that we may understand the different significance that luck has for a man pursuing an avocation—an irresponsible dilettante, as it were—and for a responsible professional man, though both occupy statuses that have relatively high incidences of unexpected success.

Serendipity and the Problem of Good Luck

The problem of unexpected evil and the problem of unexpected good luck both have a certain symmetry and asymmetry: Both raise the question of responsibility, but while in the case of evil there is a need for exculpation and, perhaps, compensation, in the case of good luck there is often a need to prove oneself worthy of it. If evil is to be accepted, it must be related to some purpose; luck, on the other hand, must, under some conditions, be related to merit. For the rest of this discussion we shall be concerned exclusively with the problem of good luck.

The problem of happy accident is sufficiently central to many moral, intellectual, and practical concerns that all manner of other problems become connected with it. People located in many different areas of society link up with the problem of happy accident various questions of policy and of evaluation. The word serendipity has tended to be diffused into certain social arenas that have a structured receptivity to a word related to the occurrence of happy accidents (see Chapter 7), and along with this structured receptivity may go a structured sensitivity to the meaning of such accidents. Happy accidents in general, and serendipity in particular, may easily become involved focally in the formulation of broad value preferences; indeed, serendipity may become a kind of inkblot, onto which are projected preferences only more and less directly related to the making of happy accidental discoveries. Associated with these preferences are implicit basic answers to the problem of luck, such as we described earlier, and *working answers* to the same problem. It is with these working answers that we are, on the whole, concerned, and they deal not with ultimate questions of responsibility, but rather with the implications of happy accidents for the fulfillment of "everyday" goals. They answer questions such as: Do happy accidents (happy accidental discoveries) *really* happen? If they may be said to happen, *how* do they happen? Is it *right* that they happen? Or can their occurrence be justified?

We find people in different social circumstances, and even within the same social circumstances, giving different answers to these questions. Although, often, their conclusions have been based on substantially the same data, they vary considerably. In part, this variety of conclusions reflects the incomplete character of the data, in part it reflects the values and attitudes of those who are interpreting the known facts. In the remainder of this book, we shall examine the analyses and interpretations of happy accidents, and especially of serendipity, in a number of domains. Because several of the individuals whose interpretations we shall discuss will not fit neatly into any one of these, we shall have to permit ourselves a certain untidiness in our categorization.

Chapter 9

THE DIVERSE
SIGNIFICANCE OF
SERENDIPITY IN SCIENCE

*A*mong the many conditions necessary for the flourishing of science in a society,[1] is the prevalence of certain beliefs and assumptions. For science to flourish people must believe, for example, in empirical rationality and they must assume the regularity and determinacy of empirical phenomena. They must believe in the possibility and support the desirability of understanding invariable relationships of cause and effect in the material universe. But commitment to these beliefs permits a variety of subordinate views as to the necessary conditions for the advancement of science and the best procedures for promoting science within a set of limiting conditions. These views also vary as to the kind of contribution the individual scientist can make or should make to the development of science, and what qualifications the scientist should have for making maximal contributions.

Underlying these views may be more and less easily discernible orientations toward such broad questions as the causes of social change, freedom and control, and the significance of the individual in the solving of social problems. (We shall find on occasion that people who would seem to take opposite sides on these questions may come to the same conclusions about science, while people on the same side come to different conclusions.) But it is not only commitments to certain values that lead

---

[1] Described in detail by Robert K. Merton in *Social Theory and Social Structure* (New York: Free Press, 1949); and by Bernard Barber in *Science and the Social Order* (New York: Free Press, 1952).

people to different opinions—their judgments may be based on the facts as they see them, facts that suggest the "best" or "only" way to advance science and the "best" people to do the job.

Since it is the special task of scientists to make discoveries, they themselves have often been concerned to understand the conditions under which discoveries are made and to use that knowledge to further the making of discoveries. Some scientists seem to have been aware of the fact that the elegance and parsimony prescribed for the presentation of the results of scientific work tend to falsify retrospectively the actual process by which the results were obtained. Others have thought that the dramatic fact of the applicability of certain discoveries to immediately comprehensible social ends has obscured the process of scientific work without which they could not have occurred. And, most significant from our point of view, scientists have often held that the role of accident in discovery has not been placed in perspective. Whether by the retrospective streamlining of accounts of discovery or because of the "human interest" inherent in the intrusion of accident into the realm of scientific rationality, the factor of accident has not been given its due. Many scientists, as well as historians and philosophers of science, have thought that the accidental component has been either underestimated, or exaggerated, or simply not understood, and that this limited understanding of the accidental component has led to an inadequate understanding both of the nature of science and of the qualities of scientists.

The intended audiences for this kind of enlightenment have varied, and often they cannot easily be determined. Such audiences may be large and heterogeneous, or relatively homogeneous. A mature scientist may try to give a group of neophytes the benefit of his experience, and so we find medical students, for example, being lectured on the significance of accidental discoveries by one of their professors. Sometimes an "elder statesman" of science may reflect autobiographically for the benefit of his less leisured peers and a few interested laymen. Or a historian of science may try to explain to a large, "educated" lay audience how, in his opinion, science progresses, not because the audience will have any occasion to use these insights directly, but because it may help it to understand the kind of moral and practical support necessary for continued scientific progress.

Observations on Unanticipated Discoveries

Almost from the time the natural sciences first became important disciplines, workers in those sciences have been aware of the existence of unanticipated discoveries and have tried to "make sense" out of their occurrence. Both Robert Hooke in the seventeenth century and Joseph Priestly in the eighteenth century drew far-reaching conclusions from their appraisals of the relative importance of chance discoveries and of planned research for the maximal development of science. It may be that these early natural scientists had especially cogent reasons for being sensitive to the problem, for it has been suggested that it is in the early stages of any particular science that such discoveries occur with unusual frequency. The English political economist Stanley Jevons, believed this was so.

> Sufficient investigation would probably show that almost every branch of art and science had an accidental beginning. . . . With the progress of any branch of science, the element of chance becomes much reduced. Not only are laws discovered which enable results to be predicted . . . but the systematic examination of phenomena and substances often leads to important and novel discoveries which can in no sense be accidental. . . . If we must attempt to draw any conclusions concerning the part that chance plays in scientific discovery, it must be allowed that it more or less affects the success of all inductive investigation, but becomes less important with the progress of any particular branch of science.[2]

Ernst Mach, the well-known German scientist and philosopher of science, contended some twenty years later that the importance of accidents in the process of discovery was substantially underrated, and he produced good reasons why under such cultural conditions as obtain in the early stages of science accidents are bound to play an especially significant part:

> It is by accidental circumstances, that is, by such as lie without his purpose, foresight, and power, that man is gradually led to the acquaintance of improved means of satisfying his wants. Let the reader picture to himself the genius of a man who could have foreseen without the help of accident that clay handled in the ordinary manner would produce a useful cooking utensil. The majority of inventions made in the early stages of civilization . . . could not have been the product of deliberate methodical reflection for the simple reason that no idea of their value and significance could have been had except from their practical use.[3]

Some fifty years later, the house organ of the A. D. Little company, a company that specializes in pioneering industrial research, came to the same conclusion about the frequency of accidental discoveries. In an

---

[2] W. Stanley Jevons, *Principles of Science: A Treatise on Logic and Scientific Method* (London, [1874] 1877), pp. 531–532.

[3] Ernst Mach, *Popular Scientific Lectures* (La Salle, Ill.: Open Court Publishing, [1895] 1943), p. 264.

anonymously authored article in the company's *Industrial Bulletin* titled "Serendipity," there is a statement to the effect that "Serendipity is more characteristic of the pioneer than of the later worker in any line, but can always occur. Our well-known earth now perhaps affords few opportunities for chance discoveries in a geographical way. However, there should continue to be many occasions for the manifestation of serendipity in the developing world of science and industry for a long time to come."[4]

These opinions, plausible as they may seem, are only opinions. No studies have ever been made of differential rates of accidental discovery in the more and less advanced stages of a science, and we have at least one opinion, that of the Standard Oil trade publication *The Lamp*, that conflicts with those presented above. (We shall discuss the article in *The Lamp* in another context.) In any case, the two notable early English scientists we have mentioned attributed a very important role to accident in the process of discovery. As long ago as 1679, Robert Hooke virtually made the serendipity pattern his philosophy of research, for in the preface to his *Lectiones culterianae* he explained that

> there is scarce one Subject of millions that may be pitched upon, but to write an exact and compleat History thereof, would require the whole time and attention of a man's life, and some thousands of Inventions and Observations to accomplish it. So on the other side no man is able to say that he will compleat this or that Inquiry, whatever it be. (The greatest part of Invention being but a luckey bitt of chance, for the most part not in our own power, and like the wind, the Spirit of Invention bloweth where and when it listeth, and we scarce know whence it came or whither 'tis gone.) 'Twill be much better therefore to *embrace the influence of Providence*, and to be diligent in the inquiry of everything we meet with. For we shall quickly find that the number of considerable observations and Inventions this way collected will a hundred fold out-strip those that are found by Design.[5]

Hooke obviously felt that the subject matter of scientific investigation was so infinitely ramified that any one subject of study quickly became inexhaustible. Because carefully planned, exhaustive study would be frustrated by the multitude of unexpected facets that would surely be encountered, it is better to follow one's scientific nose, as it were, or, in seventeenth-century language, to "embrace the influence of Providence" and make the most of the data that happen to come one's way. Most discoveries, Hooke thought, were both unexpected and unpredictable, and important "observations" were, therefore, far more likely to come

---

[4] "Serendipity," *Industrial Bulletin of Arthur D. Little, Inc.* 160 (July 1940).
[5] Robert Hooke, *Lectiones cultlerianae* (1679), in R. T. Gunther, *Early Science in Oxford* (vol. 10 is in part dedicated to *The Life and Work of Robert Hooke*) (Oxford: Oxford University Press, 1935).

about through "diligent" but unplanned work than by "Design." Hooke was, of course, by temperament incapable of sticking to any one project for any length of time, and to a certain extent he was surely simply generalizing his own experience in scientific research. He was to some extent identifying his own needs with the ways of Providence, while following the guidance of the Spirit of Invention.

Joseph Priestley, writing about one hundred years later than Hooke (in 1775), came to substantially the same conclusions about the relative importance of chance and design in the making of discoveries. Priestley attached great importance to the "truth of a remark that I have more than once made in my philosophical writings," namely, "that more is owing to what we call *chance*, that is, philosophically speaking, to the observation of *events arising from unknown causes*, than to any proper *design*, or preconceived *theory* in this business. This does not appear in the works of those who write *synthetically* upon these subjects; but would, I doubt not, appear very strikingly in those who are the most celebrated for their philosophical acumen, did they write *analytically* and ingeniously." If the most competent scientist-philosophers would pause to analyze the ways in which they arrive at scientific results, they would appreciate the extent to which they move without plan from observations that "happen" to attract their attention to the formulation of explanations of those observations, and finally to some experimental testing of those explanations.

If we cross the Channel from England to France and advance about another hundred years, we find French scientists, outstanding biologists in this case, commenting on the way accidents contribute to the advancement of science. All of them attribute significance to chance, but they see the factor of chance playing very different parts. One of the most famous remarks of all time on the role of chance was made in 1854 by Louis Pasteur, in his opening speech as dean of the new Faculté des Sciences at Lille. Addressing a student body composed, according to René Vallery-Radot, of "the sons of manufacturers," boys in whom it was presumably difficult to kindle any interest in intellectual pursuits for their own sake, Pasteur made an eloquent plea for the importance of the development of scientific theory: "Without theory, practice is but routine born of habit. Theory alone can bring forth and develop the spirit of invention. It is to you specially that it will belong not to share the opinion of those narrow minds who disdain everything in science which has not an immediate application."[6] Useful inventions, such as the telegraph, go back to theo-

<hr/>

[6] René Vallery-Radot, *The Life of Pasteur* (London: Constable and Company, [1901] 1920), p. 76.

retical advances, and although in these advances chance may play a part, it is a part subordinate to that of theoretical sophistication:

> Do you know when it first saw the light, this electric telegraph, one of the most marvellous applications of modern science: It was in that memorable year, 1822: Oersted, a Danish physicist, held in his hands a piece of copper wire, joined by its extremities to the two poles of a Volta pile. On his table was a magnetized needle on its pivot, and *he suddenly saw (by chance you will say, but chance only favours the mind which is prepared)* the needle move and take up a position quite different from the one assigned to it by terrestrial magnetism.[7]

Almost incidentally, as it were, Pasteur included the ability to take advantage of the chance occurrence among the supremely important consequences for science of developing theory.

Claude Bernard, Pasteur's contemporary, is somewhat more inclined to emphasize the way in which loose devotion to theory (not the abandonment of theory) may be an asset in furthering science in important but unexpected ways. Bernard describes his conception of the scientist in quest of discoveries:

> Experimental ideas are often born by chance, with the help of some casual observation. Nothing is more common; and this is really the simplest way of beginning a piece of scientific work. We take a walk, so to speak, in the realm of science, and we pursue what happens to present itself to our eyes. Bacon compares scientific investigation with hunting; the observations that present themselves are the game. Keeping the same simile, we may add that, if the game presents itself when we are looking for it, it may also present itself when we are not looking for it, or when we are looking for game of another kind.[8]

This is almost the generic formula for accidental discoveries, and the figure of speech Bernard uses cannot but recall to us Walpole's derivation of the word serendipity: "as their Highnesses [the three Princes of Serendip] travelled, they were always making discoveries, by accidents and sagacity, of things which they were not in quest of." Bernard describes an experiment he had performed, having to do with blood sugar, which produced results quite different from those he had expected on the basis of existing theory. His limited commitment to that existing theory stood him in good stead: "So I noted a new fact, unforeseen in theory, which men had not noticed, doubtless because they were under the influence of contrary theories which they had too confidently accepted. I therefore

---

[7] Ibid. Emphasis added.
[8] Claude Bernard, *An Introduction to the Study of Experimental Medicine* (London: Macmillan, [1865] 1927), pp. 151–152.

abandoned my hypothesis on the spot so as to pursue the unexpected result which has since become the fertile origin of a new path of investigation." Finally, Bernard integrates this example of an unanticipated discovery into a "set of principles of the experimental method that we have established," principles that make specific provision for the occurrence of the unexpected. The relevant principle is stated thus: "that in the presence of a well-noted, new fact which contradicts a theory, instead of keeping the theory and abandoning the fact, I should keep and study the fact, and [hasten] to give up the theory, thus conforming to the precept which we proposed in the second chapter: 'When we meet a fact which contradicts prevailing theory, we must accept the fact and abandon the theory even when the theory is supported by great names and generally accepted.'"[9] Clinging to preconceived ideas is, according to Bernard, one of the "great stumbling blocks of the experimental method."[10]

Our third great French biologist, Charles Richet, writing somewhat later than Pasteur and Bernard, attributes a larger role to chance in discovery than either of them did. Richet does not expatiate on the relative importance of theory or of "principles of experimental science" in the making of discoveries; what impresses him is the extent to which the scientist's eyes are opened, willy-nilly, by unexpected occurrences. "Indeed, I want to show how, in any discovery, whether slight or important, our personal role counts for very little, so little that it amounts to nothing. It will be a rather humiliating profession of faith, since I attribute a considerable role to chance. Nevertheless it will be recognized that chance was aided by perseverance."[11]

Richet's account of the discovery of zomotherapy illustrates his profession of faith. "It was chance alone that permitted Hericourt and myself to discover zomotherapy, that is to say, the treatment of tuberculosis by the ingestion of the juice of raw meat: our perspicacity had no great share in it." He describes his experiments with dogs who had been injected with tubercular virus, one of which unaccountably survived the injections. This dog, one of sixteen, happened to have been fed on raw meat.

> I thought then that we had made a mistake and that it had not been inoculated with the tubercular virus. But no! we could see that it had on its paw the little scar, proof of injection.
>
> Happily, I recommenced the experiment. But it was—I admit—without great conviction, for our blindness is such that unforeseen facts *have to force our hand* to make themselves accepted. . . .

[9] Ibid., pp. 163–164.
[10] Ibid., p. 23.
[11] Charles Richet, *The Natural History of a Savant* (New York: George H. Doran, 1927), p. 105.

Then, but after the event, I understood. I ought to have foreseen it. But neither I, nor anyone, had dreamt of it, until the moment when the experiment, far superior to our poor imagination, began to speak so as to make us understand.

In part, Richet's account is motivated by modesty, modesty that has to do with that relationship between chance and merit of which we shall say more presently. But since claiming or disclaiming credit does not seem to be a salient feature of Richet's description of the role of chance, we shall pay no further attention to it for now. Furthermore, his disclaimer seems to have little to do with the institutionalized humility of science, that humility which demands that scientists see their own contributions as only small accretions to the cumulative body of scientific knowledge, impossible without the preexistence of such knowledge. Rather than either personal modesty or institutionalized humility, Richet appears to have a rather pessimistic view of scientists' conceptual powers. When he speaks of "our poor imagination," he is not referring to the limited imagination of Hericourt and himself, but of the whole scientific fraternity. Rather like Hooke, Richet is suggesting that scientists cannot plan the substance of their discoveries, they cannot know what form they will take. Their foresight is extremely limited, though their hindsight is comprehensive. Scientists are not alert travelers or hunters with definite objectives and incidental rewards, but dull and persevering toilers who are occasionally rewarded for their labors by bumping into something that it would be hard, indeed, to miss. If this is something of an overstatement, it does bring out the contrasting views of a Pasteur or a Bernard, on the one hand, and a Richet or a Hooke, on the other.

Recently a view akin to the views of Hooke and Richet has been expressed by the Nobel prize–winning American physicist P. W. Bridgman. Bridgman reflects, dryly and impersonally, "how seldom the course of scientific development has been the logical course. . . . Much more often the course of development is determined by factors which are quite adventitious as far as any connection goes with immediate human purpose."

The problem of the role of accident in discovery evidently generates ideas about the nature of science and of scientists. The proponents of the various views that we have discussed so far have not (with the exceptions of Pasteur and Mach, to which we shall revert later) been visibly concerned with pleading causes or prescribing for the conduct of others. They have reflected on their own experience in scientific work or on their acquaintance with the history of science not so much to convince contrary-minded people that their views are superior but simply to put at the disposal of others the benefit of their insights. The same is true of nineteenth century historians of science such as William Whewell and George

Gore. Whewell's thought ran along the same lines as Pasteur's on the subject of accident in scientific discovery, while Gore was more inclined to try to specify the conditions for the differential predictability of different kinds of discoveries (quantitative discoveries, he thought, were more predictable than qualitative ones).

Certain "cases" have, to be sure, been implicit in some of the views we have examined: a case *for* theory *as against* empiricism or vice versa; a case *for* planned *as against* unplanned research; a case *for* the scientist's control over the development of science *as against* his impotence with regard to the substance of scientific discoveries; or, finally, a case *for* the merit of the beneficiary of luck *as against* his lack of merit or vice versa. These cases have all been of secondary importance. Even Pasteur's plea to the embryonic businessmen to forego immediately useful rewards could only be perceived as a result of the context supplied by his biographer, and Mach's plea cannot be divined from the explicit content of his paper, it can only be inferred from other evidence. Mach's polemical orientation against the Marxists may be inferred from the extremely hostile response that his philosophy of science aroused in the writings of Lenin. Lenin put "Machists" in the same category as "opportunists" and declared formal war on all of them in 1908.

To orthodox Marxists the suggestion that discoveries could be occasioned by accidents rather than by the inexorable development of the material base of society was anathema. Since the Marxists believe that all social and physical phenomena are rigidly determined, inventions are, in principle, predictable, and the job of the historian or philosopher of science is to work out ways of predicting them.

This commitment to predictability is shared by others who are not Marxists. The historian of science S. C. Gilfillan, who believes in complete determinism, has been working on just this problem of the prediction of inventions for several decades.[12] Joseph Rossman, the author of *The Psychology of the Inventor*, and who is an admirer of Gilfillan's, is also basically unsympathetic to the notion that chance plays an important part in discovery, though he does concede that accidental discoveries happen. Rossman thinks that stories about the importance of accident have been much exaggerated because they appeal to the popular imagination:

> As a matter of fact, chance or accident plays a very small part in inventing today. [This seems to leave the door open to the suggestion that it had once played a larger part.] We have seen . . . that the inventors in this study employ deliberate and systematic methods in making their inventions in which chance has a very small part. Only 75 out of 259 inventors who were asked

[12] S. C. Gilfillan, *The Sociology of Invention* (Chicago: Follett Publishing, 1935).

whether chance or accident played any part in their inventing replied in the affirmative . . . [!!]

Many impressive and highly colored stories have been told of accidental discoveries and inventions. It is natural, of course, that such stories should appeal to the popular imagination. A careful study of these stories of acciden-tal invention, however, will reveal the fact that lucky accidents only happen to those who deserve them.

In nearly all cases we find that the accident happens only after a persistent and carefully conducted search for what is wanted.[13]

Rossman's inventors reach their carefully specified goals after a due out-put of effort, and, as he sees it, the unexpected discoveries they make along the way are of no great significance. He does admit that they oc-cur: "It frequently happens that while an inventor is engaged in making an invention he will stumble over some facts for which he was not look-ing. It is well known that this often happens to workers in many different fields. This lucky find is termed 'serendipity' after the Persian god of chance [sic]. It is really 'accidental sagacity.'"[14]

It might be possible to conclude from Rossman's study that it is be-cause he is concerned with inventors as compared to scientists engaged in "basic" research that the factor of accident in their experience is deemed relatively unimportant. Inventors, after all, may have more specific goals than research scientists. But there is at least one very distinguished inven-tor who, if he had been asked, would certainly have been among those seventy-five who answered Rossman to the effect that accidents were im-portant. He is Frank Rieber, who was described in a brief profile in *For-tune* as "one of the most successful inventors in the U.S."[15] After enu-merating some of the important inventions Rieber has to his credit, the piece goes on to describe some of "the things that he still wants to tackle: 'talking books' for the blind, recorded on thin, plastic strips that play for hours without adjustment and cost no more than ordinary books; a complete system of robot weather stations, . . . [etc.]" Apparently, however, Rieber does not take the solution of these problems for granted, nor is he sure that his efforts will not be deflected to others: "'I'm like those three mythical princes of Serendip, he muses. 'Everywhere along the way the princes encountered new and useful things for which they had not been searching at the time. The word for that is "serendipity" and I guess it describes the experience of most inventors.'"

In the course of minimizing the importance of accidental discoveries

[13] Joseph Rossman, *The Psychology of the Inventor* (Washington, D.C.: W. F. Roberts Co., 1931), pp. 117–118.

[14] Ibid., p. 120.

[15] "Shorts and Faces," *Fortune* (Fall 1946).

Rossman has raised another issue that others have been concerned with, that of the public image of the role of accident. It has been suggested several times that the lay public *likes* to think that discoveries are made by accident, that is, that the public attributes too large a role to the factor of accident. (We shall say more later about all-or-nothing views of the role of accident.) Rossman thinks it is "natural" that the notion of accidental discovery should appeal to the popular imagination—but he does not explain why it should be natural. Otto Glasser, in his biography of Roentgen, also alleges that a certain version of Roentgen's discovery that is both false and attributes too much importance to accident ("In the widely circulated fable of the book and the key, Middleton states . . .") "has had a wide appeal to the imagination of the general public."[16] Again, the image of the lay public's "imagination" is left vague and undefined. The same is true of the remark made by the historian of science William Cecil Dampier-Whetham, in connection, incidentally, with Roentgen's invention: "Great discoveries are made accidentally less often than the populace likes to think."[17]

In some allegations about the lay public's predilection for the idea that discoveries are made by accident we do get clues as to the state of mind that those who identify with scientists project onto the public. In an article on "The Happy Accident," Franklin C. McLean suggests that people like to believe that discoveries happen by chance because this view is congenial with the lazy man's simple outlook on life.[18] Even Thomas Jefferson, no lazy layman, presumably, was too much inclined, according to McLean, to attribute discoveries to accident: "In a letter to Dr. Caspar Wistar in 1807, Thomas Jefferson refers to 'the slow hand of accident' transferring mysteries to the table of sober fact, as though a majority of the advances in science made up to that time had come by some such vicarious route." If Jefferson could err, how much more prone is "common human nature" to misunderstand science! "It is common human nature to embrace such a happy accident. Who would not desire also to stumble upon some such chance fame and fortune? The dramas appeals to every basic mentality. It is ever the short cut, the easy way, the avoidance of good hard honest sustained labor that is always so enthusiastically welcomed by the mass mind." As McLean sees it, the public likes the idea of accidental discoveries because they represent the fulfillment of its pervasive wishful thought about easy success.

---

[16] Otto Glasser, *Wilhelm Conrad Roentgen and the Early History of the Roentgen Rays* (Springfield, Ill., and Baltimore, Md.: Charles C. Thomas, 1934), pp. 14–15.

[17] William Cecil Dampier-Whetham, *A History of Science and Its Relations with Philosophy and Religion* (Cambridge: Cambridge University Press, 1929), p. 383.

[18] Franklin C. McLean, "The Happy Accident," *Scientific Monthly* 53 (1941): 61–70.

McLean thinks that the public likes to "embrace [the] happy accident" not only because it makes discoveries seem easy, but also because it tends to pull the scientist off his pedestal and to bring his activities down to earth. The public's bewilderment and resentment in the face of abstract intellectual activity that it cannot hope to understand is soothed by diminishing the stature of the scientist, by putting him on the same level as "everyone else." To quote McLean: "Probably the prevailing conception of scientific discovery is most popularly exemplified by Charles Goodyear's 'accidental' discovery of the vulcanization of rubber. Everyone knows and can visualize the picture of Goodyear puttering around in his humble kitchen with his fuming mixtures and clumsily upsetting his pan of India rubber and sulphur; the ensuing sputtering on the hot lid of the stove, and, lo, the great scientific discovery!" Obviously, this sort of thing might happen to anybody, and that is, presumably, a gratifying notion.

A similar conception of the public's desire to diminish the stature of the scientist is found in a piece of institutional advertising by the Pfizer Company in the *Journal of the American Medical Association*, in which it is alleged that "it is a favorite thesis of people who do not understand (or entirely like) science that everything important has come about through sheerest accident. In this way they imply that if they had been there under a tree when the apple fell, they would have done at least as well as Newton."

The assumption of some hostility on the part of the public toward science, hostility that may find a convenient focus in the occurrence of accidental discoveries, is not merely a projection of the scientists' uneasiness about apparently accidental success. The existence of such hostility is exemplified by the concluding remarks the columnist James S. Pooler makes at the end of one of his folksy columns ("Sunny Side") in the *Detroit Free Press*. After summarizing the article on "Serendipity" that appeared in *The Lamp*, Pooler says, "We were telling all this to an old, old man we know. And he said—'Serendipity, that's a nice, fancy word—chance and smart fellows.' I hope it grows up to mean God and God-given intelligence." Apart from the ambivalence that may be implied here toward a "fancy" word, there can be little doubt that the mixture of "chance and smart fellows" is looked upon by Pooler as threatening. "Smart fellows" are only too likely to discover something that will have harmful consequences for the public, and in their irresponsible way, these godless fellows would be heedless of these consequences. Pooler is not discounting the positive contributions of science (the column tells of such useful accidental discoveries as penicillin, X-rays, and synthetic dyes), but he is expressing the common apprehension about the possible disruptive consequences of further scientific advances. Science is bound to bring about change, and in the hands of that alien breed, the intellectuals, change might do more harm than good. As Bernard Barber has put it,

> Rationality, wherever manifested, has the same effect of producing changes and of undermining established social routines. Social instability is in part, then, the price we pay for our institutionalization of rationality.
>
> Social instability and its consequences are not something to be taken lightly. It is not strange that they should be the cause of that ambivalence toward rationality in general and science in particular which seems to be widespread, though usually latent, in our society. . . . The standing routines and vested interests of every member of society are many times attacked and overthrown by "the process of rationalization." No doubt the feelings of hostility and uneasiness which result are usually counterbalanced and even outweighed by the favorable consequences of science that all of us also experience. But there remains in each of us a residue of ambivalence.[19]

To the extent, then, that the uninformed public really does believe that accidents play a part of predominant important in the process of discovery, latent antirationalism and dreams of easy success may be factors contributing to this image of science. But the prevalence and importance of these attitudes should not be exaggerated. Though the laity, which shares only in part the values and the knowledge that permit science to flourish, may have good reasons to underrate and to distrust scientists, scientists, for their part, have reasons for distorting the sentiments of the laity.

With this discussion of the public image of accidental discovery we have moved gradually out of the realm of problems having primarily to do with the *place* of accident in the development of science (as discussed, for example, by Hooke, Bernard, and Bridgman) and into the realm of problems having to do with the *legitimacy* of the occurrence of accidents. We have had a foretaste of the "cases" we mentioned earlier, and which we shall now take up more thoroughly: cases made for and against accidental discoveries, arguing about their significance for the merit of scientists and about their proper contribution to the conduct of scientific work.

Serendipity: Merit or Luck?

In the world of science, the imputation of luck in general and of serendipity in particular is generally connected with a judgment about an individual's merit. To be considered lucky is undesirable—it implies that achievements are really undeserved and that the lucky individual cannot

---

[19] Barber, *Science and the Social Order*, p. 211.

be counted on to perform reliably. (If he is just lucky, after all, luck might easily desert him.) Having serendipity, on the other hand, may be judged to be either a good or a bad thing. When having serendipity is considered meritorious, the component of luck in serendipity is minimized; when serendipity is regarded as discreditable, the factor of luck is thought to be of paramount importance. To put it another way, when serendipity is used to enhance the reputation of an individual, the component of luck is made dependent on qualities that are unambiguously admired. Luck or chance, according to these formulations, does not favor people at random; rather, it is prepared minds who are able to benefit from luck, and to preparedness may be linked such qualities as alertness, flexibility, courage, and assiduity. Only the able and virtuous are lucky in the field of discovery, just as on the battlefield fortune favors only the brave. When, on the other hand, serendipity is used to detract from the individual's reputation, the factor of luck is stressed, and it is coupled with such qualities as passivity, irresponsibility, pretension, and unreliability. If the individual succeeded, it is alleged, it was by luck alone, luck which he had no reason to expect, which he has no right to take credit for, and which will not come his way again.

The most widely known advocate of the merit attached to serendipity is Dr. Walter Cannon. Cannon restricts the meaning of the word serendipity itself to the "accidental leads to fresh insights." Such "advantages from happy chance" he feels are "almost certain to be encountered" "in the life of an investigator whose researches range extensively." Examples of such accidental leads can be found in the scientific biographies of Oersted, Bernard, Richet, and Pasteur, to name only a few of the greatest scientists. Of his own experience, Cannon says, "During nearly five decades of scientific experimenting, instances of serendipity have several times been my good fortune."

Cannon, however, has not exhausted the subject of serendipity by telling anecdotes about accidental leads; nor is he satisfied by describing the "surprise" and "astonishment" that the great scientists experienced when confronted with an unexpected observation or insight. "The unforeseen contingency," he says, may occasion scientific advance because of the serious problem it presents." This observation seems to constitute the bridge (not an easily observed bridge in his pleasantly rambling text) to "the next point, which is quite as important as serendipity itself. I refer to the presence of the prepared mind." Cannon mentions the three well-known legends of Archimedes' bath, of Newton and the apple, and of James Watt and the steam kettle, and he draws the following conclusion from them: "Many a man floated in water before Archimedes; apples fell from trees as long ago as the Garden of Eden (exact date uncertain!) and the onrush of steam against resistance could have been noted at any time since the discovery of fire and its use under a covered pot of water. In all these cases it was eons before the significance of these events was perceived. Obviously a chance discovery involves both the phenomenon to be observed and the appropriate, intelligent observer."[20]

In describing Fleming's discovery of penicillin, he tells of the "pregnant hint" that was given to Fleming when the culture he was working with underwent dissolution by accidental contamination with a mold— "A careless worker might have thrown the culture away"—but Fleming took the hint. Finally, Cannon quotes the dicta of Pasteur and of Joseph Henry ("Dans les champs de l'observation, le hasard ne favorise que les esprits preparés," and "The seeds of great discoveries are constantly floating around us, but they only take root in minds well prepared to receive them") to clinch his argument that, in effect, chance discoveries are made by scientists who deserve them. Such scientists deserve them because they have in high degree those qualities that are essential in a good scientist.

Thus, when Cannon is evaluating the significance of luck in the achievements of other scientists, he strongly emphasizes that luck plays a subordinate role to their great talents. We might be willing to accept this as his dispassionate appraisal of the significance of luck for the evaluation of scientists in general were it not for the fact that in evaluating the significance of serendipity in his own experience he just as emphatically reverses the relative weight to be given to accident and ability. The problem of merit, which is never explicitly raised in connection with others, turns out to be a touchy matter after all; the calm of his description of serendipity in the lives of others becomes ruffled when he describes his own brushes with serendipity: "Perhaps you will be tolerant . . . if I recount to you an example of serendipity that fell to my lot. After all, an investigator is not taking undue credit to himself when he calls attention to the fact that results which he has obtained in his researches have depended on a fortuitous incident and not on his own intelligence and insight! Of course, the case which I shall cite is not to be compared with the great discoveries of Bernard, Richet, and Pasteur."[21]

It was Cannon's modesty that tripped him up: It led him in effect (one gathers, quite unintentionally) to make a rather strange distinction between the great scientists, who make accidental discoveries because of their great ability, and the more humble ones, who "depend on a fortuitous incident" for their success. One suspects that Cannon would have chortled at allowing such an inconsistency in his scientific work to derive

[20] Walter B. Cannon, *The Way of an Investigator* (New York: W. W. Norton, 1945), p. 75.

[21] Walter B. Cannon, "The Role of Chance in Discovery," *Scientific Monthly* (19 March 1940): 208.

from an otherwise admirable sentiment such as modesty. His tacit distinction reveals Cannon's latent sentiment that to protect the standing of scientists it is essential to stress that basic relationship between serendipity and competence (or better, great ability); only the truly competent make scientific gains by serendipity—although, it appears, there is at least one exception.

Willis Whitney, for many years director of research for General Electric, independently conceives of the relationship between individual merit and serendipity in much the same way as Cannon. "As one of the users and appreciators of the term *serendipity*," Whitney was asked, as we have seen, to comment on a brief article titled "Serendipity" that appeared in the *Industrial Bulletin of Arthur D. Little, Inc.*, the industrial consulting firm. His commentary plainly indicated his conviction that serendipity was not an experience that happened to just anyone; rather, individuals had to have made active efforts in order for them to recognize the fruits of the unexpected: "In every individual's stock of knowledge (his conscious and subconscious assets) there lie the peculiar items or records of his former thoughts. Some of them may 'pop out' or 'come to mind' when a novel or unexpected event crosses his mental threshold. Some sort of catalysis has taken place. This all indicates dependence of the gift of serendipity upon the total (even forgotten) knowledge and training of the individual. This gives us all a continuing reason for learning more wherever we are."[22] Thus the case for knowledge of whatever sort, for without a substantial stock of knowledge, individuals lack the requirements for "some sort of catalysis" to take place. Indeed, the "gift of serendipity" *depends* on such knowledge. Moreover, the possibility that the accumulation of knowledge may lead to serendipity should motivate individuals never to cease learning.

It is this same kind of unceasing preparation that leads eventually to inspiration that has been described by Gilbert Murray:

> Poetry needs ecstasy or inspiration; true, but the inspiration will not merely be imperfect, it will simply not come, except to a mind that has by some long process of thinking and feeling been prepared for it. The preparation need not be conscious or specialized. When Paul had his vision on the road to Damascus, or Augustine heard the words, "Tolle, lege," or Plotinus and St. Francis were uplifted into their special ecstasies, they had not, of course, been practising or rehearsing those ecstasies, but they had been living the kind of life and concentrating upon the kind of thought or effort to which that inspired hour was a natural crown. A poet who tosses off some exquisite lyric, apparently on the spur of the moment, has almost certainly been so

[22] Willis Whitney, in *Industrial Bulletin of Arthur D. Little, Inc.* (July 1940).

living as, first, to be exquisitely prepared for that particular mood or emotion, and, secondly, to have developed the technical skill which enables him to write what he wants to write. It is, if one thinks of it soberly, absurd to suppose that inspiration falls like rain equally on him who lives among poetical thoughts and him who thinks only of his digestion and his bank balance.[23]

Murray is speaking of poetry rather than of scholarship or science, but like Whitney and Cannon he is analyzing the quality of preparedness that is the prerequisite for great achievement.

While both Cannon and Whitney consider the occurrence of serendipity to signify that a superior scientist is at work, they can point only rather vaguely to the ways in which such scientists "[find] unforeseen evidence of [their] ideas, or which surprise come upon new objects or ideations which were not being sought." (This, it will be remembered, is Cannon's description of what is found by serendipity.) David Seegal, in his article "Chance and the Prepared Mind," joins in the notion that it requires special preparation to make accidental discoveries:

> As the individual proceeds through his student-hood (which should never cease), he or she learns first with some surprise and then with satisfaction that many of the great advances in medical science have come by simple means and often by *chance*. It would seem as if Providence were exercising wit and playfulness in hiding the missing piece of the scientific puzzle behind a nearby elm tree, while the search went on in a distant and exotic forest. But the rewarding chance observation may be missed even when the investigator finds the elm tree unless he has had a sound training in his chosen field. He may lack the receptors characteristic of the trained mind to take advantage of the chance observation.[24]

Dr. Seegal can quote Joseph Henry, Pasteur, and Dr. Cannon himself on the subject of the prepared mind but he goes beyond them in trying to specify the conditions under which serendipity occurs: "the conditions or climates which . . . more effectively prepare the mind for the chance observations we will experience many times during our lives."

Granting the importance of the acquisition of factual knowledge, Dr. Seegal wants to "turn your attention to certain *other* aspects of learning which are useful for the hyper-sensitization of the mind in order that the chance seed of a discovery may find a more fertile soil." First, he maintains, students should improve their capacity for free association: "Undoubtedly this . . . quality of low synaptic resistance (free flow of ideas)

---

[23] Gilbert Murray, *The Classical Tradition in Poetry* (London: Humphrey Milford, Oxford University Press, 1927), pp. 46–47.

[24] David Seegal, "Chance and the Prepared Mind." [Editor's note: We have been unable to locate this article.]

must depend largely upon inborn characteristics, but we have been impressed through the years by the capacity of many students to sharpen their intellectual reflexes and increase the degree of useful associations of ideas." Second, he recommends self-confidence, dedication, and "pride of work performed." "If this pride is at a low ebb, the chance observation will have little opportunity of being recognized or acted upon." And third, he believes that exposure to a trained investigator may transform students from "potential routinists to thoughtful, imaginative, and creative physicians."

Dr. Seegal then proceeds to explore what "differentiates the routine from the productive mind in our profession." One such differentiating quality is the response to exceptions: "Variations from the expected theme may be a nuisance to the routinist, but to the trained and prepared mind they may be the stimulus for a new symphony." Also, the agile and "prepared mind will be struck by *similarities* of reaction in different biological states." Finally, able investigators will try to provide themselves with a certain amount of leisure set aside for reflection:

> I trust that you will not feel that I am making a plea for a full return to medieval times when I suggest that the prepared mind might be the better fitted to recognize the chance observation if more opportunities were presented for the investigator to place his feet on the desk. In this position it is possible to doze and it is surprising [*sic*] that such an apparent period of unproductivity is sometimes a prelude to the entrance of an idea. This uncluttered mental state offers an opportunity for rumination of your mental cud. These periods of rest and chance for receptivity will not be given to you gratuitously. They must be earned, worked for, and treasured.

Dr. Seegal adds the further, tacitly sociological, suggestion that such time for reflection may be especially productive "when shared by others of similar interests." "The chance observation might often have been overlooked if there had been but one mind to perceive it. Furthermore, those who have learned to share their scientific bread have often found it returned many times enriched."

Here, then, is the offering of quite an explicit set of prescriptions for the securing of chance's favors—without question, these favors don't just happen; they must be earned, and the ability to earn them is somewhat limited by an individual's natural endowment. To bring his message home, Dr. Seegal illustrates each of his points with examples of well-known discoveries in which chance played a part. He describes Edward Jenner's "capacity to observe and associate the apparently unrelated facts" of the similarity between smallpox and cowpox and the apparent immunity to smallpox of those milkers who had contracted cowpox. He tells how Pasteur tried unsuccessfully to infect some hens with an aged cholera culture, and several days later was equally unsuccessful when he tried to infect them with a young and virulent strain of the organism: "This was outside of his previous experience, but Pasteur did not become irritated and disinterested [*sic*] by this vagary of biological experimentation." He repeated his experiment and discovered that the aged culture could produce immunity. "Here was a basic principle which was established from the pursuit of a chance observation." Another discovery of Pasteur's shows that "Serendip need not necessarily be a laboratory—that a meadow might serve as well." This was his discovery of the transmission of anthrax spores by earthworms: Pasteur noticed the profusion of earthworm castings at the sites of recently buried sick sheep, and followed up his guess that these worms might be spreading the disease.

Dr. Seegal also appreciates the fact that the missing of a chance discovery highlights the qualities required for making it, and he uses the discovery of penicillin as an illustration:

> Many examples might be cited of occasions when the scientist had just missed taking advantage of a chance observation. However, there are probably *scores* of bacteriologists who feel that the opportunity to discover penicillin might have been theirs. Many of these workers had observed that when two different organisms were growing side by side, the growth of one organism might inhibit the growth of the other. However, it was necessary to wait for Dr. Fleming's thoughtful evaluation of this phenomenon and Dr. Florey's fine contribution to the chemical side of the problem before penicillin was developed as a powerful therapeutic agent.

The "pregnant hint" may come from a patient as well as from a set of plates. When Leslie Gay and his associates at Johns Hopkins were working on the alleviation of the symptoms of allergic diseases by the use of antihistamines, a patient reported after a dose of Dramamine that her symptoms had not been relieved but that "she was delighted to discover that the car sickness to which she was subject did not occur on her journey home [by a street car]." The doctors followed up this chance lead, and "extensive studies since that time have shown that the chance observation of the patient and the appreciation of the possibilities by her physicians have given the medical profession a useful drug for the treatment of sea sickness and associated conditions. Indeed, suggestive evidence is at hand to indicate that the drug may relieve some of the distressing symptoms of the vomiting [stemming from] pregnancy and Ménière's disease."

At the conclusion of his paper, Dr. Seegal reiterates that the experience of serendipity is there for any scientific investigator, whether student or full-fledged physician:

> Are these anecdotes concerning chance and the prepared mind to be looked upon as mere contributions of a few? Or do they indicate tools, methods,

and points of view which are available to all of you who are prepared to work, observe, and correlate? I have chosen some of the more important examples of the rewards which are offered to chance and the prepared mind, but there are innumerable less spectacular advances in medicine which have been made by individuals like yourselves who have looked and acted. . . . It is never too soon to be alert and to question all rules as well as all exceptions.

According to Dr. Seegal then, students should cultivate in themselves the requisites for serendipity. Serendipity must be earned, and it can be earned by attending to certain specific personal qualities. Moreover, by reverse implication, the making of many accidental discoveries itself becomes a mark of the possession of those excellent qualities. Thus, like Cannon, Seegal feels that to attribute serendipity to individuals is to say something flattering about them. They link serendipity firmly with qualities eminently desirable in scientists, qualities which Dr. Seegal is, in fact, commending to neophyte scientists. Dr. Seegal realizes that neophytes in any field are likely to lack especially that flexibility which permits the mature expert to combine disciplined work with imaginative forays beyond the anticipated limits of that work. Serendipity is, therefore, no threat to the reputation of a scientist; rather, the ability to take advantage of the unexpected is taken to be a mark of maturity and distinction.

For some, however, there always lurks behind a word that has any connection whatever with luck the danger that it may be identified wholly with luck. The imputation of serendipity can carry the implication that the discoverer was "*just* lucky" and this, in turn, calls his ability seriously into question. The imputation of "mere luck" may not actually be intended, but where there exists some special sensitivity about the recognition of merit, the very possibility of such an imputation is resented. So it is that Milton Rosenau appears to be somewhat sensitive to aspersions on the reputation of distinguished scientists as well as on his own reputation, with the result that his paper titled "Serendipity" shifts, it would seem unwittingly, between praise for serendipity and the dismissal of serendipity. The key to this oscillation lies in the degree of importance assigned to the component of accident in discovery. Dr. Rosenau is glad to allow accident a subordinate role but he will not tolerate any suggestion that accident is more important than other components. With an apt metaphor Rosenau describes the process of making unanticipated discoveries, but the description is permeated by an uninhibited defensive tone and contains cautions against ascribing too much to accident:

Many a scientific adventurer sails the uncharted seas and sets his course for a certain objective, only to find unknown land and unsuspected ports in strange parts. To reach such harbors, he must ship and sail, do and dare; he must quest and question. These chance discoveries are called "accidental" but there is nothing fortuitous about them, for laggards drift by a haven that may be a heaven. They pass by ports of opportunity. Only the determined sailor, who is not afraid to seek, to work, to try, who is inquisitive and alert to find, will come back to his home port with discovery in his cargo.[25]

Chance discoveries, then, depend on an impressive list of estimable qualities in a scientist: enterprise, courage, curiosity, imagination, determination, assiduity, and alertness: There is "nothing fortuitous" in so-called accidental discoveries.

Rosenau shifts to eloquence to defend Pasteur against charges that accident had a large role in his discoveries. "There have been those who would belittle the work of our patron saint Pasteur, by insinuating that like a Prince of Serendip, his results were achieved more by accident than by sagacity. What a distorted viewpoint of a genius whose unremitting toil and zeal gave so much happiness to the world and the security of life!"[26] And in discussing the same traditional anecdotes as Cannon did, Rosenau uses a much more aggressive tone:

> There are many absurd stories about the discoveries that seem to come about through accidental blunders in the sense of Serendipity. It is said that Archimedes got the idea of specific gravity by noting the buoyancy of his body in water. How can anyone believe that Isaac Newton drew such deep deductions as the law of gravity and its universality by noting the fall of an apple as he lay musing on the grass in an orchard? Another one of these intriguing legends is told of James Watt, the teapot and the steam engine. Such things do not happen to Tom, Dick and Harry; yet Tom takes a bath, Dick sees an apple fall, and Harry observes the steam lift the cover of the teapot.[27]

In every field, Rosenau insists vehemently, hunches and chance discoveries come only to those of proven competence, competence achieved by training and proved by achievements that had nothing whatsoever to do with chance. In science as in art, "inspiration implies perspiration. The master word has always been and always will be 'work.' Chance and accidental discoveries in scientific research come only through preparation by toil and thought."[28]

Several discoveries are described by Rosenau in which, as he sees it, "There was much of sagacity . . . and little of accident"[29]: H. C. J. Gram's discovery of Gram-positive and Gram-negative bacteria, which he

---

[25] Milton J. Rosenau, "Serendipity," *Journal of Bacteriology* 29 (1935): 92.
[26] Ibid.
[27] Ibid., p. 93.
[28] Ibid.
[29] Ibid., p. 94.

found while looking for something else; Jenner's discovery of vaccination; Roentgen's discovery of X-rays. And he defends even his own discoveries against slur. He opens the story of his own experiences: "You may think that the pioneer work on anaphylaxis started in the reign of Serendipity. If you will pardon a personal allusion, it was something like this." Then he sets the record straight and tells how he and his assistant at the Hygienic Laboratory in Washington, D.C., discovered "the well-known phenomenon of seric anaphylaxis." He concludes the story by repeating: "You may think we stumbled into this in the reign of Serendipity, but I am sure that if my good friend and colleague John F. Anderson were here, he would insist with me that the results were obtained through many years of work, and there was joy in the work."[30]

Dr. Rosenau's objection to the allegation that accident plays a significant role in the process of discovery is evident when he puts the word *accident* in quotation marks, as in his description of the work of Ronald Ross on malaria. Not only does he think that science is not significantly advanced by accidents (of which more later), but also that it is ungenerous and even unfair to the investigator to attribute his success to them. Rosenau is so sensitive on the subject that it often appears as if to him any mention of accident carries the meaning of "only accident."

In another passage, however, Rosenau turns around and sees merit in the "blunderer from Serendip": "Right here I should like to put in a word to beginners in research concerning the unexpected that is constantly cropping up in exploring the unknown. The natural tendency of the unprepared mind is to discard the unusual. It is dismissed because not wanted—it does not conform to the preconceived plan. The unusual and unexpected may be significant and perhaps have to wait for a blunderer from Serendip on a journey of adventure in quest of discovery." If it be granted, then, that it is the very mark of competence to recognize the significance of the unusual or unexpected, then Dr. Rosenau is quite ready to admit this kind of "accident" as part of the process of discovery. Like Dr. Cannon, he can see the limitations of plans so carefully preconceived that the exploration of unexpected exciting possibilities is precluded. The voyager from Serendip—now called a "blunderer" in order, modestly, to understate his undoubted abilities—will not ignore the unusual because it interferes with a preestablished conception of regularity: Exceptions do not prove rules; rather, they call for the formulation of new ones.

Just as Rosenau tends to identify the suggestion of an accidental component in the making of a discovery with the imputation that accident, and accident alone, was responsible for the discovery, so do two histo-

[30] Ibid., p. 95.

rians of science seek to defend the subjects of their studies against similar "aspersions": Otto Glasser's two biographies of Roentgen include such a defense, as does R. C. Stauffer's article on Oersted. Glasser corrects those who would belittle Roentgen's achievement, and he castigates them by impugning their motives;

> Although the discovery of the X-rays made a tremendous impression upon all scientists of his time, it is noteworthy that later the credit due to Roentgen often was minimized by lesser men. Many unfounded rumours based on envy and jealousy termed the great achievement a purely [N.B.] accidental circumstance. These envious persons did not know—and even now it is not generally known—that Roentgen would have been one of the greatest scientists of the nineteenth century even without his discovery of the roentgen rays. Roentgen had taken up the work with cathode rays because he had a feeling that there were still many unsolved problems connected with them, in spite of the large number of valuable observations which had been made by other scientists. Like every research scientist he was looking for new fields when he discovered a "new kind of rays" and brought to such a brilliant culmination the investigations of his illustrious predecessors from Von Guericke to Lenard. The discovery was the logical and perhaps necessary conclusion to a long series of investigations and the *only* [our italics] accidental circumstance connected with it was that Roentgen happened to note the fluorescence on a screen at the particular time that the phenomenon was evident. It is very probable that if the observation had not been made on November 8, it would have been made later. Other workers with cathode rays had made similar observations, but their significance was fully recognized only by Roentgen.[31]

In Glasser's defense of Roentgen it is implied that the intrusion of any accidental factor casts doubt on Roentgen's greatness, and he tries to reduce the role of accident to an absolute minimum. But at least one authority whom he cites to support his defense is more inclined than he to admit that an accidental factor in success is no disgrace, and that it is a token of Roentgen's very greatness that he saw the significance of the accident. As Glasser puts it, "Münsterberg, the great Harvard philosopher, silenced some of [the] voices of envy with: 'Suppose chance helped. There were many galvanic effects in the world before Galvani saw by chance the contraction of a frog's leg on an iron gate. The world is always full of such chances, and only the Galvani's and Roentgens are few.'" While Glasser, bending over backwards to do his hero justice, shares Rosenau's sensitivity to any interpretations of accident, Münster-

---

[31] Glasser, *Wilhelm Conrad Roentgen*, pp. 71–72.

berg's view might be put as follows: If accident is not all, neither is it nothing.

The article by Stauffer attempts to set the record straight concerning "Persistent Errors Regarding Oersted's Discovery of Electromagnetism."[32] The persistent errors are of two kinds: errors in dates and "unwarranted emphasis on the role of chance." We are not interested in the errors in dating Oersted's work; as for the second kind of errors, what is significant for our purpose is the explicit assumption by Stauffer, reflecting Oersted's own view, that any suggestion of accident is tantamount to an unwarranted emphasis on the role of chance and a depreciation of Oersted's achievement. Here is Stauffer's account of the history of the erroneous version of Oersted's discovery and of the attempts to refute it:

> The first German announcement of Oersted's discovery was also the first appearance of the accident version. In his introduction to his translation of the Latin paper by Oersted which first reported the discovery, Ludwig Wilhelm Gilbert, the editor of the *Annalen der Physik,* said that, "What every search and effort had not produced, came to Professor Oersted in Copenhagen by an accident during his lectures on electricity and magnetism in the past winter." Although Gilbert admitted he had difficulty understanding Oersted's Latin account, this does not explain his unwillingness to give Oersted any credit for the discovery. As Schelling observed, the accident was a purely Gilbertian embellishment. This grudging account did serve, however, to evoke a specific refutation of Gilbert's version by Oersted himself. It also aroused the objections of J. L. G. Meinecke and S. T. v. Sommering, expressed in contemporary letters to Oersted. The accident version was again denied after Oersted's death by Johan Georg Forchhammer, the geologist and chemist, in a commemorative address. Here Forchhammer, who had been Oersted's amanuensis in 1818, stated: "Oersted was *searching* for this connection between those two great forces of nature.'
>
> Of the many subsequent repetitions of the accident version only two are worthy of note. For Louis Pasteur the story served as the occasion for one his best known epigrams: *dans les champs de l'observation le hasard ne favorise que les esprits préparés.*

The second instance of its repetition that is worthy of note to Stauffer is that of Christopher Hansteen, which Stauffer establishes as being mere hearsay instead of an authentic eyewitness account. In the "Summary" at the end of the article, Stauffer denies once more categorically that accident played *any* part in Oersted's discovery: "It was *Naturphilosophis*, not chance, that led to the discovery of electromagnetism." It may very well

---

[32] R. C. Stauffer, "Persistent Errors Regarding Oersted's Discovery of Electromagnetism," *Isis* 44 (December 1953): 307–308.

be true beyond doubt that chance had nothing whatever to do with Oersted's discovery, but it is also true that Stauffer is hypersensitive to the "accident version." Perhaps Ludwig Gilbert intended his remarks about the accident to be derogatory, and perhaps not; it does seem ironic that Stauffer does not see in Pasteur's "version" of the story a tribute to Oersted, but just another repetition of the obnoxious error.

If Dr. Rosenau tempers his view that the imputation of serendipity is damaging to a scientist's reputation, George Burch of the Tulane Medical School has no doubt but that serendipity is a stigma. In a pamphlet titled *Of Research People,* Burch contrasts "the true investigator" with the "fashionable laboratory worker" as follows:

> The ability to make discoveries is the fundamental and peculiar attribute of the *true investigator.* He knows what, where, and how to search, the number of discoveries being limited only by the time available for work. His investigations are well planned to yield significant data. The immediate or relative importance of the discoveries is comparatively inconsequential. Importance defined in terms of immediate applicability, blurs the significance of discoveries. Those of immediate practicability, regardless of magnitude, arouse acclamation, even if engendered by fundamental but less exciting discoveries of scientists relegated to little attention or fame. The feat of introducing an entirely new concept usually occurs only once in an investigator's lifetime, but an able investigator continues to make significant discoveries. He is, by nature, restless and is not content to reap laurels from one discovery. Serendipity is possible for anyone, but such discoverers may be merely fashionable laboratory workers and not necessarily investigators. To gamble upon serendipity is the practice of many laboratory workers of variable ability, as well as of some benefactors. Remember, the degree of acclaim by one's fellow man is not an accurate index of research ability and accomplishments.[33]

Dr. Burch glibly identifies everything he scorns with serendipity. His "fashionable laboratory worker" or pseudoscientist works without research design or plan, he sacrifices truly significant pure science to speciously important applied science, he looks to the lay public for applause instead of hoping for the approval of his peers, and, once successful, he rests on his false laurels instead of steadfastly pursuing his researches. Incompetent that he is, he must gamble on serendipity (accident) for his success, and because his success creates so much noise and fury, he can even find benefactors to subsidize this perversion of true science. For Dr. Burch, the difference between the qualities of the true investigator and the beneficiary of serendipity is that between the meritorious and the

---

[33] George Burch, *Of Research and People* (New York: Grune and Stratton, 1955), pp. 6–8.

meretricious, between the sound and the specious. It is the difference between steady, disciplined scientific work, whose quality is appreciated and rewarded by the scientist's peers, and casual, random efforts that happen to pay off in results that the lay public understands (again the image of the benighted lay public!) and praises extravagantly.

The different views about the legitimacy of accidental discovery that we have examined so far have been concerned primarily with the import accidental discoveries have for the *merit* of the scientist who makes them. Cannon, Rosenau, Stauffer, and Burch are in substantial agreement about the qualities a good scientist should have (indeed, there is very general agreement on this subject), but they disagree to some extent about the compatibility of the occurrence of accidental discoveries—and by the same token, of serendipity—with these qualities. Those who interpret accidental discoveries as being only in *subordinate* measure accidental find the making of such discoveries entirely compatible with qualities esteemed in a scientist: alertness, training, toil and perseverance, and the like. However, those who interpret serendipity as granting *overriding* importance to the component of accident in the process of discovery use it as a summary label for all the qualities unbecoming to a scientist.

The Legitimacy of Serendipity in Scientific Work

The question of the legitimacy of accidental discoveries may also be debated without any judgment of the quality of the scientist whose work is involved. The question may be raised, instead, in order to establish what the "best" procedures are for the conduct of scientific investigation. The issue, then, is not *whether or not* accidental discoveries play a part in the development of science but, rather, *how large* a part is compatible with the *maximal* development of science. Various people concerned with science have presented different prescriptions for the making of scientific progress, and in these prescriptions the legitimacy of accidental discoveries generally has a minor but illuminating place. If, for example, serendipity is frowned upon, the prescription is likely to rely heavily on the development of scientific theory; if, on the other hand, serendipity is applauded, the prescription is likely to place great stock in empirical investigation. Although prescriptions tending toward these two poles of "theory" and "empiricism" are, perhaps, most frequently encountered, there are several variations of this theme, as well as attempts to synthesize the two poles.

The English scientist Norman Pirie is one who is doubtful of the usefulness of unstructured research. He believes that the most efficient way to promote science is not simply by observation but rather by focused inquiry:

> Some people think that the philosophy a scientist accepts is not of very much importance; his job is to observe phenomena. This is a gross oversimplification and it involves the subsidiary hypothesis that all scientists are fully equipped with serendipity.* [f.n. *A pleasant word coined by Horace Walpole for the faculty of often finding something more interesting than the thing you set out to look for.] A sensible philosophy controlled by a relevant set of concepts saves so much research time that it can nearly act as a substitute for genius; it may be that this is what we mean by genius.[34]

The misapprehension that Pirie is trying to correct is that just by looking, as it were, scientists are bound to see something more valuable beyond any expectation. Instead, he thinks that any scientist with training in theory (a "relevant set of concepts") is almost as well equipped as possible; he may, indeed, even be in possession of that equipment which distinguishes the genius. In any case, whether it be the near-equivalent of genius or the actual equivalent, "a sensible philosophy controlled by a relevant set of concepts" is the sine qua non for Pirie of satisfactory scientific progress.

Bradford N. Craver, a pharmacologist associated with the Squibb Institute for Medical Research (New Brunswick, New Jersey), is, likewise, on the warpath against the alleged claim that serendipity, and serendipity alone, will advance pharmacological science. In his introductory remarks to a monograph on *Experimental Methods for the Evaluation of Drugs in Various Disease States,* Craver proclaims: "The conference on which this monograph is based was arranged in the conviction that serendipity alone is an insufficient provider of better therapeutic agents. Greater reliance can be placed on the systematic study of the experimental techniques employed in pharmacological research for selecting types of physiological activity that may have clinical usefulness."[35] Craver implies that there are some who are relying on "serendipity alone," who are entirely dependent on chance for their positive results. Exaggerated though this picture may be, Craver is making his plea for systematic study in the context of the large amount of hit-or-miss research that is, actually, current in pharmacology.

Perhaps it is because the social sciences, like pharmacology, are relatively undeveloped as far as theory is concerned that several social scientists seem to share Craver's opinion that there is too much serendipity in

---

[34] Norman Pirie, "Concepts Out of Context: The Pied Piper of Science," *British Journal of the Philosophy of Science* 2 (February 1952): 269–280.

[35] Bradford N. Craver, in *Experimental Methods for the Evaluation of Drugs in Various Disease States,* ed. Bradford N. Craver (New York: New York Academy of Sciences, 1956).

their fields. (Let us repeat again that they are not so much casting aspersions on the caliber of individual social scientists as deploring a certain approach to scientific work.) Daniel Glaser, for example, thinks his sociological specialty, criminology, has been hurt by too much serendipity, because it has led to a kind of theoretical eclecticism: "Serendipity—the influence of 'unanticipated, anomalous and strategic' observations [the quote is from Merton's discussion of the serendipity pattern, of which more later]—in causing us to revise theory, has usually resulted in patchwork eclecticism rather than the systematic revision of criminality theory. Where one behavioral image does not fit, we skip to another. Most textbooks in criminology present a cluster of disparate atomistic theories as 'the theory of multiple causation.' These should be regarded as temporary expedients in the course of revising theory."[36] In criminology, as Glaser sees it, accidental discoveries have not been integrated into a body of systematic theory, but have engendered an inchoate collection of ad hoc theories. Serendipity is only incidentally to blame for this "patchwork eclecticism": It stimulated criminologists to a revision of theory, but it did not have to lead to an unsatisfactory revision.

Other social scientists have been more generally opposed to the influence of serendipity on their science. Edward Suchman, in his review of Herbert Hyman's *Survey Design and Procedures*,[37] lists his "Likes" and "Dislikes" where the book is concerned, and under "Dislikes," point 2 is: "The accent on 'secondary' analysis and serendipity, rather than initial hypothesis formation." Suchman takes it for granted, it seems, that the reader will understand what the harmful consequences of secondary analysis and serendipity are.

In another book review, Otis Durant Duncan dismisses Walter and Jean K. Boek's *Society and Health* with the comment: "Whatever increment of knowledge or refinement of method it may confer upon the social scientist will be only 'serendipitous.'"[38] The implication is that such serendipitous increments to knowledge are not worth very much. (In this case there is also a reflection on the competence of the authors: Duncan is suggesting that whatever merit the book may have it has by chance rather than by the authors' design.)

Svend Riemer has yet another unsympathetic view of the place of serendipity in the development of science. As far as Riemer is concerned, serendipity is not the will o' the wisp that leads those astray who do not understand the need for systematization in science; rather, it is the *last*

[36] Daniel Glaser, "Criminality Theories and Behavioral Images," *American Journal of Sociology* 61 (March 1956): 437.

[37] *American Sociological Review* 21 (April 1956): 233.

[38] *Rural Sociology* 21 (June 1956): 210.

resort of social scientists for whom situational factors have made systematization impossible. In a generally critical article titled "Empirical Training and Sociological Thought," Riemer deplores the way social science is organized. (We shall deal with this aspect of his argument later, when we discuss the relationship between different conceptions of the proper organization of scientific activity and serendipity.) However, one of the consequences of this malorganization, as Riemer sees it, is relevant to the problem of the legitimacy of accidental discoveries in the optimal conduct of scientific work. Riemer thinks that because social science is "overorganized," "the development of systematic theory is left to chance. Discipline unfolds along lines of accidental methodological improvements. Inasmuch as improvements remain uncoordinated and unrelated to broad sociological perspectives, sociological theory must make its intermittent appearance and try to explain what is really unexplainable: trends in the growth of a science that does not grow by premeditated plan but by fortuitous circumstance. We learn to rely increasingly on serendipity for the advancement of our science."[39] The dependence on serendipity in current social science comes faute de mieux. It testifies to the aimless character of that science, which grows in a haphazard fashion instead of in the streamlined way Riemer thinks is the only proper one. Riemer is addicted to the all-or-nothing formulation (it has great formal similarity to identifications of serendipity with "mere" luck): Either scientists are exclusively concerned with systematization or the growth of theory is left to chance; either science grows by premeditated plan *or* by fortuitous circumstance. The history of science shows, of course, that science has grown both by plan and by accident, but Riemer thinks that to benefit from ("rely on") accident is a measure of desperation.

It would appear, then, that those who are committed to the idea that science, natural or social, progresses in the measure that scientists are properly concerned with the development of theory ("a relevant set of concepts," or "initial hypothesis formation," or "systematization" refer, roughly speaking, to theory) view accidental discoveries in general and serendipity in particular with skepticism, dismay, and even alarm. The occurrence of accidental discoveries or serendipity tends to be interpreted as "dependence" on chance for the advancement of a science, and serendipity has even become a kind of symptom of the parlous state of that science.

But this dim view of the significance of accidental discoveries is not the only one. Others have taken the occurrence of accidental discoveries to be a token not of the languishing of a science but of its flourishing. Far

---

[39] Svend Riemer, "Empirical Training and Sociological Thought," *American Journal of Sociology* 59 (September 1953): 112.

from cautioning scientists against serendipity, they would be more inclined to predict a sad future for a science that had to get along without it. Those in this camp are not opposed to the elaboration of general concepts, but they are concerned to give proper recognition to the way empirical data may play an independent and unpredictable role in this elaboration. They would rephrase Charles Richet's remark that "unforeseen facts *have to force our* hand" to read instead, "unforeseen facts *should be permitted to guide our hand.*"

But how is this to be done? As Edgar Allan Poe saw it in the middle of the nineteenth century, the problem had already been solved by fiat, as it were. In the words of Poe's scientific detective, Auguste Dupin, the benefits of accident had become so obvious that only one policy was both desirable and possible for scientists:

> Experience has shown, and a true philosophy will always show, that a vast, perhaps the larger portion of truth, arises from the seemingly irrelevant. It is through the spirit of this principle, if not precisely through its letter, that modern science has resolved to *calculate upon the unforeseen.* But perhaps you do not comprehend me. The history of human knowledge has so uninterruptedly shown that to collateral, or incidental, or accidental events we are indebted for the most numerous and most valuable discoveries, that it has at length become necessary, in any prospective view of improvement, to make not only large, but the largest allowance for inventions that shall arise by chance, and quite out of the range of ordinary expectation. It is no longer philosophical to base, upon what has been, a vision of what is to be. *Accident* is admitted as a portion of the substructure. We make chance a matter of absolute calculation. We subject the unlooked for and unimagined to the mathematical formulae of the schools.[40]

"Modern science," as we have seen, is actually far less unanimous on the subject of accident than Poe thought. Apart from that, however, even those who agree that it is the "seemingly irrelevant" that contains important clues to truth, have succeeded only partially in establishing ways of "calculating upon the unforeseen." Accident production, as we might call it, has been approved by many in principle, and, like those who would prevent accidents, they have sought in the history of past accidents the causes of future ones. But specifying the auspicious conditions for accidental discoveries is at least as difficult as finding the nefarious conditions of automobile accidents. Making room for the crucial element of surprise is as difficult as preventing it, and much more difficult than those who deplore the prevalence of serendipity in science seem to think. Mere carelessness is not enough, for indiscriminate carelessness would produce

[40] Edgar Allan Poe, "The Mystery of Marie Roget."

"bad" accidents as well as possible "good" ones. The accidents that are desired must be of a particular kind. They must have implications for a body of scientific knowledge; in fact, they must add to that body of knowledge. Furthermore, such accidents must be capable of repetition—while they come (by definition) unexpectedly, their contribution to science depends on their subjection to planned experimentation or observation. Such accidents don't just grow on trees, and among those people concerned with the development of science who rather wish they did, we find a variety of prescriptions for their cultivation.

Only a few of the explanations of the occurrence of happy accidental discoveries do *not* aim to tell the reader how such discoveries should be made. These are descriptions of the occurrence of accidental discoveries for the benefit of laymen, who are presumed to know little about this aspect of scientific work (cf. the image of the lay public that likes to think all discoveries are accidental). In a collection of popularized essays about atomic energy, for example, Harlow Shapley in an article titled "It's an Old Story with the Stars" explains:

> The fossil-pickers and chemists working with radioactivity in the rocks had found unknowingly something greater than they knew. It has been called serendipity—this faculty or fact of accidentally finding a result of superior significance while searching for something else. It occurs frequently, but seldom has it led to such a magnificent revelation as the exposure of the secret of the stars. Serendipity operated while the geochemists measured the ratio of lead to uranium in various old strata of rocks and while the paleontologists pondered the structure of fossil plants. For they found results that forced the astrophysicists . . . to make a speculative invasion of nuclear physics, and to find there a clue to the coming of a terrestrial atomic era.[41]

As Shapley describes it, one kind of scientific work turned up unexpected results that "forced" scientists in another field to make further investigations, and these further researches were fruitful beyond any expectation. That is how these things frequently happen.

Or, again, Edmund W. Sinnott, a biologist and the former dean of the Graduate School at Yale, expounds the theme of the endless possibilities for scientific exploration to a lay audience, and he remarks that "scientific discoveries . . . are often quite unpredictable. Planned and organized research, directed to a given end, is often highly successful, but frequently some strange maverick of fact may present itself, some curious exception to what was confidently predicted, that may be far more significant than everything else. . . . Serendipity, the happy faculty of finding things that

---

[41] Harlow Shapley, "It's an Old Story with the Stars," in *One World or None*, ed. Dexter Masters and Katherine Way (New York: Whittlesey House, McGraw-Hill, 1946), p. 8.

one did not set out to seek, is an important trait of every scientist."[42] If all scientists have serendipity, then acquiring it, or fostering the occurrence of "strange mavericks of fact," presents no problem. As in Shapley's description of these events to the layman, they just happen.

Finally, we might cite, in this vein, the assertion by Earl Ubell, the science editor of the *New York Herald Tribune*, that "among scientists it is now almost proverbial that they make their *best* discoveries while searching for something else. Horace Walpole called that phenomenon serendipity." Ubell, again, suggests that serendipity simply befalls those who are engaged in research. However, his statement surely needs modification: Although some scientists may accept such a bit of folk-wisdom without further question, many others are anxious to establish somewhat more precisely the determinate conditions of serendipity.

How Serendipity Happens in Science

There have been two chief approaches to the problem of *how* accidents happen in science. One has been concerned more and less directly with the issue of *planned* as against *unplanned* research in relation to accidental discoveries. (This approach is closely related to that which considers the bearing of the organization of research on accident production, of which more later.) The second approach has focused on the nature of the *interplay between fact and theory* in accident production.

The scientist who has perhaps considered more carefully than anyone else the question of the bearing of planning on the occurrence of accidents is Irving Langmuir. Langmuir has been concerned with the furthering of science rather than with the making of scientists, with the conditions that permit good scientists to be creative, rather than with the ways in which they come to be good scientists. In defining these auspicious conditions, Langmuir draws on his own research experience in the G.E. labs, and he tries to formulate some general principles for the conduct of research. These general principles constitute a more explicit version of the practical wisdom of Willis Whitney, who was director of the labs when Langmuir first went to work there in 1905, and from whom he himself took over the directorship later. Speaking to the National Academy of Sciences (November 17, 1947), Dr. Langmuir made some general remarks about the conditions of serendipity, before coming to the substance of his paper on "The Growth of Particles in Smokes and Clouds and the Production of Snow from Supercooled Clouds."[43]

---

[42] Edmund W. Sinnott, *Think* (November 1951).

[43] Irving Langmuir, speech to National Academy of Sciences, published as "The Growth

In these introductory remarks, Dr. Langmuir wants to leave his audience with "certain threads of thought" as a result of telling about his work with smoke and clouds:

> These threads are of many kinds. In the first place, the researches grew out of one another and yet there was no obvious connection. It was always an accidental connection. [For "obvious" and "accidental" read, perhaps, "premeditated" and "unplanned."]
>
> Several years ago, Dr. Whitney introduced into the laboratory in Schenectady the word "Serendipity," and it is very appropriate for certain kinds of scientific research. The word was invented or coined by Sir Horace Walpole about two hundred years ago. . . . [There follows a not too accurate version of the coinage of the word.]
>
> Now I would like to give my own definition of serendipity and see if you can find its application in the activities about which I shall tell you: It is the art of profiting from unexpected occurrences. It has a great deal to do with the planning of research.

Here we come to the heart of Langmuir's argument, to his case *against* the planners, who may be very successful or very unsuccessful, depending on the merit of their overall plan, and *for* unplanned research, which, he almost says, is bound to be very satisfactory: "The American public likes to think of supermen. There are supermen of all kinds: in Wall Street, in labor unions, in our Government. There are economic planners who can plan all things, and given time enough they would make everything very good or very bad. They are supposed to know how to plan. Now the examples I want to give you are all examples of unplanned research where things happened in a way that nobody could have planned and arranged and as a result we got results which were most satisfactory." Langmuir thinks that the American public, in its anxiety to bet on a sure thing, is likely to bet on the wrong thing. He thinks that if, instead, the public were willing to accept more uncertainty, it would be more certain to be rewarded. Reluctance to accept uncertainty limits the scope of available rewards, and may preclude rewards altogether. In his eagerness to undermine the validity of planning, Langmuir goes almost as far as Hooke in putting the scientist in the hands of Providence: "[The satisfactory results] occurred as though we were just drifting with the wind. These things came about by accident, and yet you will see that a series of rather fortunate accidents started us on a research problem just about the time we were finishing another, in a way which gave a continuity which has turned out to be very useful." Though Langmuir does not deny that

---

of Particles in Smokes and Clouds and the Production of Snow from Supercooled Clouds," *Proceedings of the American Philosophical Society* 92 (July 1948), n. 3.

there is some scope for planning, his description of the proper conditions for such planning would make it almost impossible to do at the "frontiers of science":

> Another thread that goes through it all is this very idea of planning; as to whether planning is always successful in accomplishing results; as to whether or not there are definite relations of cause and effect. If there are definite relations of cause and effect and you can see the effect produced by a given cause, then of course we should go ahead and plan.
>
> But suppose in some cases there is no relation of cause and effect. Then planning does not get us very far. All we can do is, like serendipity, put ourselves into a favorable position to profit by unexpected occurrences.

If planning is possible only when relations between cause and effect have already been established, this really amounts to saying that planning is impossible for the researcher—it may be all right for the engineer.

Planning, then, is inappropriate to most research, but what is the opposite of planning? How does the research scientist put himself in a "favorable position" that will permit him "to profit from unexpected occurrences"? Langmuir describes this favorable position in two addresses. It consists both of a set of assumptions about scientific phenomena held by the researcher, and a corresponding definition of the nature and aims of research on the part of the administrator of a research program. The attitudes and aims of those responsible for organizing research will concern us later. As far as the assumptions of the scientists are concerned, these have to do with the determinacy of physical phenomena. Scientists have established the fact that in certain areas relationships can only be stated in terms of probability rather than causality. Langmuir describes this dualism in science as the difference between "convergent" and "divergent" phenomena:

> We must recognize *convergent phenomena* where the behavior of the system can be determined from the average behavior of its component parts. The fluctuating details of the individual atoms average out giving a result that converges to a definite state corresponding to a natural law.
>
> There are also phenomena in which a single discontinuous event, which may originate from a single quantum change, becomes magnified in its effect so that the whole behavior of the system does depend upon a chain reaction that started from a very small beginning. These we may call *divergent phenomena*.[44]

[44] Irving Langmuir, "Freedom, the Opportunity to Profit from the Unexpected," Commencement Address delivered at the Pratt Institute (1 June 1956) and reprinted in the *G.E. Research Laboratory Bulletin* (fall 1956).

These divergent phenomena are of great importance in nuclear physics, in genetics, in psychology,[45] and in biology. It is the prevalence of these chain reactions that makes any detailed planning of research on the part of the scientist impossible: "Life seems to teem with divergencies. When you realize that the divergent phenomena frequently affect human lives, you must recognize that complete planning is not justified. Plans should be made but often only of a general kind."[46]

The art of profiting from the unexpected, then, that art, to return to Poe's phrase, of "calculating upon the unforeseen," is an art of general planning. The scientist lays out a general area of problems in which he is interested, but he steers clear of formulating specific problems too far in advance. One thing leads to another, and who can tell what will turn up? It is foolish as well as presumptuous to make precise plans for future work. For it is Langmuir's conviction that "calculation upon the unforeseen" is not only consonant with scientific principle but that it will also pay off in other ways that are highly desirable. Calculation upon divergent phenomena will produce results that are both unexpected *and* important: "You don't know all the things that are going to happen. Too many of them are unexpected. And it is many of these unexpected things that are going to be the most profitable, the most useful things you do."[47]

The crowning reward of such general planning is that it will preserve a freedom of inquiry, a freedom of opportunity, that is not only rational and efficient but is also part of a good way of life. "Freedom of opportunity as developed by democracy is the best human reaction to divergent phenomena. We may, in fact, define 'freedom' as 'the opportunity to profit from the unexpected.'"[48] It is in the best interests of science and of democratic society therefore that serendipity be held in high esteem.

Professor Salvador Luria of the University of Illinois has arrived at a somewhat different way of describing the kind of research process that combines planning in general with indefiniteness as to specific problems. In his account of "The T2 Mystery," he explains how "controlled sloppiness" permits the occurrence of fruitful accidents:

> Our story has as its critical episode one of those coincidences that show how discovery often depends on chance, or rather on what has been called "serendipity"—the chance observation falling on a receptive eye. The episode is a good illustration of the principle of "controlled sloppiness," which states that

---

[45] "We are continually confronted with situations where we must make a choice and this choice sometimes alters the future course of our lives."

[46] Ibid.

[47] Irving Langmuir, address to the G.E. Research Colloquium (December 1951), partly paraphrased and partly quoted in *G.E. Shareholders' Quarterly* (25 July 1952).

[48] Langmuir, "Freedom."

it often pays to do somewhat untidy experiments, provided one is aware of the element of untidiness. In this way unexpected results, sometimes real discoveries, have a chance to come up. When they do, we can trace their cause to the untidy, but known, features of the experiment.[49]

The assumption behind the principle of "controlled sloppiness" is that scientific work is not cut and dried, that, instead, there are always loose ends, and that it pays to be aware of those loose ends. In the absence of a rigid plan it is possible to pay attention to the untidy ends, and they may turn out to be of considerable importance. Compulsive tidiness in experimentation is even more crippling than in other areas of life: It permits the scientist neatly to verify or disconfirm hypotheses, but it keeps him from leaving himself open to fruitful surprises. Since science is by nature dynamic, compulsive tidiness goes against the grain of science. That does not mean that the scientist should abandon systematic experimentation: It means only that he should try to answer one question at a time, even though, inevitably, unanswered ones may "lie around" for a while. The unanswered questions may depend for their solution on the scientist's work on antecedent problems. In any case, if he keeps them for their potential worth, rather than discarding them because they clutter things up, the chances for the occurrence of serendipity will be much greater. Luria is urging a limited untidiness as part of a procedure for work rather than as the component of a psychological state. Such untidiness is, perhaps, the patterned counterpart to the subjective flexibility that Dr. Seegal was speaking of when he exhorted his students to develop "a capacity for free association."

The patterned occurrence of unexpected discoveries in science—which Langmuir explains elaborately by the divergence of phenomena, and Luria puts down, less analytically, to the efficacy of "controlled sloppiness"— has also been discussed by scientists in terms of a figure of speech drawn from industry: the byproduct. The making of byproducts in industry is the efficient utilization of the waste that the making of the primary product entails. On occasion the obsolescence of the primary product has made the erstwhile byproduct assume first importance as a source of income for the company. In its inception, however, the byproduct is an unanticipated result of the production of a specified commodity. In the same way, planned scientific work may have its byproducts. The products and the byproducts of research are, in effect, Claude Bernard's two kinds of game: that "[which presents] itself when we are looking for it, [and that which] may also present itself when we are not looking for it or when we are looking for game of another kind."

[49] Salvador E. Luria, "The T2 Mystery" [with biographical sketch], *Scientific American* 192 (April 1955): 22, 93–94. T-2 is a bacteriophage.

Ellice McDonald is concerned that recognition be given to the by-products of research, and, like Poe's detective Auguste Dupin, he stresses the importance of those phenomena that appear to be incidental or peripheral to the scientist's focus of interest:

> In the conception [of new problems] the incidental suggestion or accidental discovery should by no means be neglected. Often the byproduct is more important than the stated effort. If research laboratories could be blessed with serendipity, their lot would be happy. This strange word was coined by Horace Walpole in 1754, and the new Oxford dictionary gives its meaning as "the faculty of making happy and unexpected discoveries by accident." Fortunate indeed is the laboratory which has serendipity. But discovery comes to the prepared mind, so that serendipity may be cultivated, for the mind is a barren soil and soon exhausted unless it be continually enriched with new ideas and new aspects of old ideas.[50]

For McDonald, serendipity is the ability to see the possibility of making byproducts, and this ability comes both from the constant acquisition of new ideas (McDonald's conception of this constant enrichment of the mind in the interest of serendipity is very similar to Willis Whitney's) and from the general awareness that valuable byproducts may at all times be lurking just off the beaten track. Milton Rosenau places the same great value on byproducts as McDonald, but with a caveat: "what may be called the byproducts in research are often more important than the primary aim of the study. The byproducts of industry are often its most useful and profitable products; but we cannot have byproducts unless we make products." Rosenau is anxious to remind scientists that they must not neglect intensive, planned activity in their quest of the more elusive byproducts of such activity, for it would literally be nonsense to make the incidental reward the essential aim of their efforts.

Rosenau seems to be more inclined than Langmuir to believe that if you take care of the planned products the byproducts will take care of themselves. The byproducts are not so much carefully nurtured accidents as inevitable consequences of planned effort; their substance may be unpredictable, but their occurrence is not. This does not mean that Rosenau sells short the importance of the byproducts of research. Indeed, both in science and in society generally, he thinks byproducts may be of very great importance, and the importance he attaches to the extrascientific byproducts enhances that of the scientific ones. The social byproduct that Rosenau is especially interested in is the effect of the scientific method of thinking on other areas of thought:

---

[50] Ellice McDonald, "Annual Report," *Journal of the Franklin Institute* (January 1939).

The secondary effects of science itself may be even more valuable than the actual discoveries that are acclaimed. One of these secondary effects is the infiltration of the scientific method into the mode and thought and attitude of herd psychology. It has taught a frank facing of realities calmly and impersonally. It has shattered shackles and traditions and superstitions that have long chained our attitudes and methods of thought. . . . It is no longer impious to ask questions, to doubt. . . . This attitude has had a profound influence upon society, its morality, its ethics, its thinking. These adventures into the realm of reality are not accidental blunders in the sense of Serendipity, but inevitable by-products of the scientific method.[51]

Unlike Cannon, Rosenau does not think of the personal qualities that make for accidental discoveries in science as having a more general usefulness in the solving of social problems; rather, he believes that properly conducted, science is bound to have good secondary consequences— for science itself, and for society.

Conceptual Steps in Research That Lead to Accidental Discoveries

The first approach to accident production has focused on the relationship between planned and unanticipated consequences in science, between the nature and scope of planning and the occurrence of accidental discoveries. Scientists have expressed their opinions on the rationale of having accidents happen and about the way scientific activities should be carried on to make them happen. This is the macroscopic view of the "how" and "why" of accidents in science: the analysis of the way science should be furthered by taking the unpredictable into account, and the reasons why the prescriptions offered are good ones. Now we come to the microscopic view: the close analysis of the conceptual steps in the process of investigation that lead to the making of an accidental discovery. Again, it may be well to emphasize that this analysis is not concerned with the personal qualities necessary for making the unanticipated discover—it is concerned only with the pattern of scientific thinking involved.

Robert K. Merton, in a close-up look at the relationship between fact and theory in accident production,[52] shows how research may help to initiate, reformulate, deflect, and clarify theory, as well serving to confirm or refute hypotheses. It is by the "serendipity pattern" in research, which

---

[51] Rosenau, "Serendipity."

[52] Robert K. Merton, "The Bearing of Empirical Research on Sociological Theory," *American Sociological Review* 13 (October 1948); reprinted in Merton, *Social Theory and Social Structure*, pp. 254–278.

we noted earlier, that empirical facts aid in the *initiation* of theory: "The serendipity pattern refers to the fairly common experience of observing an *unanticipated, anomalous and strategic* datum which becomes the occasion for developing a new theory or for extending an existing theory." All three elements in the pattern are essential to an accidental discovery, that is, to the unanticipated development or extension of theory—the observation in and of itself is no discovery. The elements, furthermore, can be described in more detail:

> The datum [that exerts a pressure for initiating theory] is, first of all, unanticipated. A research directed toward the test of one hypothesis yields a fortuitous by-product, an unexpected observation which bears upon theories not in question when the research was begun.
>
> Secondly, the observation is anomalous, surprising, either because it seems inconsistent with prevailing theory or with other established facts. In either case, the seeming inconsistency provokes curiosity; it stimulates the investigator to "make sense of the datum," to fit it into a broader frame of knowledge. . . .
>
> And thirdly, in noting that the unexpected fact must be "strategic," i.e., that it must permit of implications which bear upon generalized theory, we are, of course, referring rather to what the observer brings to the datum than to the datum itself. For it obviously requires a theoretically sensitized observer to detect the universal in the particular.

The focus here is not on explaining how byproducts are best obtained, or how the scientist should allow for surprises, or how he should prepare himself to recognize the anomalous; instead, it is on a set of social-psychological and intellectual conditions that make for accidental discoveries. Given research set up for a certain purpose, some unexpected, puzzling data, and a scientist capable of being puzzled—given all of these, an accidental discovery will occur, because the relationship between fact and theory in science is such that it must occur.

In *Science and the Social Order*, Bernard Barber takes up the social and psychological factors in the process of invention in general; he is interested in accidental invention (or discovery) because it illuminates the general process. The role of the serendipity pattern in invention, and also of hunches, he thinks, shows that "the role of the individual in scientific research, however much his function and his particular problem may have been conditioned by society, is still an active one."[53] Barber brings out the extent to which the "individual creative imagination" is at work in happy accidental discoveries by pointing out that the "'unexpected' occurrences in the pattern of serendipity have been passively seen by other scientists; they are actively noticed only by the discoverer. They are ac-

[53] Barber, *Science and the Social Order*.

tively noticed, that is, by the scientist who has carefully studied his problem over a long time and is thereby ready, if he can create some anticipatory ideas, to take advantage of the 'unexpected' occurrence."[54] Barber tends to take for granted that an accidental discovery depends on the observation of unexpected and anomalous data: His emphasis is on the strategic character of the data.

Accidental discoveries also receive passing attention in a textbook on *Scientific Social Surveys and Research*. There the importance of occurrences that are extrinsic to the "problem at hand" is noted: "We should not overlook the fact that chance findings can be very fruitful. . . . Many great scientific successes are often prefaced by *seeming failures* and chance occurrences which at first glance *may appear utterly futile* to the problem at hand. However, such devious scientific paths have led to the discovery of the X-ray, radium, penicillin—the modern instances of serendipity. Scientists cultivate serendipity by being constantly alert for chance occurrences that may lead to new explanations and discoveries."[55] Further on in the book, the effect unanticipated data may have on the formulation of hypotheses and theories again is remarked on: "The observer must be sensitive to, and take cognizance of, wholly unanticipated and chance observations which may come to play a very important role in the research process in both finding strategic data and initiating new significant hypotheses and theories."[56] Like Riemer, however, though in lesser degree, the authors, Pauline V. Young and Calvin Fisher, appear to be wary of theoretical eclecticism as a consequence of serendipity. After explaining how scientists cultivate serendipity, they add what might seem to be a non sequitur if we had not encountered the same suspicion of serendipity before: "Planning, reflection, and hypotheses give direction to the inquiry. Without them, the inquiry may result in a mass of unrelated facts, mere 'intellectual scraps,' which fail to explain or throw light on the problems at hand."[57]

Among those interested in interpreting the significance of serendipity for research, broadly conceived, we have, finally, Roger Hilsman, who finds it opportune to discuss this question in his book on *Strategic Intelligence*. Hilsman states very clearly in what sense serendipity should, and in what sense it should not, be thought to enter into the research process. As he sees it, happy accidental discoveries, or serendipity, occur when a trained observer encounters unexpected and unfamiliar data; indeed, the fruitful interaction that occurs when the trained mind meets

[54] Ibid.
[55] Pauline V. Young and Calvin Fisher, *Scientific Social Surveys and Research: An Introduction to the Background, Content, Methods, Principles, and Analysis of Social Studies* (New York: Prentice-Hall, 1956), p. 107. Emphasis added.
[56] Ibid., pp. 156–157.
[57] Ibid., p. 107.

this kind of data is one of Hilsman's proofs of the importance of training: The trained investigator "is . . . in a better position to make the most of lucky accidents and to evolve meaningful hypotheses. Being familiar with theory, he not only knows what is unknown territory, but also what needs to be accounted for in altering prevailing theory as well as how to go about doing these things."[58] Hilsman adds a footnote to this statement to tell a little of the history of the word serendipity, and to support his interpretation with an allusion to Merton's discussion of the serendipity pattern. He has, evidently, read Merton's discussion somewhat selectively, with particular attention to Merton's use of the concept of a "strategic datum":

> The word [serendipity] has recently come into fashion (See Robert K. Merton, *Social Theory and Social Structure* . . .) to describe happy accidents in research and to emphasize the role of the theoretically prepared mind in making the most of such happy accidents. For example, when some growing bacteria were killed by a mold that had come into the laboratory through a window left open in carelessness, an untrained mind might have merely thrown out the contaminated culture and started over. But to the trained mind of Sir Alexander Fleming, this accident was highly significant and led eventually to the discovery of penicillin and the whole family of antibiotics.

Hilsman is less interested in what constitutes an accident in science than in the impact of this (unanalyzed) accident on the theoretically sophisticated investigator. To conclude his footnote, Hilsman provides a warning somewhat reminiscent of Dr. Rosenau's warning that the finding of by-products depends on the making of products. He cautions the reader against a noxious misinterpretation of serendipity that would make the word synonymous with a mindless, intuitive, and passive dependence on "inspiration" for the answer to scientific or strategic problems: "The word has unfortunately been sometimes used carelessly to describe the flash of inspiration that leads to a final, significant hypothesis after research has been going on for some time. The danger of this use of the word is the confusion it would cause if it was understood to mean that in research one could plunge into data, seize the hypothesis abounding there, and never have to think at all." It is not quite clear from this whether Hilsman would prefer to ignore the occurrence of those flashes of inspiration that Cannon called "hunches." But he certainly does not want people to equate serendipity with inspiration *alone* (again, the all-or-nothing formulation), or to think that accidental discoveries are easily come by, without thought or toil.

[58] Roger Hilsman, *Strategic Intelligence and National Decisions* (New York: Free Press, 1956), pp. 70–71.

Chapter 10
SERENDIPITY AS
IDEOLOGY AND POLITICS
OF SCIENCE

*T*he problems that serendipity raises in connection with the organization and administration of scientific research are somewhat different from those that are raised by accidental discoveries in the work of individual scientists. As far as individual scientists are concerned, serendipity raises problems having to do with the proper definition of their tasks and the optimal performance of them. In organized scientific research, on the other hand, tasks and goals are not defined exclusively *by* the individual scientists; they are defined *for* them, to some extent, by someone with the authority to do so. The administrator of scientific research defines the expectations he has of individual scientists who are accountable to him for their results. (We shall discuss the accountability of the administrator in the section on accidental discovery in business and in government research.) It is his job to define his expectations in such a way as to "get the most out of" the scientists whose work he is directing. In some instances, important decisions about the proper administration of research—about the proper definition of the tasks of scientists in a research organization—may be revealed in attitudes toward the occurrence of accidental discoveries.

Serendipity and Strategy in Research Organizations

Research administrators may be authoritarian or they may be permissive, they may see the interests of the individual scientists as being identical with those of the organization as a whole or they may not, and such preferences for relative autonomy and independence or for relatively rigid control may be refracted through the problem of the legitimacy or desirability of serendipity.

The occurrence of accidental discoveries may be extolled as a token of the proper degree of autonomy of the individual scientists in an organization, or of the proper flexibility of the organization as a whole, or it may be deplored as a sign of too rigid an organization. (Because the scientific community is individualistic in its values, it is not surprising that there are no instances of the charge that serendipity reflects too little formal organization.)

It is congruent with Irving Langmuir's views about the harmfulness of too much planning in science that he should also believe that an administrator of scientific research should define the expectations he has of those working under his direction as generally as possible. Such permissiveness on his part and freedom on theirs, Langmuir believes, satisfies the needs both of the scientists and of the research organization. The scientists are content because their individualistic values are not being violated and they are able to go, within wide limits, wherever their quest for fundamental scientific concepts may lead them, and the needs of the organization are being met because the scientists are productive under this liberal regime. Here is Langmuir's description of his own happy experience as a scientist in the G.E. laboratories under Willis Whitney's direction: When he first came to work for the company,

> Dr. Whitney . . . invited him to look around at the work that was being done there and to choose whatever kind of research he would most like to do.
>
> "In a purely selfish way," says Dr. Langmuir, "I decided that I'd like to work on high vacuum and on tungsten filaments, because here was a field where the laboratory had two things that I had never heard of before, two things in which they certainly had excelled anything anywhere else in the world. So I tried that for three years, having a lot of fun. Dr. Whitney used to come around every day and see every man in the laboratory and ask him 'Are you having fun today?' And they all were having fun.
>
> "After three years or so, I was telling Dr. Whitney one time: 'I'm having a lot of fun, but I really don't know what good this is to the General Electric Company.' 'That's not your worry,' he said. 'That's mine. As long as you are doing something, finding out about high vacuums and tungsten filaments and things that are related to the work the company is doing, why, we want to see that work go on.'"[1]

[1] Irving Langmuir, address to the G.E. Research Colloquium (December 1951), partly paraphrased and partly quoted in *G.E. Shareholders' Quarterly* (25 July 1952).

Whitney's permissiveness paid off in some very significant unanticipated discoveries by Langmuir, with valuable applications for G.E. (We shall discuss the question of the extrascientific aims of research later.)

Langmuir has, of course, in his subsequent tenure as director of the labs, made every effort to reproduce the social conditions created by Whitney. He has tried to do this by very general planning, that is, by a vague and unspecific definition of expectations from his staff, by forgoing work directed toward specific discoveries and thereby making discoveries possible. As he puts it, "You can't plan to make discoveries. But you can plan work that will probably lead to discoveries." Or again, "you can organize a laboratory so as to increase the probabilities that things will happen there. And in doing so, keep the flexibility, keep the freedom. That's what freedom is for. All of us in this Research are interested in freedom. We know that in true freedom we can do things that could never be done under planning."[2] Though Langmuir does not say so explicitly, it is evident that he is willing to pay one part of the price of freedom—uncertainty. The social arrangements that he believes to be morally and practically best will *probably*, not certainly, bring concrete rewards. But the risk is, after all, almost a mandatory one if it is recognized that if the risks of freedom are avoided the likelihood of failure is magnified.

Langmuir moves out of the laboratory and onto the stage of world history to demonstrate this point:

> Stalin believes that everything can be planned. Marx believed that everything could be planned. That's the trouble with all dictators. They think they can run the world by planning from above. What did Mussolini try to do? What did Hitler try to do? They had plans for conquering the world and they knew just how to do it. They failed. They failed for many reasons, but one of the reasons is that you can't run things that way. And I think that no matter how far you go in dictatorship, no matter how far it may succeed, it will ultimately fail because of the impossibility of planning on a world-wide scale.[3]

To put Langmuir's argument another way, dictatorship in the laboratory and in the political sphere is as impractical as it is morally repugnant; it runs counter to what men of the eighteenth century would have called the "natural laws" of society in general and the world of science in particular. The policy of leaving nothing to chance is inherently doomed to failure: It flies in the face of human nature, and especially the nature of rational, independent scientists.

The views expressed by Langmuir on the necessity of freedom for the

[2] Ibid.
[3] Ibid.

research scientist have been echoed by others who have come more or less independently to the same conclusions. Recently, for example, we received a private communication from DeWitt Stetten, now at the National Institutes of Health and formerly at the Public Health Research Institute of the City of New York. Apropos of our inquiries about his interest in serendipity, Dr. Stetten wrote:

> In the course of setting up our division in the Public Health Research Institute of the City of New York, it was my good fortune to have as an assistant, Mr. Frank Rennie. . . . Mr. Rennie's official duties related to the operation and maintenance of our mass spectrometer and a small instrument shop. He is a man of unusual perception and very diverse interests and skills and I observed, coming from our mass spectrometer room, a continuous stream of ideas based upon accidental discoveries unrelated to the primary function of that room. Because these events seemed to be well described by the word, serendipity, we labeled the room the "serendipity room."

After describing this episode in his career as an administrator, Dr. Stetten went on to make some general remarks about the significance of the occurrence of serendipity in the laboratory:

> The idea implicit in serendipity is of vast significance in the development of science. This idea has been variously expressed and one such expression I keep framed on the wall of my office: "Only a small part of scientific progress has resulted from planned research for specific objectives. A much more important part has been made possible by the freedom of the scientist to follow his own curiosity in search of truth!" This quotation comes from Dr. Irving Langmuir. . . . I believe that this statement will meet with concordance by the majority of laboratory workers.

It seems "obvious" that it would have been regrettable if Mr. Rennie, and all like him, had been prevented from exercising curiosity freely, and yet neither Stetten nor Langmuir takes it completely for granted that Rennie, or the young Langmuir, should be cut loose in this way. The quotation on Dr. Stetten's wall seems to be both a manifesto and a reminder that the scientists on his staff be given maximal freedom consonant with the general purpose of the organization he is administrating.

Again, in an article that appeared in the Standard Oil trade publication *The Lamp* (an article that we shall have occasion to discuss more extensively below), Langmuir's ideas are both quoted and rephrased:

> Cultivating serendipity is, essentially, a matter or being constantly on the lookout for the chance reaction that may lead to a discovery. Dr. Irving Langmuir . . . deliberately nurtures serendipity by *never setting himself a specific goal*. As he puts it, he just "has fun in the laboratory."

"Discovery cannot be planned," Langmuir said recently. "But we can plan work that will lead to discoveries."

In other words, research directors can help create an atmosphere in which the muse of serendipity is most likely to be wooed and won. They often do this by planning programs *broad enough* to *allow* their researchers the freedom to follow leads they chance upon.[4]

The research director, then, is empowered to make a decision about the desirability of having serendipity occur in his laboratory, and if he decides in favor of serendipity, he can manipulate the structure of his organization accordingly. He can create a suitable atmosphere for the making of accidental discoveries by giving his researchers, within broad limits, the autonomy to decide what is "interesting," that is, to decide whether they will follow up an expected line of investigation or divert their energies to an unexpected one that arouses their scientific curiosity.

Although Langmuir or Stetten might be said to take the occurrence of accidental discoveries to be a sign of a well-run research organization—one in which expectations are sufficiently loosely defined—they would surely not go so far as to suggest that *any* expectations imposed on the scientist from without would handicap him intolerably in the making of significant discoveries, anticipated and unanticipated alike. Theodore Kopanyi, professor and chairman of the Department of Pharmacology at Georgetown University, however, takes this extreme position in his attack on the Magnuson-Kilgore bill to create a National Science Foundation. He contends that such a foundation would impose restrictions on scientists that would throttle the future development of American science: "A disruption of serious proportions in American science is threatened by a bill pending in the Congress which provides for the regulation of scientific research. . . . The effect of the passage of this bill would, in my opinion, be as destructive in the field of science as the atomic bombs were destructive of the Japanese cities of Hiroshima and Nagasaki."[5] The scientific mind, according to Kopanyi, can develop only in very few milieus, for it needs "unhampered, unrestricted, unregimented freedom to explore." Universities and research institutions have provided such conditions fairly satisfactorily: "The university scientist does not have to make elaborate petitions for grants and contracts with detailed blueprints; he merely has to get a modest sum of money for the purchase of some necessary instruments, chemicals, glassware, or laboratory animals and then start on his research, *responsible only to his own conscience*."[6] A certain

---

[4] "Serendipity," *The Lamp* (September 1953). Emphasis added.

[5] Theodore Kopanyi, in *Bulletin, American Association of University Professors* (1945): 681–696.

[6] Ibid. Emphasis added.

inferior kind of scientific research may, to be sure, be "organized and directed"—this is the "spadework or routine type" of work that is necessary when "the fundamental discovery has already been made, the goal is already in sight and all that is needed is to corroborate, to extend, and to apply the basic observations made." Industry can be depended on to further such research, without harm to basic science.

The role of the government in science, however, Kopanyi fears will be utterly nefarious because it will stifle basic research:

> Basic discoveries in science may be the result of accidents, are often made by the scientist working on a fallacious hypothesis or on the basis of correct theory but the discovery made on the correct basis was not anticipated and turned out to be far more important that the original goal. . . .
>
> Every scientist knows that some of the most important discoveries in the history of science have been made by accident or on the basis of false assumptions. Walter B. Cannon called it "serendipity." . . .
>
> Is "serendipity" by act of Congress feasible?

Kopanyi's answer, of course, is "no!" because the National Science Foundation will award contracts only for specific discoveries, like a test for cancer, and the scientist who has this "assignment" "plugs on at the task for which he has contracted, and what might well have been the most important outcome of his research falls by the wayside." Kopanyi implies that accidental discoveries in science are likely to be the most important ones, and that a National Science Foundation will make them impossible by regimenting the work of scientists. The article goes on to describe the travails of American science as it becomes dominated and corrupted by government money, but we are interested here only in the place accidental discoveries have in Kopanyi's argument against the National Science Foundation.

Svend Riemer, whose complaint about the lack of systematization in contemporary sociology we discussed earlier, thinks that this lack of systematization is a consequence of the bureaucratization of sociology and that, in turn, the advancement of sociological science by serendipity alone is the consequence of the absence of systematic theory. The prominence of accidental discoveries, then, is for Riemer a symptom of *too much* organization in science, not of freedom. Riemer sees the dangers of over-organization much as Langmuir would: The bureaucratized scientists are expected to turn in reliable, routine performances; they are unfitted for creative scholarship; they become engineers or craftsmen. But whereas Langmuir would suppose that any scientific advances, let alone unanticipated ones, would be stifled by such a bureaucracy (whether real or mythical), Riemer thinks that in this unhappy situation the only way for science to make any progress is by serendipity:

Initiative is submerged in specialized bureaucratic performance or allocated to a narrow range of leadership positions, access to which is obtained actually through the renunciation of initiative.

Room for initiative remains in the phrasing of questionnaires and in the analytical procedures involved in the evaluation of collected materials. But tasks of this nature are so restricted in scope that concerns with general problems are rarely maintained. Thus, the development of systematic science is left to chance. . . . We learn to rely increasingly upon serendipity for the advancement of our science.[7]

Serendipity for Riemer is a bare flicker of life in a crippled science, not a sign of health.

Administrators of research organizations vary not only in the ways they define a desirable performance on the part of their subordinates, but also in their ability to shift the major focus of interest of the organization as a whole from one problem to another. M. Brewster Smith, a professor of psychology at New York University, suggests that flexibility in shifting the work of the organization to significant problems as they arise makes for important contributions to science. In his review of *When Prophecy Fails* by Leon Festinger, Henry W. Riecken, and Stanley Schachter, which describes the study of a short-lived prophetic movement in the Midwest,[8] Smith observes that the book is a "noteworthy venture" in that it represents "an exemplary instance in which alert social psychologists with a theory to test were able to see the relevance of a passing event, and to respond to it in time and in sufficient force to capture the pertinent data." It is the response "in time and in sufficient force" that is of particular interest here, for it is as members of a flexible research organization that the authors were able to take scientific advantage of a social movement that *happened* at a certain time, without any regard for the current plans and projects of the research institute. Here is the way Smith puts it (not quite coherently, perhaps, but the main import is clear): "How better to advance a science of social phenomena than to capture elusive social events and put to them questions of theory forged and sharpened in the laboratory? Luck is required to turn the passing event to scientific advantage, but serendipity can be nurtured by imagination and institutional flexibility. Perhaps research institutes like the Minnesota one can provide for a planful flexibility of operations that will make such studies less of a rarity than they are at present." To speak of the flexibility of a research institute is, of course, tantamount to speaking of the flexibility of

---

[7] Svend Riemer, "Empirical Training and Sociological Thought," *American Journal of Sociology* 59 (September 1953): 112.

[8] M. Brewster Smith, review of *When Prophesy Fails, Contemporary Psychology* (April 1957).

the individual or individuals responsible for decisions about the activities of the members (or part of the members) of the institute. What Smith is suggesting, therefore, is that serendipity can be nurtured not merely by the capacity of the individual scientist to shift his attention to a strategic, unexpected phenomenon, but also by the capacity of a director of research to take such a step.

Allen Wallis, an official of the Fort Foundation, has a perspective on the relationship between serendipity and the administration of research somewhat different from those we have examined so far.[9] Langmuir, Kopanyi, Riemer, and Smith have decided in advance, as it were, that accidental discoveries do or do not play an important part in a flourishing science, and they believe that the organization of scientific activities may play a part in facilitating or inhibiting the occurrence of such discoveries. Accordingly, since Langmuir, for example, believes that serendipity *is* important for the advancement of science, he wishes to organize scientific research in such a way as to encourage serendipity; and Riemer, who thinks accidental discoveries are of small significance, wishes to organize science in such a way as to encourage other patterns of discovery. Wallis, however, raises the prior question as to whether science is significantly served by serendipity or whether important scientific advances are more often made in other ways. In asking and answering this question, he suggests implications for a theory of the strategy of scientific advancement:

> Small differences in the distribution [of people of great talent among the various sciences] are exceedingly important *if the progress of a science depends on the appearance of exceedingly rare individuals who make revolutionary break-throughs* that innumerable others exploit for years; for relatively small changes in, say, the median of a typical distribution result in great changes in the frequency of rare individuals. *If,* on the other hand, *the progress of science is more continuous and mostly the product of serendipity*—the knack of spotting and exploiting good things encountered while searching for something else—then it is important that substantial quantities of effort be applied; indeed, under the extreme assumption that *scientific advances are pure chance,* like royal flushes and quintuplets, sheer volume of effort is what determines progress. Unquestionably, scientific progress springs from both sources (and all intermediate combinations) but in either case the conclusion is clear that a large amount of high quality application to the behavioral sciences is one of the most important requirements for their advance.

---

[9] As shown in his report titled, in the unpublished version, "The 1953–54 Program of University Surveys of the Behavioral Sciences."

In other words, if it is the giants of the scientific community who are chiefly responsible for progress, it is they who should be encouraged by whatever support is possible; if it is all those scientists capable of serendipity who do the most for science, all such scientists are worthy of maximal support; and, finally, if anyone busy with science may make discoveries (like the monkeys with the typewriters writing the books in the British Museum), then the more such people who are kept busy, the more discoveries will probably be made. Since, actually, "revolutionary breakthroughs" by great scientists, accidental or otherwise, *and* smaller advances by men of lesser stature, again accidental or otherwise, are involved in the progress of science, the best strategy is to encourage as much "high quality application" as possible. This conclusion is, perhaps, not a startling one, but the reasoning by which Wallis arrives at it is unusual and illuminating.

Wallis sees discovery production as a problem of administration, but for him it is not a problem involving the degree of autonomy of scientists within a more or less formally structured organization; rather it is the problem of whom to support within a scientific community (in this case, the community is that of all those engaged in work in the social or behavioral sciences). For Kopanyi, to be sure, the two problems are almost identical: Any selective support *is* restrictive of freedom, and is, therefore, harmful to science. Selective support constitutes dictation to the scientists and it perverts their work by defining success and allotting rewards in a way that is necessarily incompatible with the proper fruition of science. But though Wallis is concerned, in a sense, with defining expectations, he is not (by intention, at least) telling scientists what kind of performance he expects of them. Instead, he is appraising their maximal potentialities as a group, with a view to realizing these potentialities as nearly as possible. Wallis's appraisal has the aim that any scientist must share: the maximal development of science as a body of knowledge, without exercising substantive discrimination within that body of knowledge.

Scientific Research in Business and in the Government

Scientific research may be carried on simply for the sake of advancing scientific knowledge, or it may be carried on for other ends. So far we have discussed interpretations of the significance of accidental discoveries in the work of scientists or in the administration of scientific activities with only peripheral attention to the ways in which the scientific or extrascientific ends of the work affected these interpretations. We have deliberately postponed consideration of the effects of extrascientific ends on interpretations of serendipity; now we shall refocus our attention to examine the significance serendipity has in activities that are being carried on for ends defined by business or by the government.

Research that is carried on under the auspices of business or of the government is not aimed primarily at benefiting science in general, though it may do so incidentally, but to serve a certain business or a certain government objective. The scientists' efforts, therefore, are vindicated not by the judgment of their peers as to the value of their contribution to science, but rather by profits for the company that employs the scientists or by some kind of public profit if it is the government that is their employer. In this context, accidental discoveries may raise questions concerning their effect on the earning power of the company or the capacity for public service of the government. Research in general and accidental discoveries in particular should be justified by useful results. Accidental discoveries may come under special scrutiny because they may appear to call into doubt the reliability of the research organization that produced them. To keep public confidence, therefore, it is sometimes argued that accidental discoveries are not a symptom of erratic performance, but that the very competence of the researchers and administrators of research makes it possible to turn even accidents to the public's advantage.

The concern felt for the interests of the company by those working in industrial research was demonstrated by Langmuir in a passage we quoted (but did not highlight) earlier: Reminiscing about his early days at G.E., Langmuir recalled that he had "had fun" for three years but had wondered "what good this is to the General Electric Company." Willis Whitney had assured him that that was his (Whitney's) worry.[10] It has to be someone's worry whether ongoing scientific work will pay off—it would be irresponsibility on the part of the relevant executives *not* to worry. They may decide, as Whitney did, that it is in everyone's interest not to pass that worry on to the working scientists, or they may put pressure on these scientists to fulfill immediately the company's need for useful results. Their decisions will depend on how long the company can afford to wait for "basic" research to pay off. As Bernard Barber has observed, "In every case, however, in the not too long run . . . the 'basic' research of industry has as its purpose some application in the immediate interests of the enterprise which subsidizes it. To think otherwise would be to ignore the social purpose of industrial organizations. The directors of industrial scientific research groups are well aware that their activities are subject to the same institutional imperatives as other industrial activities, well aware that they must lead to the maximization of profit in the long run."[11]

[10] *G.E. Shareholders' Quarterly* (25 July 1952).

[11] Bernard Barber, *Science and the Social Order* (New York: Free Press, 1952), p. 97.

The worries of industrial or government research executives are not simple, however. They must obtain short- or long-run results that will benefit the company or the public, and they must, from time to time at least, convince the lay public that they are obtaining these results. It is a problem in public relations, then, as well as a problem in the administration of science. The lay public that has invested in a certain company or has voted for the government must be reassured that its trust is justified. That is why Langmuir's reminiscences about his research, including the happy ending in serendipity, were printed in the *Shareholders' Quarterly*. Eventual serendipity, indeed, may make it all the more convincing that research not only pays off, but that science works in mysterious ways—mysterious to the layman, that is—and that it is best to trust experienced experts like Langmuir to make science profitable.

The Standard Oil trade publication *The Lamp* is intended to explain some of the developments in the company to the layman, and thus, among other things, it is interested in presenting the company's scientific research in a favorable light. The article on "Serendipity" in *The Lamp* is partly based on Langmuir's remarks, partly it elaborates on them. The (anonymous) author of the article is less inclined than Langmuir is to make the occurrence of accidental discoveries an index of a well-functioning laboratory. He is willing to attribute to accident some important discoveries, but he wants to have it clearly understood that such discoveries constitute a minority. He does not say that accidental discoveries are of secondary importance qualitatively as well as quantitatively, but it is clear that he would not like it to be thought that the Standard Oil laboratories were *depending* on serendipity for their significant advances: "Scientific discoveries are usually made by painstaking research, with each great success prefaced by a thousand failures. But sometimes they seem to drop out of the blue, the gift of happy chance—a wind blowing through a laboratory window, a key tossed carelessly upon a photographic plate. . . . Of course, significant chance discoveries are the blue diamonds of laboratory searching. They are as rare as they are unpredictable. *Well-organized research along clearly defined lines* is most often the method by which modern science achieves its goals."[12]

This does not mean that serendipity is not worth cultivating. Research directors can do so by "planning programs broad enough to allow their researchers the freedom to follow any leads they chance upon." Standard Oil laboratories, the piece goes on to say, are so organized that nothing stands in the way of accidental discoveries, and serendipity has produced, for example, a valuable fungicide called SR-406. Here is how it happened: "It happened when a team of SOD chemists were working on a

[12] "Serendipity." Emphasis added.

general research program aimed at finding new uses for petroleum products on the farm. *They were free to do any experiments that interested them.* One of the chemists began to search for an insecticide." In writing down the chemical formula of his 406th experiment, he noted that its chemical structure looked perfect for an insecticide. Tests showed that the insects probably thought so too. They thrived on it. To the chemist's *surprise,* however, further tests revealed the same compound to be an extraordinary fungicide."[13]

The writer does not explain how the chemist came to try out the effects of his failure-as-an-insecticide on fungi, but he suggests, at another point in his article, that a crucial step in the making of an accidental discovery may be taken when an alert scientist asks of an unexpected phenomenon: "What is it?"

Serendipity, as our anonymous author repeats frequently, is an "art," the art of being curious at the opportune but unexpected moment. It is an art that is as old as the most primitive science, and one that accompanied both the development of alchemy and of modern science:

> All of the basic crafts—such as those of pottery, embalming, the tanning of leather—probably originated through the magic nexus of happy chance and an alert imagination, which is the essence of serendipity.
>
>   . . . It was pure serendipity that created the bridge that led from the groping of alchemy to the science of chemistry. This was the result of a discovery made by . . . [Paracelsus] . . . who happened to drop pieces of iron into what was later to be known as sulphuric acid, thereby liberating free hydrogen. Some scientific historians believe that Henry Cavendish, the father of modern chemistry, based his work on this discovery. . . .
>
>   It was not until men threw aside chimerical goals and began to devote lives to a search for truths about the real world that serendipity became a true art. The two most illuminating exhibits of serendipity—the discoveries of penicillin and the X ray—shows how much of the art as well as accident its practice requires.

Just what distinguishes the modern "artist in serendipity" from the ancient is not quite clear. Evidently, the author is less interested in pinning down the elusive quality that William Henry Perkin (the discoverer of artificial dyes), or Roentgen, or Fleming had than in assuring the public that Standard Oil scientists have that quality along with other qualities requisite for making scientific progress: training and technical skill. Given this combination of qualities, science in general and serendipity in particular are becoming ever more effective:

[13] Ibid. Emphasis added.

It is doubtful whether the atomic age will ever produce another boy prodigy like Perkin, for science has become infinitely more intricate and purposeful that it was when he made his stone-in-the pond discovery. Serendipity, however, is by no means a dying art. On the contrary, scientists today are better prepared than ever to capitalize on the unexpected, because of the accumulated knowledge and refined techniques at their command.

All research is a voyage to the edge of the unknown, and who can say what discoveries will be made there, by scientists endowed with training and skill and imagination—and blessed by happy chance?

Standard Oil wants the public to view serendipity in proper perspective: as frosting on the cake of competent and reliable research activities.

Even in a communication to university professors, who might be assumed to be better informed about and less suspicious of the nature of scientific progress, Standard Oil seems anxious to have it understood that a certain development was not the result of serendipity alone. In a letter the company sent out accompanying a booklet on the development of Butyl, it would appear, to the members of the faculties of scientific and engineering schools,[14] Standard Oil explained that though Butyl had been discovered by serendipity, much hard scientific work had been done on the product since the original accidental discovery: "The discovery of Butyl during a research project on motor oils was certainly an example of serendipity, *but* its usefulness to mankind could be proved only by extensive laboratory research. In fact, after ten years' work . . ."[15] Serendipity *alone* would put the company in a bad light, just as it was felt to detract from the merit of individual scientists, so success by virtue of serendipity is justified by subsequent effort.

The problem of the predictability of results is more than academic when scientific research becomes part of business and government activities, because the lay public is entitled to some reasonable prospective assurance that its investments are safe or that its vote of confidence is justified, as well as to retrospective vindication of its trust. Executives responsible for business and government research are concerned, therefore, to assure the public that useful scientific discoveries of some kind are bound to occur, though, partly at least, the substance of scientific progress cannot be predicted. Sometimes the public is told that the most important discoveries are the unexpected ones, and that these accidental discoveries depend on unhampered explorations by scientists. The impli-

---

[14] We know of only two of the recipients—Professor Julian Eisenstein of the College of Chemistry and Physics at Pennsylvania State University, and Professor Herbert Simon of the Graduate School of Industrial Administration at the Carnegie Institute of Technology—and the above tentative generalization may well be quite wrong.

[15] Emphasis added.

cation is that if the public will courageously tolerate uncertainty (apparent uncertainty, that is), it will surely receive tangible benefits as its reward. (To put it in a somewhat earthier way, if you are going to ask people to make an exception to the old rule about buying pigs in pokes, the least you can do is to promise them that the pigs are good ones.)

The worthwhileness of taking a chance on scientific research and serendipity was put graphically before the public a few years ago in an advertisement by the pharmaceutical company Merck and Co. It is not unlikely that the theme of the advertisement was inspired more directly than is usual by the president of the company, George W. Merck. As long ago as 1946, Merck was discussing the question of predictability in science in an article in *Chemical and Engineering News*, and expressing the opinion that "prediction of the results of scientific inquiry is not easy; neither is it safe. There are too many 'if's,'" and these uncertainties are the result of the "indulgence" of research workers in serendipity. A few years later, when *Time* magazine published a cover story on Mr. Merck, he again tried to make the public aware of the fruitfulness of expanding the boundaries of the known. As *Time* recreates the interview, "Of one thing [George W. Merck] is confident: there is more of the unknown ahead than the scientists have left behind. And there is nothing [he] enjoys more than the thought of unexpected adventures in the offing. 'For one thing,' he says, 'there's always serendipity. Remember the story of the Three Princes of Serendip who went out looking for treasure? They didn't find what they were looking for, but they kept finding other things just as valuable. That's serendipity, and our business is full of it.'" Finally, in 1954, in a full-page advertisement for Merck and Co., the message is really brought home to the public. The advertisement shows three Arabs riding on camels with accompanying text that reads in part:

"Serendipity" in Chemical Research

No field of human endeavor illustrates better than chemistry the story of *The Three Princes of Serendip*. These princes went out searching for treasure. They never found what they were looking for, but they found other things just as valuable.

Typical of "serendipity" in chemical research is the case of the expedition Merck sent to Guatemala some years ago to investigate a tree sap used by the natives as a remedy for certain diseases. Subsequent studies in the Merck Laboratories proved that the sap, while unsuitable for medicinal use, yields an enzyme that is highly useful in connection with certain large scale fermentation processes.

The Merck advertisement is intended to impress on the public an even more fundamental point than the article in *The Lamp* is trying to make. The latter assumes the public's support for research, and is only reassuring the stockholder that the occurrence of accidental discoveries does not mean that the research is incompetently conducted. Merck is using the theme of the travels of the three princes to dramatize the point that though basic research may be unpredictable, it pays off. That is to say, a relatively unstructured kind of research, which it may be hard for the layman to connect with anything so specifically useful as pharmaceutical products, should be trusted to yield just such products. The less predictable results of basic research are likely to be just as valuable as the specifiable results of research narrowly focused on immediate practical problems.

The claims of the Merck advertisement and Mr. Merck's own remark that "his business" is "full of" serendipity are supported by an appraisal of the pharmaceutical industry made by Baker, Weeks and Co., a firm of investment consultants. In that company's *Weekly Review*, a publication giving advice to investors, the pharmaceutical industry is judged to be a promising field for investment. To be sure, Baker, Weeks thinks that way about the industry not because of, but in spite of the fact that serendipity is rampant in pharmaceutical research: "Worthwhile results from research in an ethical drug company are not solely the results of brain work. An important element of luck is unavoidable, and some of the most dramatic finds have been attributed by their discoverers mainly to good fortune. Thus, it has been well said that in drug research, *serendipity* can be more important than *sagacity*. In any case, drug research is proving so productive that fully 85% of the drugs now in common use by physicians have been introduced within the past 15 years."[16] In other words, what does it matter if those engaged in drug research have been lucky, when their lucky discoveries have so consistently turned out to be useful? Baker, Weeks is making no general case for serendipity, nor is it suggesting to investors that the relationship between luck and brainwork that it perceives in ethical drug research is a sound one in general (the introductory remarks about the inevitability of a certain amount of luck have little connection with the rest of the statement and seem to have been thrown in for trimming, as it were). The attitude of these investment consultants is an eminently practical one: Serendipity in industrial research may or may not look like a reasonably safe investment, but as of April 16, 1956, it looks like a good thing in the pharmaceutical industry.

As we saw earlier, one of the pharmaceutical companies that Baker, Weeks may have had in mind, Pfizer, resists having discoveries attributed blithely to accident. A piece of institutional advertising by that company that appeared in the *Journal of the American Medical Association* (July 17, 1954) tends, rather, to minimize discoveries that happen by accident.

[16] Baker, Weeks and Co. *Weekly Review* (16 April 1956).

Entitled "Serendipity: The Happy Accident," the article tries to show that the accidental component in discovery is more apparent than real, whatever people may think:

> It is a favorite thesis of people who do not understand (or entirely like) science that everything important has come about through sheerest accident. In this way they imply that if they had been there under a tree when the apple fell, they would have done at least as well as Newton. They are mistaken. Many other ingredients are vital to serendipity in science. They include the "prepared mind" mentioned by the late Walter B. Cannon as capable of recognizing the thing that has happened and what it may signify, Freudian "accident-proneness" extended to the happy accident as much as the adverse one, and a peculiar philosophical substance called the *Zeitgeist*, a spirit of the time. Zeitgeist is definable in this way: If Gus does not discover the wheel in 40,000 B.C., his nephew Bug is sure to do so within the century, for the makings of such a concept, and the need for it, evolve with the culture and come into being at a certain historic time. In fact, there is a question whether happy accidents of science are accidents at all.

In case after case of alleged accidental discovery, the writer of the article tries to show that the importance of the accidental factor was minimal and that the discovery would in each case have been made without the intervention of the accident. These discoveries happened to coincide in time with the occurrence of accidents, but they were bound to happen anyway. Here is the author's withering dismissal of the importance of accident in Roentgen's discovery:

> Wilhelm Konrad Roentgen seems to have had a remarkable sense of what he was about—still more remarkable since the roentgen ray was not present to his senses but was actually an undescribed phenomenon taking place invisibly in the neighborhood of an evacuated glass bulb having 2 electrodes. It so happened that a piece of cardboard, coated with barium platinocyanide crystals, glowed in the presence of this phenomenon. Roentgen was interested. He found that "solid objects" cast a shadow—incomprehensibly, because the tube was covered with black paper and was certainly unable to emit light. Then he *happened* to pick up the cardboard, and he *happened* to see the shadow of his own phalanges. Thus he *happened* to conceive of all radiology in that instant, showing more acute insight than could ever be set down to accident.

In this case, as in the case of Henri Becquerel's discovery of radioactivity by "accident" about a year later, the author holds that it was the constellation of events, not the accident, that was crucial: "Here were a phenomenon and a man equal to recognizing it, and all that accident could do was to keep them apart or bring them together at a certain instant."

The anonymous author's opinion about the importance of accident in this and other discoveries[17] closely parallels that of the late historian of science George Sarton. Writing about the discovery of X-rays, Dr. Sarton observed:

> The year 1896 might well be taken as the opening year of modern physics, and we are now beginning to see the succession of its memorable events in its true perspective.
>
> And yet there is no absolute beginning in science; or to put it otherwise, every beginning is also a climax. Indeed it is only because it is a climax—the clinching of a long series of efforts—that it can be the beginning of a new evolution on a higher level. This applies very well to Roentgen's discovery.
>
> It was in the air when it was made, and if Roentgen had not been available or had been less persistent and successful, the self-same discovery would have been made sooner or later—and probably very soon—by Lenard or someone else.[18]

If anything is lucky or accidental, then, about a discovery, it is happening to have been the one who made it, happening to have achieved priority, rather than the discovery itself, which is quasi-inevitable. Luck merely speeds up, in a socially insignificant degree, what would have been discovered in any case. But the luck of achieving priority is not undeserved; the discoverer, and here we quote Dr. Sarton again, "was undoubtedly lucky—we all need a modicum of luck to show our mettle and do our best—but he fully deserved his success, having prepared himself for it by a lifetime of ingenious and indefatigable research." (Here is independent confirmation of the argument of the Pfizer company from a historian of science who certainly had no interest whatever in establishing the competence and reliability of the research department of a particular industry.)

The Pfizer company's anonymous author uses the phenomenon of missed discoveries to show that it is all-important that the proper Zeitgeist be present and the qualified scientist at hand before some accident may precipitate an important discovery: "There can be negative as well as positive serendipity: the tragedy of the opportunity missed, the hundred and one precious remedies that are standing on shelves everywhere, known only to chemists who are busy doing something else with them . . . Some happy accidents are still badly needed." But in spite of this dim view of the likelihood that significant discoveries will be made unexpectedly, the author thinks there is some scope for the cultivation of serendipity: "It calls for the prepared mind and the favorable opportunity, but

---

[17] For example, his remark: "It is largely a question of whose hands the several ingredients (drug, patient, etc.) falls into first."

[18] George Sarton, "The Discovery of X-rays, with a Facsimile Reproduction of Roentgen's First Account of Them Published Early in 1896," *Isis* (March 1937): p. 358, n. 2.

of these something can be known in advance. The work of generations of previous workers must go into preparing the mind, which must in addition be tainted with xenophilia (Boring's name for love of the strange and novel) and must flourish in a congenial *Zeitgeist*; a modicum of genius is a useful ingredient. Given so much, an accident is awfully likely to happen." Such cultivation has nothing to do with the proper organization of research; rather it is a matter of developing the right qualities in individual scientists.

So far we have examined only discussions of the problem of accidental discovery by those connected with industrial research. In the one instance we have of such a discussion from the area of government-sponsored research, there is evidence of the same concern to assure the public that there is nothing wrong with accidental discoveries.

J. T. Seery, head of Research and Development in the Bureau of Supplies and Accounts of the Navy Department, tries to show that there is very little luck in successful research. Like the author of the article in *The Lamp*, he considers serendipity to be an art, and he goes somewhat farther than that author in trying to pin down how that art works. He may split a few hairs in the attempt, and his subtle distinctions between "interesting" and "promising" in the passage that follows may elude us, but he seems to be trying to spell out the difference between the disciplined investigation of strategic data and the sterile pursuit of a planned study:

> Serendipity, the "art of finding valuable or agreeable things not sought for," has been said to account for as much progress in research and development as careful planning. If the art could be measured, one probably would find that the successful manufacture of floating soap was 99 and 44/100 percent pure serendipity. Penicillin and DDT are two more recent examples of experimenters being alert to recognize and exploit unexpected phenomena.
>
> The real skill, of course, lies in the ability to forgo the unpromising occurrence, however interesting it may be, and relentlessly pursue that which, found but unsought, may pay great dividends in progress.[19]

Seery then goes on to describe recent unexpected discoveries made by those doing research on clothing and textiles for the Bureau of Supplies and Accounts. He concludes that "these unexpected but profitable occurrences are almost always the byproducts of planned research and development. So the future depends to a considerable extent, not on happenstance alone, but skill in serendipity." The kind of skill that Seery is demanding, which will, as far as possible, take the "happenstance" out of scientific progress by distinguishing between the "promising" and the *merely* "interesting," is the ability on the part of scientists to make the

---

[19] *Monthly Newsletter* 16, no. 8 (15 August 1952).

crucial guess as to which of the several problems confronting them at any one time will, when solved, turn out to be more important either for science in general, or a particular applied branch of science. With such an interpretation of serendipity, as a skill that permits discrimination between the merely "interesting" and the genuinely significant, Seery has insured that discoveries in which serendipity plays a part *must* be important ones. Given research scientists with such a skill, discoveries may be unexpected, but they cannot be either accidental or trivial. Chance has been well harnessed—it can lead scientists with serendipity only to success.

In the business world serendipity has been used most frequently to demonstrate the valid function of research in industry. The occurrence of accidental discoveries provides an opportunity to explain and justify the usefulness of research—a good opportunity, because accidental discoveries both have "human interest" and go to the heart of the nature of science. Norman Stabler, the financial editor of the *New York Herald Tribune*, has quite a different slant on serendipity. Inspired by the article on serendipity in *The Lamp*, Stabler uses serendipity to express his scorn of those presumptuous, incompetent (and possibly venal) economists who think they know more about the stock market than observers closely connected with the market; serendipity, in effect, nullifies their feeble efforts at predicting the financial future. "If serendipity plays an important role in the exact sciences of chemistry and physics, how much more must it make itself felt in the inexact domain of economics? . . . The article in 'The Lamp' was concerned only with the scientific applications of luck. How much more does the experimenter in a field that leans heavily on that unknown quantity, public psychology, need the help of Lady Luck!"[20] In spite of all the efforts of economists over the years to establish laws governing "the ups and downs of the business trend and the stock market," confusion reigns: "We are in a situation right now when businessmen and investors are more inclined to ask 'What is it?' than to proclaim they have the answer." Certain economists, to be sure, claim to have an answer, but it is hardly a trustworthy one: "The press is full of statements from leading business and political leaders that conditions are fundamentally sound, and one wonders why there is a necessity to belabor the point. In ancient alchemy, research was more accident than art, and possibly our modern economic scientists are in that same relative position now. Theirs is no more an exact science now than was that of their ancient forebears, whose only reason for research was to discover a method of transmuting base metals into gold." Other economists, wise after the event of a stock market decline, are just as little use

---

[20] C. Norman Stabler, in *New York Herald Tribune* (24 September 1953).

to the investor: "According to some students of stock market ways, we have just recently had it confirmed that we are in a bear market. Holders of about half of the issues of stock traded in on our major exchanges would say that they personally have known it for two years and they don't need any serendipity. All they need is the last syllable [Stabler means the last two syllables] of the word." When one group of economists predicts one thing and the other group predicts the opposite, and neither group really knows what it is talking about, one group may, in the end, be right by chance, and the other happy to find itself wrong: "If the economists who help the big industrialists write their speeches are correct, then the stock market economists are going to be wrong. All of a sudden they may come across a little serendipity in this bear market. That is, they may discover something they weren't looking for." The alchemist-economists may make gold (though Stabler appears to be skeptical), and the others may find gold that they never knew existed.

A NOTE ON SERENDIPITY AS A POLITICAL METAPHOR

*A*lthough it would be rather surprising to find an American politician explicitly making allowance in his policies for accidental success or serendipity, it would also seem to be the rule in American politics that policies are formulated in a pragmatic and piecemeal fashion. As Irving Langmuir said in his talk on planning in research, it is the totalitarian dictators who have grand and comprehensive plans, while it is the "American way" to make policies step by step, without trying to anticipate too far in advance each step's consequences. In England, the absence of general plan or principle in the political sphere was for a long time glorified as the English talent for "muddling through." Disraeli once remarked, "I suppose, to use our national motto, 'something will turn up'"—and the ironic reference is, of course, to the apparent adoption by Englishmen as a nation of the motto of Dickens's Mr. Micawber. Mrs. Micawber, to be sure, had a different approach to life's problems, which she voiced rather tentatively: "Now I am convinced myself, and this I have pointed out to Mr. Micawber several times of late, that things cannot be expected to turn up of themselves. We must, in a measure, assist to turn them up. I may be wrong, but I have formed that opinion."[1] Two observers of the American political scene have commented on the likelihood of good things turning up unassisted in Washington: One has held that good things did so turn up, and were the more delightful for being unexpected, while the other has deplored the disinclination of certain politicians to assist in turning them up.

Gerald Johnson is the journalist-commentator who concludes that in

---

[1] Charles Dickens, *David Copperfield* (1849–1850), chapter 28.

the era of the New Deal the American public, the "common man," was perennially the beneficiary of serendipity. Johnson was an admirer of Franklin D. Roosevelt and of the New Deal, and approved of the broad programs that were developed in the 1930s to cope with economic problems, but he thinks, as we saw earlier, that the incidental windfalls that accompanied these programs did a great deal, too, to brighten the common man's days.

> The New Deal brought into our national life another novelty extremely difficult to describe, because it was unplanned, unexpected and unauthorized by the New Dealers, and is by its nature intangible and elusive. It is possible, indeed, that the great and important never perceived it at all, but the little fellow, the common man, the undistinguished citizen saw it vividly and it was one of the features of the regime that enchanted him.
>
> To pin it down in words one must borrow an expression, and a staggering one, from Horace Walpole. . . .
>
> It can be argued plausibly that one thing about the New Deal that fascinated the common man was its serendipity. Through it, he was always discovering interesting, or amusing, or delightful things while hunting for something quite different; and this tended to keep him keyed up and expectant. Sometimes what he discovered was a personality; sometimes it was an institution; sometimes a lifting of his mental horizon; but always it was something by which he was astonished and pleased.[2]

Without the New Deal's stress on planning, these unexpected benefits would have been less conspicuous and, perhaps, less appreciated; as it was, people felt, according to Johnson, that they were getting a bonus, a surprise present. It has been pointed out that surprise gifts have the important function of institutionalizing unexpected good luck, and that in a world that harbors for people a considerable share of unexpected bad luck, a complementary quota of good luck does much for morale.[3] Johnson seems to be suggesting that it was this unexpected good luck, or serendipity, that did much to keep up the common man's morale during the Depression.

Serendipity came in different forms. One of the unexpected consequences of the establishment of the Civilian Conservation Corps (CCC) was the increase of the value of the public domain by many millions: "We didn't undertake it for financial profit. We were looking for something else, and the profit was simply a pleasant thing that we ran across accidentally. It was so much velvet; and picking up velvet is one delightful

---

[2] Gerald W. Johnson, *Incredible Tale: The Odyssey of the Average American in the Last Half Century* (New York: Harper, 1950), pp. 190–191.

[3] Bernard Barber's analysis.

form of serendipity." Another such consequence of the CCC was the introduction to the wonders of the countryside of many boys who had known only city life:

> As the average man saw it, the original idea of the CCC was negative; it was to prevent a whole generation of boys from degenerating into bums. No doubt there were uncommon men among the planners of the scheme who saw much more in it than that; but just to avoid the production of innumerable bums was enough for the common man, and with that he would have been content.
>
> However, when you snatch a youth out of the environment to which he has been accustomed, and project him into one completely different, more happens to his mind than to his muscles. . . . Never again can he view the world as comprised in a few city blocks.

Johnson is not primarily interested in whether the New Dealers planned it this way or not—they deserve ample credit for what they did plan, and the unexpected consequences of their activities benefited the common man, too, because so many of them were positive surprises.

> There is no doubt that the commoner at home enjoyed during the six years before the government was turned into a war machine certain usufructs of the New Deal that were not strictly political, that were not expected and that usually were not planned. Aside from all other considerations, it was a magnificent show. Only a part of this was attributable to the New Dealers. Some of the reactions against them were as entertaining, and as packed with thrills, as anything they did. Every day the common man picked up his paper in the knowledge that the chances were at least three to one that he would find in it something from Washington that would give him a great kick.[4]

A magnificent show, indeed!

James Reston's view of the Republican "show" in Washington in the summer of 1956 is a much less sanguine one. Reston finds the style of political life in Washington at that time open to strong criticism, and has adapted serendipity for the purpose of that critique. Serendipity becomes the tag for a kind of shallow, falsely optimistic opportunism in political affairs—a Micawberish optimism that denies the patent realities and seeks to convert people into believing in the fantasies being supplied in place of the realities.

Reston asserts that the administration thinks it has found a "cure-all. It is a rare tonic called 'serendipity.'" Serendipity, according to Reston, was coined by Horace Walpole from a fairy tale about the Three Princes of Serendip, "who were never dismayed by misfortune, but always managed

[4] Johnson, *Incredible Tale*, pp. 190–194.

to discover, by chance or sagacity, wonderful things they had not sought. This is what is happening here now." In spite of legitimate concern about a number of problems, "the administration is pretending that they do not exist. One such problem is the President's [Eisenhower's] health." Secretary of State Dulles is "in a particularly serendipic mood,"—"downgrading" the president's illness to an "indisposition," minimizing the threat to the foreign aid appropriation, "and full of bounce and optimism about many other things."

This optimism, as Reston sees it, is not only foolish but reprehensible, since it is being used to pull the wool over the public's eyes to conceal weaknesses in the administration. Reston thinks a "conscious effort" is being made to give the public false impressions about such matters as the president's health, and that, for political purposes, "highly fanciful" interpretations are being put on the success of American foreign policy, interpretations that do not square with the facts. However, the public is really only getting what it deserves, for while the administration cannot afford to face facts, the public does not want to: "For the time being, . . . 'Serendipity' is the word, and since this coincides with the mood of the country, it seems to be a valuable addition to the Administration's political arsenal."

Serendipity is for Reston the word that catches up all the irresponsibility that he sees around him. The word is being put to use from the point of view of a critic of current political policy, and it is well on the way to becoming a *Schimpfworn*. For Reston the word has none of those positive associations—alertness, imaginativeness, activity—that it has had for many scientists, nor is he willing to let it stand for the unearned increments of government policy, as Johnson does; for him, it stands simply for deception and self-deception, for the wishful belief that "something will turn up" if need be, and that in the meantime it is as well to pretend that all goes well.

A Note on Serendipity in the Humanities

*T*he interpretations given to serendipity by humanists vary, perhaps, more than do the interpretations in the other social domains that we have discussed. In science, business, and politics, accidental discoveries raised problems of merit and of justification: Some might think that accidental discoveries were useful and justifiable while others rejected them as harmful or discreditable, but all passed some judgment on the implications of accidental discoveries. Among humanists (writers, scholars, collectors of items of literary and historical interest) some also evaluate accidental discoveries and the individuals who make them, but others look upon them as being entirely beyond the scope of the individual's responsibility.

The making of accidental discoveries by collectors and literary scholars has on several occasions been attributed to a personal trait that they happened to be endowed with. Though collectors and literary scholars, like scientists, actually have to be prepared to make accidental discoveries—they must know in a general way where to look and what to look for—their stock of knowledge does not have the systematic quality of science, and the fact of their preparedness may, consequently, be less visible. Also, the nature of the happy accidents that befall them consists, frequently, of unexpectedly locating a desired item or of the unhoped for anticipation of others in the recognition of a valuable item; the human drama of such events may serve to conceal (as it does in some accidental scientific discoveries) the knowledge and effort necessary for making the discovery. Finally, collecting is often a hobby, and the fact that browsing through bookstores or antique shops or catalogues is "fun" distracts attention from the skill and effort involved.

Horace Walpole defined serendipity as a special quality with which he happened to be equipped by nature, and other collectors and literary men have subsequently used the word in a similar way. Edward Solly called serendipity a "particular kind of natural cleverness," and he described an instance in the course of his genealogical studies when "by the aid of *Serendipity*—that is looking for one thing and finding another," he accidentally found a missing piece family history. "Vebna," a contributor to *Notes and Queries* in the 1880s, calls serendipity "a gift, or good fortune, or whatever else it may be called," which aided him in the search for a certain quotation from Cicero. Christopher Morley describes dramatically an instance of serendipity, which he thinks of as a personal "characteristic"; the instance is famous in literary circles, and interesting.

> In the meantime [while Colonel Isham was making various important discoveries] no less astounding was the adventure of Professor Claude Colleer Abbott, then Lecturer in English at the University of Aberdeen. Perhaps *he is the best example* in our time *of Walpole's serendipity*, characteristic of one who finds something he wasn't looking for. What Professor Abbott was actually looking for was material for a new life of James Beattie, an 18th century Scottish philosopher, a friend of both Boswell and Johnson, but quite forgettable now. Abbott found it necessary to examine papers preserved at Fettercairn House, not far from Aberdeen, an estate of Lord Clinton. To his *amazement*, going through a capharnaum of miscellany . . . Professor Abbott found what completely *side-tracked* him from Beattie. In bags, bundles, bean sacks, and helter-skelter through the closets and attics of the ancient house, he found some sixteen hundred letters and manuscripts and documents to, from, and by Johnson, Boswell, and their friends. Among them was the journal here printed.[1]

Though Professor Abbott was "amazed" and "side-tracked," his discovery is not attributed to his appreciation of new, significant materials, but to his *being* a case of serendipity. An instance of serendipity, then, is equivalent to one of double-jointedness or of perfect pitch.

It has been suggested that serendipity is an attribute that collectors have *as a group*, without any very clear explanation of how this happens. Whereas scientists were much concerned about ways of cultivating serendipity, some collectors and literary men, at least, seem to think that there is really nothing that can or should be done about it. Alan Walbank deems serendipity to be an acquired trait of the "junkhunter," the obverse of an occupational disease. The trait, which "works in conjunction" with "a kind of sixth sense, collector's radar you might call it, which

---

[1] Christopher Morley, "Preface," in James Boswell, *London Journal, 1762–1763* (New York: McGraw-Hill, 1950), p. xxi. Emphasis added.

'homes' him on desirable sites," is a "knack developed by the sensory equipment of the keen collector and junkhunter, that of serendipity."[2] Walbank goes on to describe some unexpected discoveries he has made by virtue of the twin traits of collector's radar and serendipity. Wilmarth Lewis also thinks that "all collectors and scholars are indebted to what Walpole called 'Serendipity.' This is the faculty for making happy accidental discoveries while looking for one thing and finding another," and, like Walbank, he connects serendipity with a mysterious world of clairvoyance and extra senses: "Serendipity leads collectors into the mysterious extrasensory world where telepathy, clairvoyance and premonition are common-place. Making all allowance for the collector's tendency to dress up his discoveries, and to forget the times when these occult impulses led to nothing, there remains a good deal that cannot be accounted for by the five senses." Lewis is aware that "the scientific world, which owes so much to Serendipity, has revived the word and given it wide currency, particularly in connection with the discovery of penicillin,"[3] but he pays little heed to the fact that the scientists' explanations of serendipity, incomplete as they are, are at odds with his own view that serendipity defies explanation.

The unexplainable, whether it be God-given or a whim of nature, is ipso facto apart from the sphere of merits and deserts. Serendipity as interpreted by Morley or Solly or Lewis does not raise questions of deserts in connection with the benefits collectors and literary scholars derive from their idiosyncratic trait. Even so, in the world of the humanities the problem of the relationship between luck in the making of discoveries and other contributing qualities has not been entirely ignored. In a specific form the problem is dealt with by John Mason Brown in his review of Wilmarth Lewis's collector's autobiography, *Collector's Progress*. Brown is concerned to dispel any notions to the effect that Lewis's success was chiefly due to luck: "In amassing the roomfuls of Walpoliana at Farmington, Mr. Lewis has had his moments of high good fortune. To use a word of Walpole's which has again come into circulation, he has had the advantage of serendipity, the faculty of making happy and accidental discoveries while looking for one thing and finding another. Mr. Lewis' persistence, however, has been greater than his luck. It has been equaled only by his endowment as a detective, his Geiger counter instinct, his scholarship, his genius for friendship, and his unflagging fervor." The many meritorious qualities that Lewis possesses, then, in addition to his special faculty, account for his success. Brown makes this judgment about the relative importance of persistence, detective skill, fervor, and so forth,

---

[2] Alan Walbank, "Joys of the Junkshop," *The Saturday Book* 14: 261–267.
[3] Wilmarth S. Lewis, *Collector's Progress* (New York: Alfred A. Knopf, 1951), p. 236.

on the one hand, and luck on the other, ad hoc as it were—he is concerned only with establishing Lewis's merit, not with general analyses of the conditions of success or with prescriptions for success.

Some humanists, however, have tried to integrate serendipity into more general prescriptions for conduct: Leslie Hotson has shown the place of serendipity in good literary scholarship, and Leonard Bacon and Samuel McChord Crothers have defined its place in a general outlook on life. In a talk to a students' literary society at Johns Hopkins University, Hotson sings the praises of literary serendipity; in a sense, he is preaching serendipity. His sermon is intended to be somewhat revolutionary in academia. As Hotson understands the special qualities of Walpole's three princes of Serendip, "they were more than three clever fellows spending their time, as Roark Bradford might say, 'jes' projeckin'.' In going out 'for to admire an' for to see' they bore minds stocked with everything but prejudices, and driven by a universal and lively curiosity which sharpened their eyes."[4] Hotson is recommending these qualities that the three princes had, in contrast to others that he implies are all too prevalent among literary historians. These desirable qualities cannot flourish when routine dominates all activities, as, he thinks, it does in the American academic world: "It would be unreasonable to expect serendipity to be greatly fostered in our schools and universities—Serendipity's chief enemy proudly calls itself 'scientific method' [in literary studies] when his true name is 'meritorious dullness.'" Hotson implies that such dullness is unwittingly encouraged by scholars who are too timid, too narrow, or too conventional to see that they are in a rut, and who think, in fact, that their conscientious progress within that rut constitutes intellectual creativity. Instead of admiring the qualities that make for unexpected discoveries—curiosity, spontaneity, imaginativeness, a sense of adventure—they frown on them.

Literary scholars who venture to depart from the beaten track may or may not make exciting, unexpected discoveries (Hotson does not quite come out and say that discoveries that are found on the beaten track are sure to be uninteresting, but the implication seems unmistakable), but their praiseworthy eccentricity is certainly of no avail unless it is accompanied by hard work. On this point Hotson is emphatic: "When all is said, the essential to bear in mind about serendipity—whether you like to call it happy accident or lucky chance—is that the Princes had to travel; and travel means labor. The searcher has to go into them thar hills, and then look about him and dig at twenty to the dozen. You don't strike

⁴ Leslie Hotson, "Literary Serendipity," *ELH: A Journal of English Literary History* 9 (July 1942): 79.

devilish good luck without weevilish hard work."[5] Hotson, to use his own folksy idiom, is laying it on the line: In literary scholarship, no more than anywhere else, do you get somethin' for nothin'.

Beyond prescribing curiosity and hard work, Hotson is not explicit about the process of literary detection. He conveys well the excitement of detective work among the documents in the Public Record Office, but he fails to explain how the researcher recognizes significant items and discards useless ones: "One can't spend a day among the records without uncovering a promising clue of some sort. Fascinating trails beckon off in every direction, and one feels like a Sherlock Holmes who has scores of mysteries pouring into his lodgings in Baker Street. The problem resolves itself into one of deciding which clues, out of the mass, are most likely to lead to the lucky spot. It is necessary to select, and that selection is dubiously made. Then the excitement, impossible to describe, of finding that you have guessed right, and the trail is growing hotter!"[6] From the point of view of the literary scholar, unlike that of the scientist, there is, even in principle, no explicit method of distinguishing between strategic data and others: Since there is no coherent body of theory to which new data may be related, only a stock of loosely organized knowledge, the literary scholar's mind is less well prepared to classify clues as "promising" or otherwise. To be sure, the scientist, too, often selects dubiously from a mass of data, but his doubts are as nothing compared to those of the literary scholar or historian.

Among humanists there are some, like Hotson, who attempt to gain insight into the infinite and subtle complexities of man's nature through the empirical study of history or literature. If they are scholars rather than critics they often do not translate their insights into general moral observations about Man; they let the facts speak for themselves, as it were. Other humanists are more interested in characterizing mankind in general terms on the basis of impressions derived from experience, and in drawing moral conclusions about the way men should best live within the limits set by God or nature or both. Leonard Bacon might be said to represent the literary scholar-cum-critic, while Samuel McChord Crothers would seem to be the erudite humanist-preacher. Bacon is anxious to bring home to people the fact that chance plays a much larger part in their lives than they are willing to admit. In an article titled "The Long Arm of Coincidence," Bacon expresses his scorn of those who shrink from the unexplainable and unpredictable (though he also despises those who fall for cheap spiritualism). He believes that people's reluctance to admit the extent to which chance rules their lives makes them unduly

[5] Ibid., p. 81.
[6] Ibid., p. 87.

critical of the novelist's use of coincidence in his stories. For Bacon the word serendipity sums up well that prevalence of unknown causes for unanticipated results which he is trying to bring to people's attention:

> Horace Walpole, never at any time such a fool as he occasionally looked, invented the word Serendipity as a name for "the capacity to make happy discoveries by chance." Evidently he must have had an eerie realization of the extent to which we are wrapped in the everlasting coincidental arms. But most of us are afraid of chance, happy or otherwise, and like to put it out of our nice airtight frame of things. We are not flattered by the thought that our successes are in large part due to it, and, as has been noticed before, we don't like to think of our failures at all. Nevertheless, chance is in the air we breathe, an incalculable wind blowing from somewhere, where nothing is or can be explained, often with a tang of irony in it which searches the marrow.[7]

Bacon thinks people are either too smug or too timid to be honest about the extent of their helplessness; their vanity or their anxiety leads them to suppose that they control their fates. Bacon is urging upon them, as an alternative to their illusion of control, not passivity, but a realistic acknowledgement of the limits of rational activity.

Though Dr. Crothers, the Unitarian minister-essayist of Cambridge, Massachusetts, in the first quarter of the twentieth century interpreted serendipity quite differently from Bacon, he, too, used it with reference to the real and supposed limits on people's control over their lives. Crothers thought that people could do *more* to overcome life's difficulties than they supposed, if they included serendipity among their virtues:

> Besides the ordinary Christian virtues I would recommend to anyone who would fit himself to live happily as well as efficiently, the cultivation of that auxiliary virtue or grace which Horace Walpole called "Serendipity." . . .
>
> I am inclined to think that in such a world as this, where our hold on all good is precarious, a man should be on the lookout for dangers. Eternal vigilance is the price we pay for all that is worth having. But when, prepared for the worst, he goes forward, his journey will be more pleasant if he has also a "serendipitaceous" mind. He will then, by a sort of accidental sagacity, discover that what he encounters is much less formidable than what he feared. Half of his enemies turn out to be friends in disguise, and half of the other half retire at his approach. After a while such words as "impracticable" and "impossible" lose their absoluteness and become synonyms for the relatively difficult.[8]

---

[7] Leonard Bacon, "The Long Arm of Coincidence," *Saturday Review of Literature* 29 (December 1946): p. 10, n. 50.

[8] Samuel McChord Crothers, *Humanly Speaking* (Boston: Houghton Mifflin, 1912), pp. xii–xiv.

Crothers is preaching a kind of calculated self-deception: People are encouraged to meet problems with a mixture of exaggerated pessimism and "serendipitaceous" optimism, and with this combination of attitudes they will be equal to almost anything. Since menaces and difficulties are quasi-deliberate distortions, normal experience will tend to appear as a series of happy accidents. Such a series of "happy accidents" will, presumably, give people an illusion of control that is useful in overcoming difficulties.

In untangling Crothers's little sermon, we have ourselves been guilty of some distortion. Crothers was really urging on people nothing more complicated than the old adage that suggests there is always a silver lining in the darkest cloud. But because he was somewhat fearful himself of the storms that twentieth-century clouds seemed to portend, he was looking for some Dutch courage and finding it in serendipity.

Afterword

AUTOBIOGRAPHIC
REFLECTIONS ON
THE TRAVELS AND ADVENTURES
OF SERENDIPITY[1]

ROBERT K. MERTON

for Elinor

Quel beau livre ne composerait-on pas en racontant la vie et
les aventures d'un mot?
　—Honoré de Balzac, *La comédie humaine* (1832)[2]

Faire l'histoire d'un mot, ce n'est jamais perdre sa peine. Bref
ou long, monotone ou varié, le voyage est toujours instructif.
　—Lucien Febvre, "Civilisation: Le mot et l'idée" (1930)[3]

*I*n long retrospect, it now seems virtually inevitable that I should have
decided to track the travels and adventures—or, put in rather more academic terms, the diffusion and reconceptualization—of the word serendipity, and then persuaded the historian of eighteenth-century France and

---

[1] As memoirs and other retrospectives sometimes do, this Afterword at times shame-
lessly paraphrases or even quotes fragments from earlier writings of mine without always
indicating that I am doing so—and without citing the specific sources. This no doubt

my longtime friend Elinor Barber to join me in that quest. "In long retrospect" because I am writing this Afterword soon after my doubly improbable ninety-first birthday. "Doubly improbable" in both the objective and subjective senses of probability. Objectively improbable, of course, because even in our time of rapidly extended life expectancy, only some 1.5 percent of my composite nationality-sex-and-complexion cohort have actually reached their nineties. And subjectively improbable because, from the time I was a young romantic haunted by the memory of Shelley, Byron, and Keats, I took it for granted that I too would not outlive my thirties. So much for the long retrospect. But in what immediate sense was the writing of the book virtually inevitable? In particular, a book manuscript that was brought to a close in the late 1950s and left abandoned on a bookshelf until Giovanna Movia and Laura Xella, my publisher friends at Il Mulino, got wind of the deserted manuscript and persuaded me to finally have it published.

My Self-Exemplifying Encounters with Serendipity

A Note on the Reflexive Concept of "Self Exemplification"

Paradoxically enough, the near inevitability of this book largely derives from the self-exemplifying character of its origins. The reflexive concept of "self-exemplification" refers to an idea (concept, hypothesis, or theory) that applies to its own content or is exemplified by its own history. As a case in point, consider the idea central to the now widely familiar aphorism that was formulated in the twelfth century by the philosopher Bernard of Chartres, "We are like dwarfs standing on the shoulders of giants and so are able to see more and see farther than the ancients," or Isaac Newton's more succinct and more personalized seventeenth-century version: "If I have seen farther, it is by standing on the shoulders of giants." In the course of thus metaphorically expressing the idea that the advance of knowledge depends largely on the prior accumulation of knowledge, Bernard in effect (happily) acknowledged that he himself stood on the shoulders of the grammarian Priscian who had formulated the gist of the idea six centuries before. Plainly, then, a self-exemplifying idea.

By way of further illustration, consider how this aboriginal but far from

---

violates fundamental norms of scholarly citation but it does have the incidental merit of somewhat reducing the excessive number of self-citations.

[2] Honoré de Balzac, *Études philosophiques et études analytiques: Louis Lambert* (Paris: Alexandre Houssiaux, 1832), p. 111.

[3] Lucien Febvre, *Civilisation: Le mot et l'idée, Premiere semaine internationale de synthèse,* 2d ed. (Paris, 1930), pp. 1–55, at p. 1.

primitive sociological idea about the dependence of current knowledge upon past knowledge has since been further specified to explain the multiple and independent occurrence of the same scientific discovery or technological invention. Ever since 1922, that idea has been associated with William F. Ogburn and Dorothy S. Thomas, who did so much to establish it in sociological thought.[4] On the basis of their study of some 150 cases of independent multiple discovery and invention, they concluded that certain innovations become virtually inevitable as certain kinds of knowledge accumulate and as innovating scientists and technologists who are focused on the same problems arrive at the same solutions.

Appropriately enough, this further idea is exemplified by its own history (almost a Shakespearean play within a play). For as observers of multiple independent discoveries have reflected on that phenomenon over the centuries, they have come to much the same explanation. Working scientists, historians and sociologists of science, anthropologists, Marxists and anti-Marxists, Comteans and anti-Comteans have called attention, time and time again though with varying degrees of perceptiveness, to both the phenomenon of independent multiple discoveries and to some of its implications for understanding how scientific knowledge develops. Again, an idea exemplified by its own history.

Consider another, and concluding, example of a self-exemplifying idea, which Laurence Sterne tucked away in his eighteenth-century literary masterwork, *Tristram Shandy*, a sportive work that broke new ground for the novel by instituting a tortuous, stream-of-consciousness style, based on Locke's notion of the association of ideas, an inner-monologue style that would not reappear until early in the twentieth century with such masterworks as James Joyce's *Ulysses* and Virginia Woolf's *Mrs. Dalloway*.[5] Here, then, is Tristram's instance of what amounts to a self-exemplifying observation, this one on the singular behavior of hypotheses: "It is in the nature of an hypothesis, when once a man has conceived it, that it assimilates everything to itself, as proper nourishment; and, from the first mo-

---

[4] W. F. Ogburn and D. S. Thomas, "Are Inventions Inevitable?" *Political Science Quarterly* 37 (1922):83–98; and W. F. Ogburn, *Social Change* (New York: Heubsch, 1922), pp. 90–122. For data on the self-exemplification of that theory of "multiples," see chapter 16 of my *Sociology of Science: Theoretical and Empirical Investigations*, ed. Norman Storer (Chicago: University of Chicago Press, 1973).

[5] I do not retract this remark (which does not imply direct influence), even though I now belatedly recall (and confirm) this entry on page 48 of Virginia Woolf's *A Writer's Diary*, ed. Leonard Woolf (New York: Harcourt Brace, 1953): "I've finished *Ulysses* and think it a mis-fire. Genius it has, I think; but of the inferior water. The book is diffuse. It is brackish. It is pretentious. It is underbred." But she goes on to confess, "I have not read it carefully; and only once: and it is very obscure; so no doubt I have scamped the virtue of it more than is fair." Nor did Woolf pause then to reflect on the Tristramesque patterns of time convolution and stream-of-consciousness in *Ulysses* any more than in her later *Mrs. Dalloway*.

ment of your begetting it, generally grows the stronger by every thing you see, hear, read, or understand."

I am also mindful that the Reverend Laurence Sterne concludes in his typically irreverent and ironic fashion: "This is of great use." But since the truth of Tristram's doctrine is exemplified by its own enunciation, any further report of the versatile uses of hypotheses undisciplined by skeptical and systematic inquiry would be altogether superfluous. What's more, these uses are recorded in chapter 19 of book 2 of *Tristram Shandy* and are thus open for inspection in every one of the uncounted editions of Tristram's masterly memoirs.

I close with this example of self-exemplifying ideas for two interlocking reasons: first, because it happens that Sterne's masterwork was promptly denounced on both literary and moral grounds by his contemporary and our neologist, Horace Walpole; and second, because this example is quoted in my own favorite and prodigal book, *On the Shoulders of Giants* (long since acronymized as *OTSOG*), the book which, it will be remembered, James Shulman persuasively hypothesizes in his imaginative and meticulous Introduction to have preempted any further outgrowth of *The Travels and Adventures of Serendipity*. Or, as he put that hypothesis more briefly, cogently, and elegantly, this book on serendipity, which was ruthlessly abandoned in the 1950s, was in effect a *preparazione OTSOGIA*.[6]

My First Serendipitous Encounter with the Word Serendipity

In saying, then, that the idea of a work on the travels and adventures of first the word and then the concept of serendipity is self-exemplifying, I only mean that it emerged and developed serendipitously. As the foregoing "time-capsule" text of 1958 hints in a terse paragraph (Chapter 7), my first consequential encounter with the word was altogether serendipitous. While in search of the history of a word beginning, I suppose, with the letters "se"—it may have been the word *sequestration* or the word *sesquipedalian*; I do not truly remember—I was riffling through the pages of volume 9 of the incomparable *Oxford English Dictionary* (*OED*)

---

[6] For the example of self-exemplification provided by *Tristram Shandy* and for diverse other examples, see R. K. Merton, *On the Shoulders of Giants: A Shandean Postscript* (New York: Free Press, 1965), pp. 29, 42, 61, 117, 174, 194. For more on the concept of "Self-exemplification," consult the index of my *Sociology of Science: An Episodic Memoir* (Carbondale: Southern Illinois University Press, 1979). Since Chapter 6 of our 1958 text and the next section of this Afterword go into some detail on the role of dictionaries in the diffusion of words and ideas, I note that, though the reflexive concept of "self-exemplification" has repeatedly appeared in print for several decades, it has yet to find its way into the incomparable and presumably all-inclusive *Oxford English Dictionary* just as the nicely translated *autoesemplificazione* in Italian versions of those varied texts has yet to find its way into the incomparable and presumably all-inclusive *Grande Dizionario della Lingua Italiana*. Perhaps this enlightening concept has not met certain criteria for inclusion.

organized on historical principles when my eye happened upon the strange-looking and melodious-sounding word serendipity. Strange-looking, since its etymology was far from obvious. It may have been that etymological obscurity which led me to pause and read the entry in detail. As is standard (if not always realizable) practice for the *OED*, the entry quoted the passage in which the word first appeared, a passage that readers of this book have come to know intimately: "A word coined by Horace Walpole, who says (Let. to Mann, 28 Jan. 1754) that he had formed it upon the title of the fairy tale, 'The Three Princes of Serendip,' the heroes of which 'were always making discoveries by accidents and sagacity, of things they were not in quest of.'"

At once serendipity became a part of my working vocabulary. This, on several counts. For one, as the 1958 text briefly notes (Chapter 7), the crucial defining phrase, "discoveries . . . they were not in quest of," instantly resonated to a theoretical fixation of mine on the sociological importance of unintended consequences of intended actions in social life generally and of unanticipated phases in the growth of knowledge. That fixated idea had taken root in my mid-1930s doctoral thesis on the unintended and rather paradoxical role played by Puritanism in advancing the cause of the new science that was emerging in seventeenth-century England[7] and was then formulated more generally in a paradigmatic essay on "The Unanticipated Consequences of Purposive Social Action."[8] The resonance of the then still esoteric word was only reinforced as I turned from the expeditiously limited entry in the *OED* to Walpole's full letter to Mann, in which he went on to reiterate most emphatically that "you must observe that *no* discovery of a thing you *are* looking for comes under this description" (emphasis in the original). Clearly, then, for its inventor the word serendipity signified a special kind of unintended and unanticipated outcome.

[7] R. K. Merton, *Science, Technology, and Society in Seventeenth-Century England* ([1938] New York: Howard Fertig, 2001).

[8] That paper was published in *American Sociological Review* 1 (1936):894–904. This immediate section of the Afterword draws heavily on an autobiographic retrospective requested by the Italian Sociological Association that examines a half-century of sociological ideas stemming from this enduring focus on the unintended and unanticipated in social life and in the growth of knowledge—such ideas as anomie-and-opportunity theory, latent function and dysfunction, the self-defeating and the self-fulfilling prophecy, and, of course, serendipity. See "Unanticipated Consequences and Kindred Sociological Ideas: A Personal Gloss," in *Robert K. Merton and Contemporary Sociology*, ed. Carlo Mongardini and Simonetta Tabboni (New Brunswick, N.J.: Transaction Publishers, 1998), pp. 295–318. For further extensions of this theme, see Raymond Boudon, *Effets pervers et ordre social* (Paris: Presses universitaires de France, [1977] 1993); Albert O. Hirschman, *Autosovversione* (Bologna: Il Mulino, 1997), pp. 64–68; and Albert O. Hirschman, *The Rhetoric of Reaction* (Cambridge, Mass.: Harvard University Press, 1991), pp. 38–42.

The word resonated all the more for me because Walpole had elected to define his "very expressive" word as referring to "discoveries, by accidents and sagacity." Now, as we know from his gossipy example of serendipity—"Lord Shaftesbury having found out the marriage of the Duke of York and Mrs Hyde by the respect with which her mother treated her at table"—Walpole was deploying the word "discovery" in a rather extended sense. He was surely not referring to *scientific* discoveries for, as we know from his vast correspondence, he hadn't the least interest in that esoteric and disciplined sphere of knowledge. In any case, he would hardly have known of "accidental discoveries" such as "animal electricity," made shortly after his altogether apt coinage when Luigi Galvani "sagaciously" happened to notice that a dissected frog leg contracted when simultaneously touched by two different metals and the physicist Alessandro Volta followed that accidental discovery with the first electric battery. (Hence, as is often the case in the reward system of science, they earned such honorific eponyms as *galvanic, galvanize, volt, voltage,* and *voltaic battery.*)

Nor, of course, as we have seen in the foregoing time-capsule text of 1958, are such "accidental discoveries" rare. They have been experienced times without end by practicing scientists and have been often noted by historians and sociologists of science who were thus primed to adopt Walpole's uniquely "expressive word" when they first encountered it. I cannot put a definite date on the serendipitous moment of my own first encounter with the word, since one doesn't ordinarily record such an episode for future reference, especially when one fails to keep a journal or a diary. But its instant resonance was plainly overdetermined. That is to say, that resonant interest had more determining factors than were necessary (and so, of course, violated the fourteenth-century maxim known as "Ockham's razor": "Entia non sunt multiplicanda præter necessitatem"). It was clearly the result of at least four long-standing and converging interests: my sociological interest in the generic phenomenon of unintended consequences of intended action, my methodological interest in the logic of theorizing, my substantive interest in the history and sociology of science, and my enduring engagement with neologisms that are needed to describe newly discovered phenomena and newly emerging ideas. Walpole had bequeathed us a word—and a protoconcept, about which more, perhaps much more, later—that happily combined elements of all four of those interests. Truly, a splendid legacy.

But if I cannot date that first serendipitous encounter with serendipity in the *OED*, I can easily date my first use of the word in print. As is briefly noted in the foregoing *Travels and Adventures*, that was in a paper published in 1945, when the prevailing positivistic outlook did not take kindly to the claim that not all consequential scientific research follows a unilinear hypothetico-deductive method. That then unpopular claim was expressed in the observation that "fruitful empirical research not only tests theoretically derived hypotheses; it also originates new hypotheses."[9] My generic interest in the unanticipated had led to a specific focus on the unanticipated (and hence, nonlinear) phases in much scientific inquiry. So it was that I soon adopted the—for me—variously encompassing word and went on to remark of an unexpected scientific finding that "this might be termed the 'serendipity' component of research, i.e., the discovery by chance or sagacity of valid results that were not sought for."[10]

That barely perceptible remark would scarcely have been enough to introduce even the word, let alone the analytical concept, of serendipity into the domain of the sociology of science. As hinted in the time-capsule text and noted in the perceptive Shulman Introduction, this evidently came about through a paper published a few years later that undertook to transform the *word* into an evocative *concept* (i.e., a general idea with successively drawn-upon-able implications). The concept of "the serendipity pattern in scientific inquiry," it was proposed, consists in the observation of "an *unanticipated, anomalous and strategic* datum that becomes the occasion for developing a new theory or for extending an existing theory."[11]

Thus, the unexpected occurs twice over in the serendipity pattern. An unanticipated observation yields an unanticipated kind of new knowledge. Or, as this was put in the 1940s: "The datum is, first of all, unanticipated. A research directed toward the test of one hypothesis yields a fortuitous by-product, an unexpected observation which bears upon theories not in question when the research was begun. Secondly, the obser-

[9] Or, as the philosopher of science, Aharon Kantorovich, and the Nobel laureate in physics, Yuval Ne'eman, would put it in their fine-grained analysis of "the principle of serendipity": "Although scientific theories cannot be logically derived from observations, the generation of theories seems to be by no means blind to the observational data. . . . [M]any creative leaps and breakthroughs in science result from *serendipitous* processes which represent the blind discoveries in science (Kantorovich and Ne'eman, 1989)." A. Kantorovich, *Scientific Discovery: Logic and Tinkering* (Albany: State University of New York Press, 1993), p. 148, citing an earlier joint paper, A. Kantorovich and Y. Ne'eman, "Serendipity as a Source of Evolutionary Progress in Science," *Studies in History and Philosophy of Science* 20 (1989):505–529.

[10] R. K. Merton, "Sociological Theory," *American Journal of Sociology* 50 (1945): 462–473; reprinted in *Social Theory and Social Structure* (New York: Free Press, 1968), pp. 157–158, nn. 4, 4c.

[11] R. K. Merton, "The Bearing of Empirical Research upon the Development of Sociological Theory," *American Sociological Review* 13 (1948):505–515 (italics in the original text). See also Maria Luisa Maniscalco, "Il concetto di 'serendipity' nell'opera di Robert K. Merton," in *L'Opera di R. K. Merton e la sociologia contemporanea*, ed. Carlo Mongardini and Simonetta Tabboni (Genoa: Edizioni Culturali Internazionali Genova, 1989), pp. 283–293.

vation is anomalous, surprising, either because it seems inconsistent with prevailing theory or with other established facts. In either case, the seeming inconsistency provokes curiosity; it stimulates the investigator to 'make sense of the datum,' to fit it into a broader frame of knowledge."

Along with these two components of the unexpected in the serendipity pattern is at least a third component having to do with the scientist who finds the datum evocative: "In noting that the unexpected fact must be strategic, i.e., that it must permit implications which bear upon theory, we are, of course, referring to what the observer brings to the datum rather than to the datum itself. For it obviously requires a theoretically sensitized observer to detect the universal in the particular." In short, the unanticipated, anomalous, and strategic datum is not simply "out there," but is in part (but only in part) a cognitive "construction" that is a function of its observer's theoretical orientation and knowledge, both explicit and tacit.

This observation on the interplay between the widely adopted practice of hypothetico-deductive inquiry and often-experienced serendipity in science deals wholly with theoretical aspects of serendipitous discovery, not with an emerging interest in the sociological history and semantics of the "outlandish" term and potential concept. However, that interest was signaled a year later by the insertion of a typically discursive footnote when the paper analyzing the "serendipity pattern" was included in the first edition of *Social Theory and Social Structure*:

Since the foregoing note was first written in 1946 [and published in 1948] the word *serendipity*, for all its etymological oddity, has diffused far beyond the limits of the academic community. The marked speed of its diffusion can be illustrated by its most recent movement among the pages of the *New York Times*. On May 22, 1949, Waldemar Kaempffert, science editor of the *Times*, had occasion to refer to serendipity in summarizing an article by the research scientist, Elice McDonald—this, in an innermost page devoted to recent developments in science. Some three weeks later, on June 14, Orville Prescott, book reviewer of the daily *Times*, has evidently become captivated by the word, for in a review of a book in which the hero has a love of outlandish words, Prescott wonders if the hero knew the word serendipity. On Independence Day of 1949, serendipity wins full social acceptance. Stripped of qualifying inverted commas and no longer needing an appositive defining phrase, serendipity appears, without apology or adornment, on the front page of the *Times*. It achieves this prominence in a news dispatch from Oklahoma City reporting an address by Sir Alexander Fleming, the discoverer of penicillin, at the dedication of the Oklahoma Medical Research Foundation. "Sir Alexander's experience, which led to the development of modern disease-killing drugs," says the dispatch under the by-line of Robert K. Plumb, "is frequently cited as an outstanding example of the importance of serendipity in science. He found penicillin by chance, but had been trained to look for significance in scientific accidents." In these travels from the esoteric page devoted to science to the less restricted columns of the book-review to the popular front-page, serendipity had become naturalized. Perhaps it would soon find its way into American unabridged dictionaries.[12]

A concluding conjecture, which was realized in the event, as we know from Chapter 6 and as we shall also see in a later section of this Afterword, focused on dictionaries as modes of linguistic diffusion. It shows how that once "outlandish" and esoteric word, which was confined to English dictionaries back in the "time-capsule" period of almost a half-century ago, has since diffused into dictionaries in many other languages.

So much, then, for this unexpectedly prolonged account of my first serendipitous and consequential encounter[13] with the word in the *OED*. But that threshold encounter had been preceded by another and, I trust, more briefly told serendipitous episode; to wit:

My Serendipitous Acquisition of the OED

As I have somehow neglected to report, though James Shulman has noted it in his observant Introduction, the copy of the *OED* in which I happened upon the word serendipity was my very own. Improbably so, since I had acquired the—for me—costly thirteen-volume edition of 1933 in the year of its publication, while I was still an impoverished graduate student at Harvard. And thereby hangs a tale that might be more informatively titled:

How FDR Incidentally Arranged for My Serendipitous Acquisition of the OED

It may be remembered that 1933 was the year in which Franklin Delano Roosevelt—soon known the world over as FDR—first took office as president of the then ailing United States. In an instant effort to cope

---

[12] Merton, *Social Theory and Social Structure* (New York: Free Press, 1949), pp. 376–377.

[13] I refer to this as my first *consequential* encounter with the word *serendipity* for a definite reason: I may possibly have seen, but if so, surely had not noticed, the word previously in a book by an inventor, Joseph Rossman, titled *The Psychology of the Inventor* (New Hyde Park, N.Y.: University Books, [1931] 1964), which, as noted in Chapter 4 and elsewhere in the 1958 text, alluded briefly (and misleadingly) to inventors' frequent "lucky finds," which are "termed 'serendipity' after the Persian god of chance." But then again, I may have come upon the Rossman book only after my serendipitous and quite definite encounter with the word in the *OED*.

with the years-long economic chaos that became known as the "Great Depression," FDR's first official act was to declare a "national banking holiday." That "holiday" or "moratorium" closed all banks in the country, which meant that Americans would not have access to their bank deposits for some unannounced duration. And rather more egocentrically, it meant that I would not have access to my meager savings at the Cambridge Savings Bank conveniently located on Harvard Square.

But all was not lost. Deprived of their customary trade, merchants on Harvard Square swiftly adapted to that condition by offering to extend credit to properly certified Harvard students. And happily for us student bibliophiles, as can be seen from the enticing advertisement in the student newspaper, the *Harvard Crimson* (Figure 1), the owner of our favorite Phillips Book Store headed the list of those rational, ingenious, risk-taking, and sympathetic merchants making that inviting offer. That led less optimistic bibliophiles, including myself, to rush to the bookshop in the thought that they might be able to convert at least some of their inaccessible and possibly irredeemable cash into books. Browsing through the bookshop, as the time-capsule text makes plain that would-be serendipitist bibliophiles often do, I came upon the ten resplendent volumes of the *OED*, that wonderfully instructive and very expensive reference work which I had been consulting in libraries since its first appearance in 1928 but which I obviously had no intention, expectation, or hope of acquiring. Whether other dictionary buffs in their equally cashless state also took special note of the incomparable *OED*, I cannot, of course, say. But I can speak for myself. Unable to resist temptation, I found myself ready to invest a sizable portion of my fellowship stipend to become the happy owner of the grandest English dictionary of them all. I am still fairly confident that I would not have dared make what was for me a huge investment back then, were it not for my pessimistic doubt that I would ever be able to salvage the whole of my scanty savings. And, of course, I became all the more eager to "close the deal" then and there when the ingenious (and, I'm inclined to think, empathetic) bookseller went on to make a handsome offer; he would accept my I.O.U. and have it apply to the purchase of the forthcoming 1933 edition—an edition announced as replete in fully thirteen volumes, rather than the first, 1928 edition of ten volumes—which would include a supplementary volume devoted to new words and to a bibliography of the sources of quotations in the entries. That further offer proved irresistible. I left the bookshop knowing that in due course I would become the blissful possessor of the newest version of the magnificent reference work that had been some seventy-five years in the making.

My pessimism about the recapturing of my savings thus turned out to be a most fortunate risk-aversive error. For, of course, FDR saw to it that

The Harvard Crimson

CAMBRIDGE, MASS. MONDAY, MARCH 6, 1933.　　PRICE FIVE CENTS

Haven Saturday by a

I. B. Stewart defeated
a time advantage of 9

L. M. Klein defeated
a time advantage of 8

towelle (Y) defeated R.
le advantage of 9 min.,

full (Y) defeated W. G.
dvantage of 4 min., 50

Whiteridge (Y) threw
body scissors and half-
n, 20 sec

Indeke (Y) defeated R.
advantage of 9 min.,

Graham (Y) defeated
a time advantage of 8

itein (Y) threw E. F.
se arm hold. Time 1

se To
re
lassed, the Freshman
defeated 47-24 by the
a meet held at New
t.

swim—Won by Win-

Harvard Square Lacks Cold Cash, Local Bankers Assert
Financial Stability---Dewing Blames Situation on West

Local Merchants in Uproar From Shortage of Change, Some Refuse To Accept Bills

While the bankers are maintaining an optimistic tone as they await developments in the present financial situation, the small traders of the vicinity have been more quickly and seriously affected by the bank holiday and are frankly worried. There is a serious shortage of coins and the question of obtaining change has become acute. This situation is further complicated by the fact that some seem to have lost faith in the federal bills and are making incredible efforts to convert their paper money into metal. One manager of a little store was seen running wildly from door to door, waving a twenty-dollar bill which he was unable to change, since he would not accept pa-

individual states.

"This is the result of a terrific drain on the gold supply which centered in New York. Much of this drain was to foreign countries. The final cure for this may lie in an embargo on gold. The immediate relief will probably come in the form of federal guarantee of private deposits. In spite of the disadvantages and the difficulties in currency shortage which this will incur I do not think the alternative, abandonment of the gold standard, either a wise or a likely step."

University Treasurer

J. W. Lowes '17, deputy treasurer of the University, declared: "No one yet knows what will be done. The University must wait and watch developments before declaring its policy. Presumably there will be some limitation on banking contracts, but these must be made definite before any line of ac-

Finance Professor Says Closing Of Banks Necessitated By Failures in Middle West

"From the Eastern point of view the bank holidays were not necessary, but to save the Western banks they were absolutely essential," declared A. S. Dewing '02, professor of Finance in the Business School, in an interview last night. "When a few banks failed in the Middle West, merchants in the districts affected, immediately sought to draw all their funds from the banks which still remained solvent, and this contagion for withdrawal spread throughout the country until moratoriums became unavoidable if whole areas were to be saved from financial ruin. New England is not one of those areas, —it is in better condition than any other part of the nation.

No. 28.　　CAMBRIDGE, MASS. WEDNESDAY, MARCH 8, 1933.

ws
om the
Houses

OP HOUSE GETS
RY OF RECORDS

in Standish Equipped With
la Records For Residents

sible by anonymous dona-
/inthrop House Committee
sed a Music Room on the
of Standish Hall for the
of all students of the House.
d with an electric victrola,
.00 records stored in the
Winthrop library.
is one recently vacated by
is furnished with chairs
In use only a week ago, it
popular, and it has been
ntinually every afternoon
up to 10 o'clock, the clos-
tudents must sign with the
r the records and the key,
is no one else waiting, may
as they wish. Otherwise,
limited to one hour. In ad-
records may be taken to
ooms, provided that fibre
sed exclusively. All equip-
broken must be replaced.

GIVES TO CHARITY

lay Dance Surplus Will Go
mbridge Relief Fund

f the surplus of the Lowell
e, which is to be held on
be turned over to the Cam-
ployment Relief Fund, ac-
statement issued by E. K.
irman of the Lowell House
ittee.
ing is the list of patrones-
. W. Brinkman, Mrs. J. L.
s. W. Y. Elliot, Mrs. D. W.
. F. Jones, Mrs. N. C. Per-
. H. Vogel, Jr., and Mrs.

BOOTLEGGERS ADAMANT IN DEMAND FOR CASH BUYING

Although almost all the merchants in Harvard Square are advertising today that they will extend credit and honor checks when presented with Bursar's card, the bootleggers in the vicinity demand honest-to-goodness money. The H.A.A. has announced that they will take checks for tickets for the hockey game with Yale tonight, but seven out of ten "importers and exporters" refused to supply beverages for checks. They seemed to oppose it on the grounds that it was not straightforward business. Three local dispensers were even willing to have charge accounts opened, although insisting on a personal interview first.

When interviewed on the subject, one of the civic-minded members of the profession stated: "What the government of the United States believes is best for the people must be correct. Every citizen should obey the instructions of the lawmakers."

RUGBY TEAM PREPARES
FOR APPROACHING GAME

Schedule Calls For Games With Yale, Princeton, and New York and French Rugby Clubs

SIXTY ALREADY PRACTICING AND FORTY MORE EXPECTED

In preparation for the coming season, the 1933 schedule of the rugby team was announced yesterday by Stanton Whitney, Jr. '34. It includes two games with Yale, one to be held here, one at New Haven. Whitney stated that he hoped to arrange a game with the Argentine champions, to be played in the Argentine some time during the coming summer.

The team is at present practicing in the Old Cage, but it is expected to be outside before next week. A Varsity team will be chosen by that time, and also a number of House teams, or, if there are not enough out to provide for these, four class teams and a Graduate School

EXTENDING CREDIT

The following merchants are extending credit, with reservations, to students who can exhibit Bursar's Card's

Phillips Book Store . . . . 1288 Mass. Avenue

George's Flower Shop . . . . Harvard Square

James F. Brine, Inc. . . . . Harvard Square

Daley's Drug Store . . . . Harvard Square

Covin, Florist . . . . Next To University Theatre

Harvard Coop . . . . Harvard Square

Hood's Creamery Lunch . . . . 11 Brattle Street

Walter & O'Rourke, Tailors . . . . Harvard Square

C. A. Stonehill, Jr. . . . . Mt. Auburn Street

Bob Slate, Student Supplies . . . . Harvard Square

Charles Ferranti Barber Shop . . . . 1302 Mass. Avenue

John Goode, Tailor, ... Dunster Street

The Harvardashery . . . . 1388 Mass. Avenue

Figure 1.

the banks were reopened in a matter of weeks.[14] But, from my acutely grateful perspective, he had also seen to it that I should come away from the national banking crisis with my very own precious set of the *OED*.[15]

That it truly was a "huge investment" at the time is evident from a simple reconstructed accounting. Not relying on my deficient memory of such matters, I have ascertained from its publishers that a set of the *OED* sold for £21 in 1933, at a time when the British pound converted into approximately $4.25. My serendipitous acquisition, then, had cost about $90. By way of further perspective (provided by the inflation calculator based on the consumer price index [CPI] of the authoritative U.S. Bureau of Labor Statistics), I also learn that an American dollar in 1933 is equivalent to about $13.20 in 2000. So it turns out that that impoverished graduate student and *OED* buff found himself dedicating the present-day equivalent of some $1,188 to his unexpected grand acquisition. That to me rather surprising fact invites a further basis for gauging the extent of both my dedication to the *OED* and my pessimistic doubt about recovering those slender savings that FDR had, first by edict and later by enjoined legislation, sequestered in the closed Cambridge Savings Bank. My financial resources were confined to the at first seemingly sumptuous annual stipend of $1,000 provided by the Robert Treat Paine fellowship and an additional stipend of $300-or-so for serving as a research assistant. But that apparently substantial sum soon dwindled to a barely livable minimum, for from that sum Harvard had deducted $400 for tuition and $140 rent for my share of a two-room suite (with just one roommate) in the graduate dormitory, Perkins Hall.[16] Since those instant

---

[14] My limited memory of "the banking crisis of 1933" turns out to be amply confirmed in the thoroughgoing book of that title by Susan Estabrook Kennedy (Lexington: University Press of Kentucky, 1973).

[15] Having recalled that FDR's first political and fiscal act as president dealt with the banking crisis, one also recalls that this was soon followed by his diplomatic and political act of visiting the retired Supreme Court justice Oliver Wendell Holmes Jr. to honor that great jurist on his ninety-second birthday. The story is told that when one of his protégés asked for his impression of the new president, Holmes promptly replied: "A second-class intellect, but a first-class temperament." Truly *se non è vero è ben trovato*. [Even as I proofread this Afterword, I receive a copy of the brilliant sociological and philosophical analysis of this apt proverb by the sociologist of knowledge Kurt H. Wolff, "Una struttura con cui giocare: aproposito del proverbio italiano '*se non è vero è ben trovato*,'" *Sociologia* (2000):33–45.] The story is well told by the historian Arthur M. Schlesinger Jr. in *The Coming of the New Deal* (Boston: Houghton, Mifflin, 1958), p. 14. It was that "temperament," of course, that found expression in his decisively closing the nation's banks, an action which in turn led, as its most trivial unexpected consequence, to my serendipitous acquisition of the *OED*.

[16] Translated into today's dollars by the CPI inflation calculator, these 1933 sums for tuition and rent would be the equivalent of about $7,500 in 2000. It should be noted, however, that the actual figures at Harvard these days are far larger, running to $25,000 for tuition and $4,810 for those very same rooms in Perkins Hall.

Only now, sometime later in the writing of this autobiographic Afterword, do I recall a

deductions left me a total of some $750 for all the rest of the academic year's expenses, it appears that I had devoted almost a third of my still remaining but inaccessible cash resources to the serendipitous purchase of the *OED*. Still, I now have only to glance at those heavily worn volumes on a nearby bookshelf to reflect on the pleasurable use to which they have been put these last seventy years and to realize yet again the impulsive wisdom of that scholarly investment.

In retrospect, then, this dictionary buff owes a great debt to FDR, to that enlightened and kindly disposed bookseller, and to all the prior others[17] who together led me to that serendipitous and only seemingly extravagant decision.

A Recent Serendipitous Encounter with Serendipity

Because my first consequential encounter with the word serendipity via the *OED* was serendipitous, it is only fitting that a quite recent encounter,

---

similar comparison of prices at Harvard and vicinity in the 1930s and today by I. Bernard Cohen, the distinguished historian of science and my colleague-at-a-distance and good friend for two-thirds of a century (who, I report with something approaching awe, has, with Anne Whitman, published the first new English translation of Newton's immortal *Principia* to appear in 270 years). Noting the subsidy of $375 provided by Harvard for publication of my doctoral dissertation, *Science, Technology, and Society in Seventeenth-Century England,* Cohen goes on to observe, "In order to determine the value of $375 in those days I use two scales. First, when I became a graduate student and young instructor at Harvard in the late 1930s and 1940s, it was common lore that a Harvard professor would receive an annual salary of $12,000. Second, in those days a nickel would buy a good cup of coffee (with cream) at any of the restaurants or cafeterias near Widener Library. On today's scale, then, this $375 would be worth between $4,750 and $5,000. If the scale used were the cost of undergraduate tuition at Harvard, or dormitory room and board, today's equivalent would be an even greater sum." This estimate is reported in footnote 72 on page 80 of I. Bernard Cohen's *Puritanism and the Rise of Modern Science: The Merton Thesis* (New Brunswick, N.J.: Rutgers University Press,1990). It is comforting to see that I.B.'s estimate and mine are of the same order of magnitude.

[17] An explanatory note on "those others": I had grown up poor in the then urban village of Philadelphia but with a great array of institutional riches close at hand. Chief among them was what amounted to a private library of some ten thousand volumes, located just a few blocks from our modest family house. Actually, it was a branch public library, which in effect had been bestowed upon me by that ultimately beneficent robber baron Andrew Carnegie. Since the neighborhood was secure enough for me to make my way to my library from the tender age of five or six, I spent countless hours there over the next decade or so, until my parents moved. That long and diligent presence led to my being adopted by the dedicated librarians, all of them women, of course. Having noted my deep and growing interest in language, they alerted me to the joys and uses of browsing in dictionaries, most particularly in the exciting early 1920s edition of the unabridged *Webster's New International Dictionary of the English Language.* So it was that I became a dictionary buff, primed to respond eagerly to the *OED* when it first appeared in 1928 and ready for its serendipitous acquisition a few years later.

occurring even as I write these reflections, should also be serendipitous and that it should also involve the *OED*, albeit through an intermediary. That encounter begins with a coincidence in which my granddaughter and her husband present me with a most knowing gift on my ninetieth birthday: a copy of the engrossing book by the English writer and adventurer Simon Winchester engagingly titled *The Professor and the Madman* and instructively subtitled *A Tale of Murder, Insanity, and the Making of the Oxford English Dictionary*.[18] Knowing that I was an *OED* buff and having learned that, nevertheless and quite inexplicably, I did not yet have a copy of the book although it was a *succès d'estime* as well as a *succès fou* which had long been on the *New York Times* bestseller list, they confidently selected it as a bound-to-be-appreciated gift, as indeed it proved to be, especially for a transcending coincidental reason.

For it happens that the author Simon Winchester has occasion to refer to Ceylon early in the book and then goes on to free-associate in terms long grown familiar to us, thus: "These days it [i.e., Ceylon] is called Sri Lanka; once the Arab traders called it Serendib, and in the eighteenth century Horace Walpole *created* a fanciful story about three princes who reigned there, and who had the enchanting habit of stumbling across wonderful things by chance. Thus was the language enriched by the word serendipity, without its inventor, who never traveled to the East, ever really knowing why" (pp. 43–45).

It is plain why I have italicized the word *created* in that passage about Walpole's coinage. For, as we know, it is not that Walpole "*created* a fanciful story about three princes" but, as he says in that now historic letter to Mann, that he had "once *read* a silly fairy tale, called *The Three Princes of Serendip*." This plain but easily corrected error begins to hold more than passing interest for us as students of cultural diffusion for it has appeared in print more than once. And that fact raises the question whether the error has been conceived independently or has diffused. Thus, the very same attribution of the "fanciful story" or "fairy tale" to Walpole also appears in volume 18 of the authoritative *Grande Dizionario della Lingua Italiana* published in 1996, two years before the Winchester book. Its unusually extensive entry for *Serendipità* concludes in this fashion: "Dall'ingl. *Serendipity*, voce coniata nel 1754 da H. Walpole nel *suo romanzo* 'Three princes of Serendip' (dove narra la storia di tre giovani che hanno il dono naturale di scoprire cose di valore senza cercale), deriv. da Serendip (to Serendib), nome dell'isola di Sri Lanka" (italics added).

---

[18] Simon Winchester, *The Professor and the Madman: A Tale of Murder, Insanity, and the Making of the Oxford English Dictionary* (New York: Harper/Collins Publishers, 1998). Initially published by Viking in England with the more sedate but less engaging title *The Surgeon of Crowthorne*.

In response to my inquiry, my newfound colleague-at-a-distance Simon Winchester confirms the guess that his mistaken attribution did not derive from the *Grande Dizionario*. He goes on to say that he cannot confidently reconstruct how he came to that fictive attribution but that he suspects it was simply a misreading of the *OED* entry on serendipity—an entry the we have ample reason to know rightly reports that Walpole formed the word "upon *the* [not *his*] fairy tale 'The Three Princes of Serendip.'" Nor can the editors of the *Grande Dizionario* reconstruct how they came to their same misattribution of what Walpole described as a "silly fairy tale." These two recent episodes lead to the distressing thought that perhaps—just perhaps—Winchester and the editors of the *Grande Dizionario* would have spared themselves and their readers this fictive attribution had I not decided to keep *The Travels and Adventures of Serendipity* from being published some forty-five years ago and had I not done so again, a dozen years ago.[19] For, as will be seen in Chapter 6 of the time-capsule text, we identify what may be the first mistaken ascription of the fairy tale to Walpole, in the edition of *Webster's New World Dictionary of the American* [*sic*] *Language* that appeared in 1954, precisely two centuries after Walpole's coinage: "serendipity (seren-dip-i-ty). n. coined by Horace Walpole (c. 1754) *after his tale* [italics added] *The Three Princes of Serendip* (i.e., Ceylon) who made such discoveries, an apparent aptitude for making fortunate discoveries accidentally."

Somewhat more likely and far more on point, had I not suppressed publication of our book back in 1958, much of the recent research on the role of serendipity in scientific discovery might have begun decades earlier. In that case, presumably we would now have a more advanced understanding of the process of serendipitous discovery in the evolution of science.[20] An intriguing, and rather distressing, counterfactual thought.

Diffusion of the Word Serendipity via Dictionaries

"Plagiarized" Definitions of Serendipity *in English Dictionaries*

As we have seen, dictionaries in general and, in anglophone cultures owing to its comprehensive coverage and historical orientation, the *OED* in particular are widely taken to legitimize newly emerging or long-suppressed words as part of the standard language. Increasingly, this is the case for the colloquial language as well. However, as lexicographers have

---

[19] It may be remembered from the preface that I was the chief culprit in suppressing prior publication.

[20] See the select bibliography in this volume.

long known, though seldom publicly emphasized, entries in their "new" dictionaries often do no more than repeat entries in prior dictionaries. Or, as it has been put by Robert Burchfield, editor of the four-volume *Supplement to the OED* (which was some twenty-five years in the making), the new dictionaries in effect often "plagiarize."[21] As a result, this practice often leads to the exclusive perpetuation of early meanings at the expense of registering semantic change.

That has evidently been the case with our word serendipity, as we see from the sampling of meanings in those thirty or so English-language dictionaries that make up Table 1. It will be noticed that the very first identified entry in the *New Century Dictionary* of 1909 and the entry in the *OED* a few years later both define serendipity in much the same way. For the one, it is primarily "the happy *faculty* or luck of finding by 'accidental sagacity' "; for the other, it is primarily "the *faculty* of making happy and unexpected discoveries by accident." (The italics are mine). In short, these archetypal entries both define the word primarily in terms of *personal attributes* of the discoverer. Only then do they go on to note that serendipity refers not only to a personal attribute but also to a particular kind of *event* or *phenomenon*: It is the "discovery of things unsought" or the experience of "looking for one thing and finding another." That secondary definition, as we have ample cause to know from my recurrent quotation of Walpole's own defining statement, echoes his insistence that his coinage referred specifically to a special kind of unanticipated event: "*no* discovery of a thing you *are* looking for comes under this description." (Here the emphasis is Walpole's, not mine.) But, as we also see from Table 1, that secondary definition of the word as also designating a special kind of unanticipated event or phenomenon soon drops from view. With only a rare exception throughout the century, the successive "new" English dictionaries follow precedent by confining their definitions to personal attributes. Serendipity is authoritatively and exclusively defined as the "faculty or ability or gift or habit or aptitude or talent" for making unexpected and felicitous discoveries. Only with the *OED Supplement* appearing in 1986 does a dictionary pause to draw attention to the further complexity of what is being designated by our word. It supplements the pioneering primary definition of the complex word, which originated in the 1912–1913 *OED*—"the *faculty* of making happy and unexpected discoveries by accident"—and its quotation from Walpole's coinage letter to Mann by expressly stating that it also refers to "the *fact or an instance* of such a discovery." That reconceptualization of the word remains unnoticed—that is, "unplagiarized"—by at least ten subsequent

---

[21] See the drastic observations on this practice by Robert Burchfield, "Dictionaries, New and Old: Who Plagiarises Whom, Why and When," *Encounter* 63 (1984):10–19.

TABLE 1

Meanings of *Serendipity* in English-Language Dictionaries, 1909–2000

Dictionary	Date	Serendipity as Personal Attribute	Serendipity as Event or Phenomenon
The Century Dictionary	1909	"the happy faculty or luck of finding by 'accidental sagacity'"	"discovery of things unsought"
The Oxford English Dictionary	1912–13	"the faculty of making happy and unexpected discoveries by accident"	[kind of discovery] 1754 let. to Mann: "this discovery, indeed, is almost of the kind which I call Serendipity" "looking for one thing and finding another"
Funk and Wagnall's New Standard Dictionary of the English Language	1913	"the ability of finding valuable things unexpectedly"	
Universal Dictionary of the English Language	1932	"the faculty of finding valuable or interesting things by chance or where one least expects them"	
Shorter Oxford English Dictionary	1933	"the faculty of making happy and unexpected discoveries by accident"	
Webster's New International Dictionary of the English Language (second edition, unabridged)	1934	"the gift of finding valuable or agreeable things not sought for"	

Source	Year	Definition
Dictionary of Word Origins	1945	"the happy faculty of finding what one did not seek"
Name into Word	1949	"the faculty or the habit of making felicitous discoveries by accident"
Concise Oxford English Dictionary	1951	"the faculty of making happy and unexpected discoveries by accident"
Concise Etymological Dictionary of the English Language	1952	"the gift for finding one thing when looking for another"
Swan's Anglo-American Dictionary	1952	"looking for one thing and finding another"
Webster's New World Dictionary of the American Language, College edition	1953	"an apparent aptitude for making fortunate discoveries accidentally"
Webster's Third New International Dictionary of the English Language Unabridged	1963	"an assumed gift for finding valuable or agreeable things not sought for"
Origins: A Short Etymological Dictionary of Modern English, fourth edition	1966	"the gift of felicitous, fortuitous discovery" "the sheer luck or accident of making a discovery by mere good fortune or when searching for something else"

TABLE 1
Continued

Dictionary	Date	Serendipity as Personal Attribute	Serendipity as Event or Phenomenon
Klein's Comprehensive Etymological Dictionary of the English Language	1971	"the gift of finding interesting things by chance"	
World Book Dictionary	1973	"the ability to find by accident"	
Morris Dictionary of Word and Phrase Origins	1977	"the faculty of making happy and unexpected discoveries by accident"	
The Oxford English Dictionary Supplement, vol. 4	1986	"the faculty of making happy and unexpected discoveries by accident"	"also, the fact or an instance of such a discovery"
The Australian Concise Oxford Dictionary	1987	"the faculty of making happy and unexpected discoveries by accident"	
The Barnhart Dictionary of Etymology	1988	"the ability to make fortunate discoveries by accident"	
Chambers English Dictionary	1990	"the faculty of making happy chance finds"	
BBC English Dictionary	1992	"the natural talent that some people have for finding interesting or valuable things by chance"	
The American Heritage Dictionary of the English Language, third edition	1992	"the faculty of making fortunate discoveries by accident"	

Random House Unabridged Dictionary, second edition, newly revised and updated	1993	"an aptitude for making desirable discoveries by accident"
Collins English Dictionary, third updated edition	1994	"the faculty of making fortunate discoveries by accident"
The Oxford Dictionary and Thesaurus, American edition	1996	"the faculty of making happy and unexpected discoveries by accident"
The Canadian Oxford Dictionary	1998	"the faculty of making happy and unexpected discoveries by accident"
Encarta World English Dictionary	1999	"a natural gift for making useful discoveries by accident"
Merriam-Webster's Collegiate Dictionary Online	1999–2000	"the faculty of finding valuable or agreeable things not sought for"
The American Heritage Dictionary of the English Language, fourth edition	2000	"the faculty of making fortunate discoveries by accident"; "the phenomenon of finding valuable or agreeable things not sought for"; "the fact or occurrence of such discoveries; an instance of making such a discovery"

dictionaries until we approach the new millennium, when *Merriam-Webster's Collegiate Dictionary Online* supplements its primary definition of the word as a "faculty" with a reference to "the *phenomenon* of finding valuable or agreeable things not sought for." And the millennium in this domain truly arrives in 2000 with the greatly expanded fourth edition of that fine desk dictionary, *The American Heritage Dictionary of the English Language*. For, as we see from the final entry in Table 1, it not only reiterates the familiar first definition—"The faculty of making fortunate discoveries by accident"—but virtually echoes the further definition of serendipity introduced some fifteen years before in the *OED Supplement* in two parts; thus: "2. The *fact or occurrence* of such discoveries" and, by way of further reconceptualization, "3. An instance of making such a discovery." This provides belated but basic recognition that "serendipity" is a concept undergoing the process of reconceptualization in the sociology and philosophy of scientific knowledge. It is no longer conceived only as a psychological attribute but also and consequentially as a more complex "*fact, instance, occurrence,* and *phenomenon*." (Because that reconceptualization will turn out to be essential to my later evolving argument about the conceptual contents of the word, the italics in those extended definitions are transparently mine.)

So much for the virtually stereotyped—that is to say, the oversimplified—meaning of serendipity as solely a personal attribute in English dictionaries of the 20th century and its most recent extension as a more complex phenomenon. But diffusion of the word did not stop at the borders of anglophone societies, as we see from the sampling of bilingual dictionaries in Table 2.

Serendipity as Niche-Word and Loanword: The Transplanting of a Foreign Idiom

We note that serendipity has spread into foreign languages in three forms: as a foreign loanword in its original English form ("serendipity"); as a slightly nativized word (thus, the Italian *serendipità*, the Spanish *serendipia*, the Welsh *serendipedd*); and as a transliteration in which the word is expressed in the corresponding characters of another alphabet (thus, the proposed Russian СЕРЕНДПНОСТ, and the approximate definitional expressions in Arabic, Chinese, Hebrew, Japanese, Persian, Tamil, and Urdu). When the word did diffuse into other languages, its definitions in bilingual dictionaries have understandably derived in large part from the antecedent English dictionaries. It is hardly surprising therefore, as we see from Table 2, that they too typically focus on personal attributes. Thus, we have the Dutch *de gave*, the French *don*, the German *die Gabe*, the Icelandic *sú gáfa*, the Italian *la capacità*, the Portuguese *faculdad*, the Serbo-Croatian *sposobnost*, the Spanish *don*, and the Turkish *kabiliyeti* corresponding to the English *gift*, *capacity*, *faculty*, and so forth.

For the moment, it is enough to call renewed attention to this virtually stereotyping practice in both English and other-language dictionaries that confines the meaning of serendipity to a personal attribute. And to note that this is no trivial matter. For, as we shall see in due course, this practice has significant conceptual and therefore theoretical consequences.[22] But now that we have somewhat sampled the virtually worldwide diffusion of the word, we must first take note that the word serendipity is, of course, far from unusual in such intercultural diffusion. It is plainly a member of the class of loanwords,[23] words that diffuse from their originating language to become variously incorporated in other languages. By way of example drawn out of the vast store of unnativized loanwords in the English language[24] without my delving into any of the dictionaries that crowd my nearby bookshelves, I am instantly put in mind of three such loanwords (no doubt because I find them especially congenial): the French *esprit de l'escalier*, the German *Schadenfreude*, and the Yiddish *chutzpah*.

Taken literally, *esprit de l'escalier* means the enigmatic "spirit of the staircase." But understood idiomatically, it refers vividly and *uniquely* to a concrete and therefore highly specific yet widely familiar kind of experience: a witty remark—in colloquial and quite inadequate English, a "wisecrack"—that comes to mind, alas, too late, only after the opportunity to deploy it has vanished. That is to say, if one is a belatedly brilliant host, one thinks of the sparkling riposte only as one is ascending the stairs to one's bedroom or, if one is a belatedly brilliant guest, one thinks of it only as one is descending the front steps from the doorway. *Esprit de l'escalier* has become a loanword in English and many another receptive

---

[22] Indeed, should "sociological semantics" ever truly emerge as a distinct subspecialty of sociolinguistics, its problematics would call for investigation of the interplay between word and thought in different societies and in different social strata and groups within the same society.

[23] In self-exemplifying style, the technical term *loanword* is itself manifestly a loanword, appearing in German as *Lehnwort* and, rather more instructively, in Italian, as *prestito linguistico*.

[24] See Mary S. Serjeantson, who opens her *History of Foreign Words in English* (New York: Barnes and Noble, 1962) with the observation that "the English language has throughout its history accepted with comparative equanimity words from other languages." This is in great contrast, say, to the struggles over "franglais" in contemporary France. I resist the temptation to indulge in a speculative digression on such questions as national differences in receptivity to alien terms that also belong in the problematics of a possibly emerging sociological semantics.

TABLE 2

Translations and Transliterations of "Serendipity" in Bilingual Dictionaries

Language	Dictionary	Date	Translated or Assimilated Term and/or Definition
Arabic	The Oxford English-Arabic Dictionary of Current Usage (Oxford: Clarendon Press)	1972	موهبة اكتشاف الاشياء النفيسة مصادفة
Chinese	Chinese-English English-Chinese Dictionary (via Dr. Dainian Fan)	2000	偶然 發現 珍貴 的 運氣 the luck to find treasure fortuitously
Czech	Anglicko- esk -Anglick Slovn k (Praha: Státni pedagogické nakladatelstv)	1994	vrozené štěstí
Danish	Engelsk-dansk Ordbog (Gyldendal)	1981	finderheld; lykketræf, (fund ved et) slumpetræf
Dutch	Cassell's English-Dutch Dutch-English Dictionary (London: Cassell Ltd.)	1978	de gave onverwachts iets goeds te ontdekken
French	The Oxford Hachette French Dictionary (Oxford University Press)	1994	don de faire des trouvailles
German	Langenscheidt's New Muret-Sanders Encyclopedic Dictionary of the English and German Languages (Berlin: Langenscheidt)	1982	die Gabe, durch Zufall glückliche u. unerwartete Entdeckungen zu machen
German	The Oxford-Duden German Dictionary (Oxford: Clarendon Press)	1994	glücklicher Zufall

Language	Dictionary	Year	סרנדיפיטי (sĕrendīpītī)
Hebrew	*The Oxford English-Hebrew Dictionary*	1996	a propensity to discover surprising things or successful solutions by chance
Indonesian	*An English-Indonesian Dictionary* (Ithaca: Cornell University Press)	1975	kesanggupan utk menemukan; keterangan dgn tak disengaja waktu mencari; yg lain
Islandic	*Ensk-íslensk Ordabok* (Örn Og Örlygur)	1984	fundsæld, fundheppni sú gáfa að uppgötva eftirsóknarverð hluti af tilviljun, slembilukka
Italian	*Dizionario Inglese-Italiano Italiano-Inglese.* Torino: Società Editrice Internazionale.	1977	la capacità di fare felici scoperte, di trovare tesori per caso
Italian	*Grande Dizionario della Lingua Italiana.* (Unione Tipografico–Editrice Torinese)	1996	serendipità Capacità di cogliere e interpretare correttamente un fatto rilevante che si presenti in mode inatteso e casuale nel corso di un indagine scientifica diversamente orientata
Japanese	*Kenkyusha's New Japanese English Dictionary*	1954	掘り出し (horidashi) dig out, pick up, find (a treasure, lucky find); to fling out じょうず (jozu) skilled, adept, expert

TABLE 2
Continued

Language	Dictionary	Date	Translated or Assimilated Term and/or Definition
Persian	*The New Unabridged English-Persian Dictionary* (Amir-Kabir Publishing and Printing Institution)	1990	(moteraghghebeh) نعجیل نعمت غیر متر ثنبه نمت غیر متر ثنبه
Polish	*The Kościuszko Foundation Dictionary: English-Polish* (London: Mouton and Co.)	1964	zdolność czynienia przypadkowo szczęśliwych a niespodziewanych odkryć
Portuguese	*Dicionário Inglês-Português* (Atualizado)	1982	faculdade de fazer, acidentalmente, descobertas felizes e inesperadas
Russian	Michael S. Macrakis, "Scattering from Fluctuating Plasmas: A Historical Perspective." (Invited lecture at the 1987 International Conference on Plasma Physics, Kiev, USSR)	1987	СЕРЕНДIПНОСТ
Russian	*The Oxford Russian Dictionary* (Oxford: Oxford University Press)	1997	**счастлйвая способность дéлать неожйданные открытия** capacity to make surprising and happy discoveries
Serbocroatian	*An English-Serbocroatian Dictionary II* (Belgrade: Prosveta)	1956	sposobnost srećnih otkrivanja slučajem
Serbocroatian	*An English-Serbocroatian Dictionary*, third edition (Cambridge University Press)	1990	sposobnost sretnog slučajnog otkrivanja

Spanish	*Collins Spanish-English English-Spanish Dictionary*, third edition (HarperCollins Publishers)	1992	serependismo
Spanish	*The Oxford Spanish Dictionary* (Oxford University Press)	1994	serendipia Don de descubrir cosas sin proponérselo
Swahili	*Tuki English-Swahili Dictionary* (Institute of Kiswahili Research, University of Dar es Salaam)	1996	kubahatisha ugunduzi
Tamil	*English-Tamil Dictionary* (University of Madras)	1965	யதேச்சையாக⊙, எதிர்பாராமல்
Turkish	*The Oxford English-Turkish Dictionary*, second edition (Oxford University Press)	1978	Iginç/değerli şeyleri rasgele bulma kabiliyeti
Urdu	*Ferozsons English Urdu Dictionary.* Ferozsons Ltd.	1961	اتفاق طور پر - غیر شریف الکتابت - کرنیق کابت -
Welsh	*The Welsh Academy English-Welsh Dictionary.* Geiriadur yr Academi.	n.d.	serendipedd

language that had no prior phrase or word to register this quite specific yet widely familiar kind of experience.

Translated literally and therefore also inadequately, *Schadenfreude* means "malicious joy."[25] But, in its idiomatic fullness, this one compounded word *uniquely* describes a far more complex and often unacknowledged human experience, to wit: the experience of taking pleasure or at least some kind of satisfaction in the observed troubles, misadventures, or mishaps of another. However, at its borders, *Schadenfreude* need not be vicious or sadistic; its several-faceted meanings can also take wide-ranging innocent form. At one extreme, there is a hint of *Schadenfreude* in the innocent pleasure a child takes in the pratfall of a clown; at another extreme, in the case of a soldier who, seeing his buddy killed in action, is filled with guilt-ridden gratitude that this ultimate bad luck has not been visited on himself.[26]

So, too, the transplanted *chutzpah* translates easily into "brazen impudence, gall, or effrontery." But again, these would-be synonyms do not convey the depth and range of its quite distinctive—indeed, its *unique*—meaning. As that witty master of the Yiddish lexicon Leo Rosten recognized, the fullest understanding of *chutzpah* is better achieved by a classic example than by explicit definition: "*Chutzpah* is that quality enshrined in a man who, having killed his mother and father, throws himself on the mercy of the court because he is an orphan."[27]

All these are loanwords firmly imbedded in English and many another language. They have become qualified as loanwords through their common functional properties as niche-words. They express, with unique precision, a familiar kind of human experience that transcends national and other cultural and linguistic boundaries. And at times they do so wittily or otherwise memorably. In short, these loanwords can be described as "niche-words": They uniquely fill semantic niches in both the originating and adoptive languages (which have no exact equivalent of their own) to identify a distinct but recurrent experience, observation, idea, or object.[28]

---

[25] In a exceedingly truncated case such as this, it is truly *tradutorre, traditore* (to put it in an amiable loanphrase). But see Scott L. Montgomery, *Science in Translation: Movements of Knowledge through Cultures and Time* (Chicago: University of Chicago Press, 2000), for an instructive account of the "crucial role" played by translation "in the history of scientific knowledge."

[26] On second thought, this last example of "guilt-ridden gratitude" over the ultimate misfortune of a comrade falls beyond the outermost limits of *Schadenfreude*. It contains no trace of malice or other perverse satisfaction. Thinking of the fine agglutinative quality of the German language that brings us such a fine niche-word as *Schadenfreude*, I'm inclined to turn to it for another suitable compound. My colleague-at-a-distance Alfred Nordmann suggests *Glücksschuldbewusstsein* or, if this considerable polysyllabic won't travel, perhaps *Glückreue*.

[27] Leo Rosten, *The Joys of Yiddish* (New York: McGraw-Hill, 1968), p. 93.

[28] Which brings us to the implied category of *potential* loanwords: unique niche-words

And so it is, and from the start has been, with the word serendipity. It, too, is a niche-word that first began to fill a semantic niche in English as it emerged from its limbo in Walpole's private correspondence. Once the singular word was detected some time after the publication of that vast correspondence, it came slowly into use as it filled a semantic niche first and, for a time, exclusively among those devoted to search and research—the book collectors, antiquarians, and, a good bit later, the scientists. Only then did it diffuse into the vernacular, also at first slowly, but recently quite riotously. As we shall see in some detail, serendipity has thus been subject to a paradoxical semantic process in the course of this accelerated diffusion. It has both acquired new differentiated meanings as it was put to work by those specialists in discovery, thus transforming it into an ever more exact, unique, and implicative concept in those special quarters while losing much of its earlier specific and unique meaning in the course of becoming a vogue word in the vernacular.

Serendipity as Individual Disposition: A Psychological Conceptualization

What, then, have we learned from the overview of the meanings ascribed to our word in dictionaries? Chiefly this: that the archetypal English and the derivative bilingual dictionaries alike largely retain a prime and almost exclusive definition of serendipity as a specific individual disposition. It is said to be a faculty, capacity, gift, or talent for making felicitous discoveries by chance. By thus conceptualizing serendipity as a personal disposition of the discoverer, such definitions tacitly claim that psychology can exhaustively and exclusively account for the complex phenomenon of serendipitous discovery. In short, these recurrent definitions lead to a reduc-

---

in one language that have yet to be adopted in other languages (for reasons which, in the absence of systematic inquiry, can only be conjectured). I think of *statunitense* as one such potential loanword. After all, there is a widespread ambivalence and hostility toward the United States of America that perceives it as an imperialistic nation engaged in establishing cultural, economic, and political hegemony over other nations. Yet, as we know, linguistic equivalents to the all-embracing and tacitly imperial terms *American* and *Americanization* are practically universal. This, even though such usage in effect symbolically expropriates the entire continent of South America and the Canadian northern half of North America. And, there, quietly waiting offstage as a potential loanword in place of *americano* and its other nativized derivatives (such as *nordamericano*) is the precise and connotationally neutral term *statunitense*, which could be easily nativized in the United States itself as both adjective and noun, thus: *statunitensean* (just as, for much the same reasons, one would deploy *etatsuienne* for Francophone audiences and *estadounidense* for Spanish-speaking audiences). Borges was evidently chafing at the bit when he referred to "Americans of the North" and "we Americans of the South," thus virtually implying the need for such potential loanwords to find their way into English idiom. See Jorge Luis Borges, *Selected Non-Fictions* (Harmondsworth: Penguin Books, 1999), pp. 309, 412.

tive reconceptualization. They lead by strong implication *only* to psychological questions that are then boiled down to this one: "What are the special psychological attributes of the *individuals* who make serendipitous discoveries?"

And, as though to push this psychological reductionism to its limit, two dictionary entries in the 1990s go on to reconceptualize serendipity as "the *natural talent* that some people have for finding interesting or valuable things by chance" and as "a *natural gift* for making useful discoveries by accident." They thus achieve a further biological reductionism in which serendipity is *nothing but* a "natural" capacity. In doing so, of course, dictionaries both reflect and perpetuate this extreme reductionism.[29] They reflect the focus of cumulating research on "creativity" (which includes research on scientific discovery in general and by inclusion serendipitous discovery as well). Thus, in a recent analysis of "the most comprehensive research bibliography [of creativity] yet published,[30] encompassing nearly 10,000 scientific and scholarly writings, only 10 entries are to be found under the heading *sociology*, only 50 under *group processes*, 50 under *environmental factors*, and only 59 under the heading *cultural factors*. The dispositional approach, by contrast, makes a dramatically stronger showing: . . . 222 [entries] were under *intelligence*, 243 under *personality*, 367 under *creative thinking* and *teaching creative thinking*, 624 under *genius* and *talent*."[31]

So, too, this overview notes that the "field's skewed emphasis on creators' dispositions to the neglect of situations" was also found in studies of recent dissertations on creativity. One such study found that "there were almost no dissertations on the traits or products of creative cultures, and on the traits or products of creative groups. The almost total absence of studies of creative groups is particularly strange, especially when one considers the theoretical importance of creative groups like the Cavendish laboratory in physics or the impressionist coterie in art." It also noted that "in psychology, the field in which efforts to understand creativity have been most highly concentrated, 87 percent of the dissertations were focused on the individual level; not a single one focused on

---

[29] The onetime physicist and longtime philosopher of science Mario Bunge has nicely elucidated "the power and limits of reduction" in an article by that title in *The Problem of Reductionism in Science*, ed. Evandro Agazzi (Dordrecht and Boston: Kluver, 1991), pp. 31–49, and in several of his recent books, such as *The Sociology-Philosophy Connection* (New Brunswick, N.J.: Transaction Publishers, 1999) and *Philosophy in Crisis* (Amherst, N.Y.: Prometheus Books, 2001).

[30] A. Rothenberg and B. Greenberg, *The Index of Scientific Writings on Creativity: 1566–1974* (Hamden, Conn.: Archon Books, 1976).

[31] Joseph Kasof, "Explaining Creativity: The Attributional Perspective," *Creativity Research Journal* 8 (1995):211–366, at 311–312.

groups or culture."[32] And the overview concludes, "Despite the massive effort devoted to it by talented scientists, the quest for dispositional causes of creativity [including serendipitous discoveries in science and technology] has been at something of an intellectual standstill for about 3 decades."[33]

That abiding and exclusive question about an imputed disposition for serendipitous discoveries is at times "answered" by a barely concealed reiteration. Thus, the chemist Royston Roberts introduces his book on "accidental discoveries in science" by promptly quoting "dictionary definitions"—" 'the *gift* of finding valuable or agreeable things not sought for' or 'the *faculty* of making fortunate and unexpected discoveries by accident.' " And he too moves from this psychological reductionism to a virtual biological reductionism by declaring in an epilogue, "I feel sure that there is an inborn ability or talent for discovery in many of those who benefit from serendipity, persons like Priestley, Pasteur, and Perkin."[34] Such would-be explanations of serendipitous discoveries *exclusively* in terms of imputed "inborn talent or ability" inevitably brings back to mind Molière's sardonic explanation that "opium causes sleep because of its dormitive power."[35]

We also recall Pasteur's reverberating maxim that "chance favors the prepared mind." But apt and memorable though it may be, the psychological black box of the "prepared mind" cannot itself explain the complexities of serendipitous discovery. For that black box focuses on the question of how it is that certain scientists rather than others take consequential note of anomalies (or "accidents") to arrive at unanticipated discoveries; it does not even begin to address the question of the relative frequency with which such consequential anomalies (or accidents) turn up in the first place. At the least, the psychological perspective needs to be integrated with a sociological perspective. For if chance favors prepared minds, it particularly favors those at work in microenvironments that make for unanticipated sociocognitive interactions between those

---

[32] Ibid., drawing upon L. Wehner, M. Csikszentmihalyi, and I. Maguari Beck, "Current Approaches Used in Studying Creativity," *Creativity Research Journal* 4 (1991):261–271.

[33] Kasof, "Explaining Creativity," pp. 312–313.

[34] Royston M. Roberts, *Serendipity: Accidental Discoveries in Science* (New York: John Wiley and Sons, 1989), pp. ix, 244.

[35] As will be recalled, Molière expresses this emptiest of tautologies in Latin (in his *La malade imaginaire*): *Opium facit dormire . . . quia est in eo virtus dormitiva*, thus rendering the tautology lexically vivid by juxtaposing *dormire* and *dormitiva*. That classic tautology is more instantly apparent than those explanations that account for repeated serendipitous discoveries by individuals as resulting from their "faculty" or "aptitude" or, most gratuitously, from their "inborn talent" for making fortunate discoveries by chance.

prepared minds. These may be described as serendipitous sociocognitive microenvironments.

Serendipitous Sociocognitive Microenvironments: A Sociological Reconceptualization

So much, then, for the subtext of those standardized entries on serendipity in dictionaries of every stripe that reflect a reductionist psychological perspective found in much research on scientific discovery. But, as I have noted rather emphatically, the 1986 *OED* did take the lead, followed some years later by a few successor dictionaries, in recording a significantly enlarged definition of serendipity. There, the word is no longer conceived solely in psychological terms as referring to a "faculty, ability, gift, habit, aptitude, or talent" for making "happy and unexpected discoveries by accident"; it is also taken to be "the fact or an instance" of such discoveries. This most recent *OED* entry thus finally does register by implication the slow but definite reconceptualization of the phenomenon of serendipity in science as a distinctive and complex cognitive process, an evolving concept now belatedly finding its place in the sociology and philosophy of science. Still, I confess to a bit of egocentric disappointment that the recent *OED* quotations that are taken to exemplify usages of the word as a complex phenomenon rather than only a psychological attribute of the serendipitist[36] do not include its first analytical usage in that sense which, it will be remembered from my frequent allusion to it, was suggested back in 1948: "The serendipity pattern refers to the fairly common experience of observing an *unanticipated, anomalous, and strategic* datum which becomes the occasion for developing a new or extending an existing theory."[37]

---

[36] Because the *Oxford English Dictionary Online* did not exist back in the 1950s, Elinor Barber and I failed to identify the first usage of *serendipitist* in James Joyce's *Finnegan's Wake* (London: Faber and Faber Limited, [May 4] 1939), p. 191. Nor had I taken note of that then-incomprehensible word in what ultimately emerged as *Finnegan's Wake* when it appeared in fragmentary portions as a "Work in Progress" in such avant-garde reviews as *Criterion* and *Transatlantic Review* (which turned up regularly in an avant-garde bookstore—its name now unhappily forgotten—off Harvard Square). It went unnoted for the best of reasons. I had not yet stumbled serendipitously upon the word *serendipity* in my 1933 edition of the *OED* and so had no reason to pause at that one incomprehensible word *serendipitist* afloat in a sea of other incomprehensible words. After all, it would have taken a lifetime to decipher every exotic Joycean neologism in that richladen vessel of portmanteau neologisms.

[37] As noted in Chapter 7 of the 1958 text and reiterated in this Afterword, the proposed analytical concept of "the serendipity pattern" first appeared in the *American Sociological Review* in 1948 and was then often reprinted in successive editions of *Social Theory and*

Although this early notion of "the serendipity pattern" fairly begs for further sociological and epistemological elucidation (which, as we shall see, it has since received), it does draw attention through its explicit reference to the "anomalous," to the workings of "accident" as well as the workings of "sagacity" in serendipitous scientific discoveries. In doing so, it argues, somewhat anachronistically, that Walpole's intuitive coinage of serendipity as "accidental sagacity" had in effect, if unwittingly, implied the sociological importance of the specific and distinctive component of "accident" along with the psychological individual attribute of the generic component of "sagacity." (*Sagacity* was then generally taken to mean "acuteness of discovery," if we are to judge from Samuel Johnson's monumental *Dictionary of the English Language* published just a year after Walpole's coinage in 1754.)[38] There is no question, of course, of great individual variability in the trained capacity for making scientific discoveries. Still, when translated into later contexts, "sagacity" is a (still ill-defined) psychological attribute that is involved in *any* sort of discovery, and not *specifically* of the serendipitous sort. It is the other component, "accident," the unanticipated circumstance that provides the opportunity for putting that sagacity to work, that gives "serendipity" its altogether distinctive meaning. And that brings us to the sociological perspective. The sociological analysis of serendipity presents us with a special case of the generic sociological problem of how differing social environments increase or decrease the probability of particular behavioral outcomes. (That is known in the sociological trade as the "agent-social structure problem.") More specifically here, it brings us to the sociological question of how certain types of sociocognitive microenvironments make for higher frequency of producing the "accidental component" in serendipitous scientific discoveries.

I shall address that question by examining serendipitous moments leading to key ideas that shaped Thomas S. Kuhn's famous and consequential work *The Structure of Scientific Revolutions.*

Social Structure. I freely confess to guilty recognition that the *OED Supplement* would probably have included that concept some forty years after its first appearance had I not held off on publishing our book on *The Travels and Adventures of Serendipity* until now.

[38] To exemplify that meaning, the sagacious Samuel Johnson wisely quotes the seventeenth-century philosopher John Locke, thus: "*Sagacity* finds out the intermediate ideas, to discover what connection there is in each link of the chain, whereby the extremes are held together." That usage, it might be said, has the term virtually assuming a psychosocial cognitive mechanism (along lines that must gladden today's adherents of "theories of the middle range" in sociology). As evidentiary context for this last, seemingly extravagant, statement, see Peter Hedström and Richard Swedberg, eds., *Social Mechanisms: An Analytical Approach to Social Theory* (New York: Cambridge University Press, 1998), passim.

Kuhn's Serendipitous Microenvironments[39]

From Thomas Kuhn's fragmentary intellectual memoirs and records of our longtime friendship, I can readily identify some of his serendipitous moments in two sociocognitive microenvironments: first as a junior fellow in the Harvard Society of Fellows from 1948 to 1951, and then as a fellow in the Center for Advanced Study in the Behavioral Sciences (Palo Alto) in 1958–1959. Although largely unnoticed in the vast library of commentary on the Kuhn oeuvre, those unanticipated encounters and observations proved to be central to basic ideas that eventually formed the framework of Kuhn's magnum opus.

The Harvard Society of Fellows and the Acquisition of a Sociological Perspective

Founded in 1933 by A. L. Lowell, the exiting president of Harvard University, the Harvard Society of Fellows was a distinctly elite institution designed to bypass the narrow specialization that typically went with graduate study. Its freedom from set rules and requirements provided instead for self-defined programs of study and research by the fellows that would broaden their interests and competence beyond the traditional confines of their disciplines of origin or would facilitate a transition from one field of science and learning to another. The young men selected as junior fellows—not at all surprisingly for the time and the institution, no serious thought was given to young women—were drawn from a national pool of nominations by select members of university faculties throughout the United States, with the lion's share coming from graduates judged to be the pick of the crop from Harvard itself.[40]

By 1948, the year in which Kuhn began his three years as a junior fellow (after a strong undergraduate degree in physics), the Society had developed a subculture that drew widely on the university's ample intellectual resources and had a thoroughly established pattern. Autobiographical snippets in Kuhn's writings provide clues to some of the immediate and delayed cognitive outcomes of working in that environment. To his mind, the institutionalized freedom provided by the Society

---

[39] This section draws substantially—at times verbatim—upon my now out of print *Sociology of Science: An Episodic Memoir*, pp. 81–109, which deals with "institutionalized serendipity" (i.e., "serendipity-prone microenvironments").

[40] George C. Homans and O. T. Bailey, *The Society of Fellows* (Cambridge, Mass.: Harvard University, 1948).

greatly facilitated the "transition to a new field of study . . . that might [otherwise] not have been achieved."[41]

The Society of Fellows thus provided a sociocognitive environment that encouraged the sort of wide-ranging exploration of developing interests and interaction with fellow scientists and scholars drawn from other disciplines (and subdisciplines) that make for serendipity in the domain of learning. Since one type of serendipity in science is the chance discovery of phenomena or ideas that were not expressly being sought, it often involves turning aside from prior cognitive interests to new ones.[42] In anticipatory fashion, Kuhn elected to turn from work on his doctoral dissertation in physics to the history of science. In that cognitively variegated environment, he ranged even further, spending much of his time "exploring fields without apparent relation to history of science but in which research now discloses problems like the ones history was bringing to my attention."[43]

From such brief autobiographical reports, we can piece together a few of the serendipitous episodes that became reflected in the development of Kuhn's thought, these episodes being facilitated by both the social microenvironment and the cognitive culture that the Society of Fellows provided in those days. One especially consequential episode had to do with Kuhn's coming upon the book that he describes, in his later opinion, perhaps overgenerously, as "an essay that anticipates many of my own ideas" and goes on to observe that only through "the sort of random exploration that the Society of Fellows permits . . . could I have encountered Ludwik Fleck's almost unknown monograph, *Entstehung und Entwicklung einer wissenschaftlichen Tatsache* (Basel, 1935)." Later, in the foreword to the English translation of Fleck's monograph (which, as an editor-translator, I had persuaded Tom Kuhn to write) he reports that his happening upon that significant monograph was one of a good many serendipitous encounters during his time in the Society of Fellows: "On that non-topic [of scientific revolutions] there was no established bibliography, and my reading was therefore exploratory, often owing much to serendipity. A footnote in R. K. Merton's *Science, Technology, and Society* led me to the work of the developmental psychologist Jean Piaget. Though Merton's book was an obvious desideratum for a prospective historian of science, Piaget's work surely was not. Even more

[41] Thomas S. Kuhn, *The Structure of Scientific Revolutions* (Chicago: University of Chicago Press, [1962] 1970), p. v.

[42] The other main type of serendipity in science involves the chance observation of an evocative phenomenon or emergence of a new idea that leads to an unexpected solution of the targeted problem.

[43] Kuhn, *Structure of Scientific Revolutions*, p. vi.

unlikely was the footnote that led me to Fleck. I found it in Hans Reich-enbach's *Experience and Prediction*."[44]

That chance encounter with Fleck was of no small moment. Rein-forced by exchanges with a sociologist in the vari-textured microenviron-ment of the Society of Fellows, it sensitized Kuhn to sociological aspects of his developing ideas. As he says: "Together with a remark from an-other Junior Fellow, Francis X. Sutton, Fleck's work made me realize that those ideas might require to be set in the sociology of the scientific com-munity."[45] Ideas reaped from serendipitous scholarship in print and from social interaction were mutually reinforcing, an episode reminiscent of the finding, long established in the sociological study of communication, that "the joining of mass media and direct personal contact" contributes to the "effective transmission of ideas."[46]

Nor were these the only consequential moments in this serendipitous microenvironment that led Kuhn to adopt a sociological orientation in his evolving work as an historian and philosopher of science. As he symp-tomatically concludes his paragraph on the Society of Fellows in the pref-ace to *The Structure of Scientific Revolutions*: "readers will find few refer-ences to these works [such as Fleck's and Piaget's] or conversations [such as those with the sociologist Sutton and the philosopher Quine] [al-though] I am indebted to them in more ways than I can now reconstruct or evaluate." (I shall have much more to say about such underreporting of serendipitous moments in research that results from the standard for-mat and expurgated content of the scientific article.)

The Center for Advanced Study in the Behavioral Sciences and the Emergence of the Kuhnian Concept of "Paradigm"

In the academic year, 1958–1959, Kuhn found his way into another mi-croenvironment, the newly established Center for Advanced Study in the Behavioral Sciences (in Palo Alto, California).[47] And this too was con-

[44] Thomas S. Kuhn, "Foreword," in Ludwik Fleck, *Genesis and Development of a Scien-tific Fact* (Chicago: University of Chicago Press, [1935] 1979), pp. viii–ix. Kuhn remarks that he "found Fleck's German extraordinarily difficult," as did Thaddeus Trenn and I as translator-editors, so much so that I threatened to withdraw when my amiable coeditor did not accept some of my revisions of the drafted translation.

[45] Kuhn, *Structure of Scientific Revolutions*, p. vii.

[46] Paul F. Lazarsfeld and Robert K. Merton, "Mass Communication, Popular Taste, and Organized Social Action," in *Communication of Ideas*, ed. Lyman Bryson (New York: Har-per and Row, 1948), pp. 95–118, at pp. 116–118.

[47] Unfortunately, a full-fledged history of the Center, detailing Paul F. Lazarsfeld's cen-tral role and my own collateral role in its origin, has yet to be written, but fortunately the historian of science Arnold Thackray has provided a preliminary sketch that documents this in part. Arnold Thackray, "CASBS: Notes toward a History," Center for Advanced Study in

ceived from the start to provide for what can be fairly described as "institutionalized serendipity." For, as we have seen, there is more to serendipity in science than the talents of individual scientists. It was thought possible to provide a microenvironment that would provide opportunity for sustained sociocognitive interaction between talents in different social science disciplines and subdisciplines that would prove to be symbiotic as talented individuals found themselves adopting new paradigmatic perspectives. This was somewhat in accord with the observation by the historian of ideas Herbert Butterfield: "Of all forms of mental activity the most difficult to induce . . . is the art of handling the same bundle of data as before, but placing them in a new system of relations with one another by giving them a different framework, all of which virtually means putting on a different thinking-cap."[48]

It was at the Center that Kuhn finally came upon an anomalous sociocognitive phenomenon that evoked a conceptual solution of the long-baffling problem that had kept him from bringing his masterwork to a conclusion. It is indeed a case in which a long "prepared mind" took imaginative and productive note of an anomaly. For that "breakthrough"[49] occurred eleven years after Kuhn had "stumbl[ed] upon the concept of a scientific revolution,"[50] seven years after he had first published some of his evolving ideas on scientific change, five years after he had received an invitation from the philosopher Rudolf Carnap and the physicist Philipp Frank to contribute a volume to the *Encyclopedia of Unified Science* (which he elected to write on "The Structure of Scientific Revolutions"), and two years after he had cleared the decks by writing the propaedeutic *Copernican Revolution*. The breakthrough that provided the solution to the problem was, of course, his distinctive concepts, first of "paradigm," and then, more dynamically, of "paradigm shifts."

In self-exemplifying style, these concepts were evoked by his sociologi-

---

the Behavioral Sciences *Annual Report* (August 1984), pp. 59–71; also "A Site for CASBS: East or West?" *Annual Report* (August 1987), pp. 63–71. Another abbreviated account will be found in my piece for the Sorbonne Symposium on Paul F. Lazarsfeld, "Working with Lazarsfeld: Notes and Contexts," in *Paul Lazarsfeld (1901–1976): La sociologie de Vienne à New York*, ed. Jacques Lautman and Bernard-Pierre Lécuyer (Paris: Editions L'Harmattan, 1998), pp. 186–188.

[48] Herbert Butterfield, *The Origins of Modern Science, 1300–1800* (London: G. Bell and Sons, 1949), p. 1.

[49] I rarely use the term *breakthrough* without it bringing to mind the jesting observation by the polymathic economist Kenneth Boulding that was intended to capture the intellectual excitement of the Center in the first year of its existence. Fellows would laughingly greet one another not with a "Good morning!" or a "Hi!" or even a "Buon giorno!" but with a simple query: "Any breakthroughs today?"

[50] Thomas S. Kuhn, *The Essential Tension: Selected Studies in Scientific Tradition and Change* (Chicago: University of Chicago Press, 1977), p. xvi.

cal and serendipitous observation of an anomaly, this one in the cognitive behavior of different kinds of scientists; that is, his observation of an unexpected departure from a previously taken-for-granted pattern of such behavior. As Kuhn explains in retrospect:

> Spending the year in a community composed predominantly of social scientists confronted me with [N.B.] unanticipated problems about the differences between such communities and those of the natural scientists among whom I had been trained. Particularly, I was struck by the number and extent of the overt disagreements between social scientists about the nature of legitimate scientific problems and methods. Both history and acquaintance made me doubt that practitioners of the natural sciences possess firmer or more permanent answers to such questions than their colleagues in social science. Yet, somehow, the practice of astronomy, physics, chemistry, or biology normally fails to evoke the controversies over fundamentals that today often seem endemic among, say, psychologists or sociologists. Attempting to discover the source of that difference led me to recognize the role in scientific research of what I have since called 'paradigms.' These I take to be universally recognized scientific achievements that for a time provide model problems and solutions to a community of practitioners. Once that piece of my puzzle fell into place, a draft of this essay emerged rapidly.[51]

Here, then, we see Kuhn's major book providing a fine specimen of "the serendipity pattern" at work in a serendipitous microenvironment, and in self-exemplifying fashion. The observation of an unanticipated, anomalous, and strategic datum becomes the occasion for identifying a new intermediate problem, which, once explored, leads to the solution of a recalcitrant basic problem. The sociologically reoriented capacity of the individual scientist synergized with sociocognitive interactions in the serendipitous microenvironment to produce a consequential discovery.

All this suggests, in short, that just as individual scientists differ in their trained capabilities for original discovery, both direct and serendipitous, so do sociocognitive microenvironments differ in the extent to which they make for discoveries, both direct and serendipitous, by the interacting scientists at work in those microenvironments.

Excursus: A Digressive Note on the Term *Paradigm*

Upon reexamining my quotation of Kuhn's own informative account of how he came serendipitously to his distinctive concept of "what I have

---

[51] Kuhn, *Structure of Scientific Revolutions*, pp. vii–viii. This self-exemplifying source of the crucial new concepts is elucidated further in a paper appearing in the same year as *Structure of Scientific Revolutions*: Thomas S. Kuhn, "The Historical Structure of Scientific Discovery," *Science* 136 (1962):760–764.

since called 'paradigms,'" I can no longer evade acknowledging that it is for me a still inexplicable fact that when he introduced that crucial concept into the final draft of his manuscript, I did not alert him to my earlier usages of that term and concept, although we had been in close touch throughout the final stages of his manuscript and he had asked me to intercede with the University of Chicago Press if they proved reluctant to publish his book independent of the *Encyclopedia of Unified Science* for which it was originally intended.[52] Nor can I explain the further fact that, in all the many years since, Tom Kuhn and I also failed to discuss what was at the least a terminological coincidence and considerably more, I believe, in its theoretical implications. That might have resulted in some interesting exchanges.

When I proposed the use of *paradigms* in sociology back in the 1940s, I soon found that many colleagues—both those nearby and those at a distance—took this to be an unusual, not to say, bizarre usage. One candid friend proceeded to inform me that, as every schoolboy and schoolgirl knew, the term *paradigm* was obviously appropriate only as an exemplar of the inflectional forms of a word taken in a declension or conjugation. In rebuking me, he of course managed to forget Plato's idea of paradigmata as well as the centuries-old usage of the English word in the extended (nongrammar) sense of exemplar or pattern. And so it was that I adopted the term *paradigm* to refer to exemplars of codified basic and often tacit assumptions, problem sets, key concepts, logic of procedure, and selectively accumulated knowledge that guide inquiry in all scientific fields.[53] Despite the manifest overlap with the concept of paradigm as it emerged in the 1962 *Structure of Scientific Revolutions*,[54] it is quite evident to me that Tom Kuhn had no idea of my usage of the term and deploying of the concept. And, of course, he put his concept to distinctive and consequential use, particularly as it led to the notion of "paradigm shifts." Where we differed was in his initial assumption that, as Imre Lakatos put it, "major fields of science are, and must be, always

[52] For a brief sample of our correspondence bearing on Kuhn's manuscript, see Jonathan R. Cole and Harriet Zuckerman, "The Emergence of a Scientific Specialty: The Self-Exemplifying Case of the Sociology of Science," in *The Idea of Social Structure*, ed. Lewis A. Coser (New York: Harcourt Brace Jovanovich, 1975), pp. 139–174, at pp. 159–160. For a more extended account of our longtime association, see Merton, *Sociology of Science: An Episodic Memoir*, pp. 71–109.

[53] By way of examples, see "Paradigm for the Sociology of Knowledge" (1945), reprinted in Merton, *The Sociology of Science: Theoretical and Empirical Investigations*, pp. 7–40, and "Paradigm for Functional Analysis in Sociology," in Merton, *Social Theory and Social Structure*, pp. 49–55 in the 1949 edition and pp. 104–109 in the 1968 edition.

[54] For the first detailed scrutiny of Kuhn's concept, see Margaret Masterman, "The Nature of a Paradigm," in *Criticism and the Growth of Knowledge*, ed. Imre Lakatos and Alan Musgrave (New York: Cambridge University Press, 1970), pp. 59–89.

dominated by one single supreme paradigm,"[55] while I assumed that scientific disciplines and specialties worked with a variety of paradigms along with shared background theoretical orientations.

This dual usage has been noted from time to time. The historian of science Yehuda Elkana has observed that though Wittgenstein used "the much-abused concept of paradigm" in his lectures, "only Merton's and Kuhn's usages became fundamental tools for a whole discipline. Thus, there is a Mertonian paradigm and a Kuhnian paradigm."[56] And the anthropologist Raymond Firth observes:

> Paradigm has become the key word for a lot of interpretation. Paradigm, a word derived from classical sources, has been in use in English since at least the seventeenth century to mean a pattern to follow, an exemplar. It also has a hint of providing the basic components underlying any variation which a phenomenon might assume. In this sense, the term was used by Robert Merton long ago when he was arguing for stricter methodology and greater awareness of the theoretical framework of sociology. . . . Modern social anthropologists use the notion of paradigm rather differently, borrowing it not from Merton but from de Saussure and Thomas Kuhn.[57]

All this holds some minor interest inasmuch as some historians and philosophers have argued that, contrary to Kuhn's own statements that his theoretical perspective on science as a social institution shared much with mine,[58] Kuhn's theoretical perspective represented a complete break with my conception of the sociology of science, while others have argued quite the opposite, maintaining that they were fundamentally much the same.[59]

---

[55] A reading of a Kuhnian paradigm that is further developed in Imre Lakatos, *The Methodology of Scientific Research Programmes* (Cambridge: Cambridge University Press, 1978), passim.

[56] Yehuda Elkana, "Culture, Cultural System, and Science," in *Essays in Memory of Imre Lakatos*, ed. R. S. Cohen, P. K. Feyerabend, and M. W. Wartofsky (Dordrecht: R. Reidel Publishing, 1976), pp. 106–107.

[57] Raymond Firth, "An Appraisal of Modern Social Anthropology," *Annual Review of Anthropology* 4 (1975):1–25, at 12, 15. For more on the usages of paradigm, see Stephen Toulmin, *Human Understanding* (Oxford: Clarendon Press, 1972), vol. 1, pp. 106–109.

[58] For an immediate example, see Kuhn, *Essential Tension*, pp. xxi–xxii, where he describes such "strident" declarations of "opposition" between his theoretical orientation and mine as "seriously misdirected."

[59] As emphatic cases in point: Ian C. Jarvie, "Explanation, Reduction, and the Sociological Turn in the Philosophy of Science, or Kuhn as Ideologue for Merton's Theory of Science," in *Centripetal Forces in the Sciences*, ed. Gerard Radnitsky (New York: Paragon House, 1988), vol. 2, pp. 299–320; T. J. Pinch, "Kuhn—The Conservative and Radical Interpretations: Are Some 'Mertonians' 'Kuhnians' and Some Kuhnians 'Mertonians'?" *Social Studies of Science* 27 ([1982] 1997):465–482; Simone P. Kropf and N. T. Lima, "Os valores e a prática institucional de ciência: as concepções de Robert Merton e Thomas Kuhn," *História, Ciências, Saude—Manguinhos* 5 (1998–1999):565–581.

I do not, of course, enter into that question in this excursus. I note only that as Kuhn came to identify the limiting constraints of paradigms as well as their functions, he independently converged on my own early observations on the dysfunctions as well as the functions of paradigms; thus:

> Since virtues readily become vices merely by being carried to excess, the sociological paradigm can be abused almost as easily as it can be used. It is a temptation to mental indolence. Equipped with a paradigm, sociologists may shut their eyes to strategic data not expressly called for by the paradigm. Thus it can be turned from a sociological field-glass into a sociological blinder. Misuse results from absolutizing the paradigm rather than using it as a tentative point of departure. But if they are recognized as provisional and changing, destined to be modified in the immediate future as they have been in the recent past, paradigms are preferable to sets of tacit assumptions.[60]

The Standard Scientific Article and Obliterated Scientific Serendipities (or SSA and OSS, as These Are Bound to Be Abbreviated in Our Age of Acronyms)[61]

In his perspicuous Introduction, James Shulman notes that "as early as 1938 [Merton] observed that the telling of a scientific finding depends on creative storytelling." That is Dr. Shulman's discerning if deliberately romantic allusion to an observation I made back then to the effect that scientific work is "presented in a rigorously logical and 'scientific' fashion (in accordance with the rules of evidence current at the time) and *not* in the order in which the theory or law was derived."[62]

And, rather to my surprise that he should have noticed both this early observation and an elucidated version of it proposed some fifty years later, Shulman goes on to observe, "As the implications of the story-telling process became even clearer to Merton (both in the subject of his research and in the process of his own work) he would write that 'the etiquette governing the writing of scientific or scholarly papers requires them to be works of vast expurgation, stripping the complex events that culminated in the published reports of everything but their delimited cognitive sub-

---

[60] Merton, *Social Theory and Social Structure*, p. 16; this and related early observations on specified functions and dysfunctions of paradigms are reprinted in R. K. Merton, *On Social Structure and Science*, ed. Piotr Sztompka (Chicago: University of Chicago Press, 1996), pp. 57–62.

[61] For observations on the age of acronyms in which the words of a title or phrase are beheaded and the decapitated initials are then collected into a unique "symbol," see Merton, *On the Shoulders of Giants*, pp. 270–272.

[62] This is quoted from my *Science, Technology, and Society in Seventeenth-Century England*.

stance.'"[63] In short, the audience demands a well-formulated argument that retrospectively imposes logical form on the romance of investigation.

But, having noted these early and late observations on the scientific article, Shulman wisely did not have his Introduction go on to anticipate me by tracking the repeated evidence of my enduring interest during the intervening decades in that divergence between public science and private scientific practice and its bearing on the infrequency with which reports of serendipitous moments that occurred in an ongoing research find their way into scientific articles.

Only now, in these autobiographic reflections, do I notice that my 1938 observation of the divergence between public and private science is immediately followed by the passage that Kuhn tells us led him seren-dipitously to the "work of the developmental psychologist Jean Piaget." That passage notes that "Piaget has described the differences between one's private way of developing one's thoughts and the order in which they are presented to others."[64] As it turns out, I was drawing upon Piaget's analysis to suggest the sociopsychological aspect of the mecha-nism making for the discrepancy between public and private science. And also only now has it become abundantly evident to me that the early observation on the rhetorical character of the scientific paper had ever since lingered in my memory. To begin with, precisely ten years later I find myself reflecting anew on some implications of the stylized model of the scientific paper which calls in effect for expurgated accounts of the actual course of the research. In that model, it is said,

> The investigator begins with a hunch or hypothesis, from this he draws var-ious inferences and these, in turn, are subjected to empirical test which con-firms or refutes the hypothesis. But this is a logical model and so fails, of course, to describe much of what actually occurs in fruitful investigation. It presents a set of [methodo]logical norms, not a description of the research experience. And, as logicians are well aware, in purifying the experience, the model may also distort it. Like other models, it abstracts from the temporal sequence of events. It exaggerates the creative role of explicit theory just as it minimizes the creative role of observation. For research is not merely logic tempered with observation. It has its psychological as well as its logical di-mension, although one would *scarcely suspect this from the logically rigorous sequence in which research is usually reported.*[65]

---

[63] An observation to which I shall soon return in more elucidated detail.

[64] The reference was to Jean Piaget, *Judgment and Reasoning in the Child* (London: K. Paul, 1929), chapters 5 and 9. Kuhn describes two of Piaget's books as "particularly impor-tant . . . since they displayed concepts and processes that also emerge directly from the history of science" (*Structure of Scientific Revolutions*, p. vi, note 2).

[65] Merton, "Bearing of Empirical Research upon the Development of Theory," pp. 505–506. I now add the italics.

And, instructively for the purpose of these Afterword reflections, this text of 1948 then proceeds directly to what is described as (the now to us all too familiar) "Serendipity Pattern: The unanticipated, anomalous, and strategic datum exerts a pressure for initiating theory" as an indication of what scientific articles typically omit from their published reports of the research.

The observation about the public record of science being misleading with regard to the actual course of the inquiry had by then manifestly become haunting. For again—as it happens, precisely another decade later—that theme of the standard scientific article (SSA hereafter) being in effect an *expurgated* version of the course taken by the research is reiterated in Chapter 9 of the 1958 time-capsule text of *Travels and Adventures of Serendipity*, thus: "the elegance and parsimony prescribed for the presentation of the results of scientific work tend to falsify retrospectively the actual process by which the results were obtained." Confined to that unpublished text, this further observation on "retrospective falsification" could of course be noted only by readers of that ongoing manuscript, as indeed it soon was, being adopted by Bernard Barber and Renée C. Fox to introduce their early and exacting case study of "serendipity gained and serendipity lost."[66]

Another ten years later—this decennial periodizing is of course mere coincidence—I find myself reverting to the theme of an expurgated SSA and retrospective falsification in more explicitly sociological terms by drawing attention to

the rock-bound difference between the finished versions of scientific work as they appear in print and the actual course of inquiry followed by the inquirer. The difference is a little like that between the textbooks of "scientific method" and the ways in which scientists actually think, feel and go about their work. The books on method present ideal patterns: how one *ought* to think, feel and act, but these tidy normative patterns, as everyone who has engaged in [scientific] inquiry knows, do not reproduce the typically untidy, opportunistic adaptations that scientists make in the course of their inquiries. Typically, the scientific paper or monograph presents an immaculate appearance which reproduces little or nothing of the intuitive leaps, false starts, mistakes, loose ends, and *happy accidents* that actually cluttered up the inquiry. The public record of science therefore fails to provide many of the source materials needed to reconstruct the actual course of scientific developments.[67]

---

[66] Appearing with the apt and engaging title "The Case of the Floppy-Eared Rabbits: An Instance of Serendipity Gained and Serendipity Lost," this then-rare case study of serendipity appeared first in the *American Journal of Sociology* 64 (1958):128–136, and was then reprinted in the more accessible anthology by Bernard Barber and Walter Hirsch, *The Sociology of Science* (Glencoe, Ill.: Free Press, 1962), pp. 525–538.

[67] Merton, *Social Theory and Social Structure*, p. 4 in the 1968. The emphatic italicized "happy accidents" is, of course, a present addition.

The theme of the expurgated SSA had plainly grown obsessive. And, as we see from the reference to "happy accidents" in the list of actual experiences omitted in the published scientific article, that theme is now linked with the further theme of obliterated scientific serendipities. This connection is reiterated once more in the preface of a new edition of *On the Shoulders of Giants: A Shandean Postscript* [*OTSOG*].[68] Interestingly enough, the book for which, it may possibly be remembered, Shulman hypothesizes in his Introduction, *The Travels and Adventures of Serendipity* was a *preparazione OTSOGIA*. And, as though this reiteration of the SSA theme were still not enough, an abbreviated version appears yet again in a paper written for the 1987 Amalfi Conference of the Associazione Italiana di Sociologia on the theme of "Unanticipated Consequences and Kindred Sociological Ideas." Here, I retrospectively come to understand my persistent interest in the canonical format and content of the standard scientific paper; it has to do, of course, with its generic cognitive consequences and implications. When mistakenly taken as an account of how the reported research actually proceeded, that representation of scientific inquiry not only leads specifically to the omission of serendipitous episodes ("happy accidents") but gives rise more generally to a misleading image of scientific work as nothing but linear hypothetico-deductive inquiry; this, presumably, in accordance with the so-called scientific method: "In short, the etiquette governing the writing of scientific (or scholarly) papers requires them to be works of vast expurgation, stripping the complex events that culminated in the published reports of everything except their delimited cognitive substance. We should perhaps note that this economic and efficient editorial norm might lead cumulatively to certain side effects—or, one might say, to certain unanticipated consequences. For it invites an imagery of scientists and scholars moving coolly, methodically, and unerringly to their reported ideas and findings."[69]

So much, I mistakenly thought, for my early and continuing interest in the SSA, the discrepancy between the actual course of a scientific inquiry and its published account. That, I thought, was something of a personal obsession; surely so, in its repeated connections with my enduring interest in serendipity. But then, in the late 1960s rather than in the mid-1930s, when I had been virtually living in seventeenth-century England, I came upon this composite observation and complaint from the pen of Francis Bacon: "That never any knowledge was delivered in the same

---

[68] "Foreword," in *On the Shoulders of Giants: A Shandean Postscript* (San Diego and New York: Harcourt Brace Jovanovich, [1965] 1985), pp. ix–xi.

[69] This is on page 297 of Mongardini and Tabboni, eds., *Robert K. Merton and Contemporary Sociology*. As will be recognized, it is this version of the expurgated SSA theme that knowingly captured the attention of James Shulman in his Introduction.

order it was invented, no not in the mathematic, though it should seem otherwise in regard that the propositions placed last do use the propositions or grants placed first for their proof and demonstration."[70]

That Baconian observation plainly provided food for further thought. After all, Bacon was writing years before the bare beginnings of "the scientific article" emerged in the *Philosophical Transactions of the Royal Society of London* from its inception in March 1664–1665 and in the *Journal des Sçavans* (three months earlier). And so it is clear that the allegedly universal practice of "never any knowledge [having] been delivered in the same order it was invented" could not have resulted *exclusively* from the institutionalized constraints placed upon the form and substance of articles in scientific journals. Still, it occurred to me that we need not lazily accept Bacon's claim of the universality of this practice, for it does seem to be the case that the historically evolving exemplar of the scientific article with its increasingly standardized style, content, and format has greatly expanded and intensified the practice of having the article report a highly stylized and expurgated version of the inquiry. I had had occasion to examine the two thousand or so papers in the *Philosophical Transactions* that appeared in the seventeenth century—in an effort to detect "foci and shifts of interest in the sciences and technology"[71]—and so I can bear witness that those accounts were generally far removed from the stylized format and typically impersonal, bland, and conventionalized idiom of the "SSA" that has evolved in the centuries since.[72] I make no effort here to indicate how remote those early modern reports in the *Philosophical Transactions* were in these respects; instead, I refer the reader to detailed analyses of this rhetorical practice by Shapin, Bazerman, and Gross.[73]

---

[70] *The Works of Francis Bacon*, collected and edited by James Spedding, Robert Leslie Ellis, and Douglas Demon Heath (Cambridge: Riverside Press, [1605] 1863), vol. 6, p. 70. The following pages on the standard scientific article draw upon and extend my discussion of "the public record" of science in the 1968 edition of *Social Theory and Social Structure*, pp. 3–8.

[71] Merton, *Science, Technology, and Society in Seventeenth-Century England*, chapter 3.

[72] For a brief account of the "institutionalization" of the scientific journal and scientific article, see Harriet Zuckerman and R. K. Merton, "Institutionalized Patterns of Evaluation in Science," *Minerva* 9 (1971):66–100, especially pp. 66–76.

[73] Steven Shapin, "Pump and Circumstance: Robert Boyle's Literary Technology," *Social Studies of Science* 14 (1984):481–520; this masterly analysis is much amplified in Steven Shapin and Simon Schaffer, *Leviathan and the Air-Pump: Hobbes, Boyle, and the Experimental Life* (Princeton, N.J.: Princeton University Press, 1985). See also Charles Bazerman, "Literate Acts and the Emergent Social Structure of Science," *Social Epistemology* 1 (1987):295–310; Charles Bazerman, *Shaping Written Knowledge: The Genre and Activity of the Experimental Article in Science* (Madison: University of Wisconsin Press, 1988); Alan G. Gross, *The Rhetoric of Science* (Cambridge, Mass.: Harvard University Press, 1990).

Observations to much the same effect as Bacon's have been made from time to time by other observant minds. Thus, a century later, Leibniz virtually approaches the subject in an off-the-record letter that eventually became part of the historical record: "It is good to study the discoveries of others in a way that discloses to us the source of the discoveries and renders them in a sort our own. And I wish that authors would give us the history of their discoveries and the steps by which they have arrived at them. When they neglect to do so, we must try to divine those steps, in order to profit from their works."[74]

And, some two centuries later, we find the physicist and philosopher Ernst Mach maintaining in almost Baconian fashion that the practice of presenting scientific and mathematical work in logically cogent fashion rather than by charting the actual paths of inquiry has persisted from the time of Euclid to the present day: "Euclid's system fascinated thinkers by its logical excellence, and its drawbacks were overlooked amid this admiration. Great inquirers, even in recent times, have been misled into following Euclid's example in the presentation of the results of their inquiries, and so into *actually concealing their methods of investigation, to the great detriment of science.*"[75]

Evidently unaware of Mach's formulation, the botanist Agnes Arber puts the practice of glossing over the actual course of inquiry in a more general and analytical frame, suggesting that "the mode of presentation of scientific work is . . . moulded by the thought prejudices of its period." Thus, she maintains that in the Euclidean period, when deduction was highly prized, the actual course of inquiry was covered over "by the artificial method of stringing propositions on an arbitrarily chosen thread of deduction," in this way obscuring its empirical aspect. Today, she argues, the scientist "being under the domination of the inductive method, even if he has in fact reached his hypothesis by analogy, his instinctive reaction is to cover his traces, and to present *all* his work—not merely his proof—in inductive form, as though it were by this process that his conclusions had actually been reached."[76]

---

[74] Gottfried Wilhelm Leibniz, *Philosophische Schriften*, ed. C. I. Gerhardt (Berlin, 1887), vol. 3, p. 568, in a letter to Louis Bourguet dated 22 March 1714.

[75] Ernst Mach, *Space and Geometry*, trans. T. J. McCormack (Chicago: Open Court Publishing, 1906), p. 118, originally published in *The Monist* (October 1903). The italics have been added.

[76] Agnes Arber, "Analogy in the History of Science," *Studies and Essays in the History of Science and Learning Offered in Homage to George Sarton*, ed. M. F. Ashley Montagu (New York: Henry Schuman, 1944), pp. 222–239, at p. 229. Arber reflects further on this matter in chapter 5 of her perceptive, subtle, and deeply informed book *The Mind and the Eye: A Study of the Biologist's Standpoint* (New York: Cambridge University Press, 1954). Of more than passing interest, as we explore the relation between the unfinished 1958 *The Travels*

The theme that the inductive form of scientific reports misrepresents the actual course of the reported inquiry has continued until the present day; not least, it seems, among scientists of the first class. Thus, to look no further, we find this to be the case for a veritable rash of Nobel laureates[77] who, as it happens, differ otherwise in various respects: in the fields of their major achievements (immunology, physics, biochemical genetics, physiology, and chemistry), the nature of their audiences, their styles of examining the pattern, and their judgments of it.

In 1963, Peter B. Medawar, the polymathic biologist who had received the Nobel Prize for having revolutionized immunological thought in ways that have led to enormous advances in organ transplantation, presented the theme of the expurgated SSA in a talk on BBC television. No doubt aiming to engage the interest of his mass audience, he adopted the title "Is the Scientific Paper a Fraud?"[78] but promptly defused that arresting, not to say sensational, title by explaining that

> by asking 'is the scientific paper a fraud?' I don't, of course, mean 'does the scientific paper misrepresent facts', and I don't mean that the interpretations you find in a scientific paper are wrong or deliberately mistaken. I mean the scientific paper may be a fraud because it misrepresents the processes of thought that accompanied or gave rise to the work that is described in the paper. That is the question, and I'll say right away that my answer to it is 'yes'. The scientific paper in its orthodox form *does* embody a mistaken conception, even a travesty, of the nature of scientific thought.

Medawar then proceeds to inform his audience, in typically ironic style, that the "traditional form" of a scientific paper divides into the "introduction" ("in which you concede, more or less graciously, that others

---

and *Adventures of Serendipity* and the thoroughly nonlinear 1965 *On the Shoulders of Giants: A Shandean Postcript*, is Arber's observation that we find efforts to record the reticular character of thought primarily in nonscientific literature, most specifically in Sterne's masterwork, *Tristram Shandy*: "Laurence Sterne and certain modern writers influenced by him [an obvious allusion to James Joyce and Virginia Woolf] have visualized and tried to convey in language, the complicated non-linear behavior of the human mind, as it darts to and fro, disregarding the shackles of temporal sequence; but few biologists [read: scientists] would dare to risk such experiments."

[77] For a sociological analysis of sociocognitive patterns among Nobel laureates and "occupants of the forty-first chair" (i.e., scientists of Nobel caliber who, like those qualified scholars who could not be included in the limited cohorts of only forty "immortals" in the French Academy, could not be granted one of the limited number of Nobel Prizes), see Harriet Zuckerman, *Scientific Elite* (New Brunswick, N.J.: Transaction Publishers [1977] 1996).

[78] Peter Medawar was also a consummate scientific essayist, and so, when this talk appeared in the 12 September 1963 issue of the BBC publication *The Listener*, readers are rightly informed that it is merely "a revised version of a broadcast talk delivered unscripted; the broadcast form and style have been preserved."

have dimly groped towards the fundamental truths that you are now about to expound"); "Methods" ("that's O.K."); "Results" (which "consists of a stream of factual information in which it's considered extremely bad form to discuss the significance of the results you are getting. You have to pretend that your mind is, so to speak, a virgin receptacle, an empty vessel, for information which floods into it from the external world for no reason which you yourself have revealed"); and "Discussion (in which "you adopt the ludicrous pretence of asking yourself if any general truths are going to emerge from the contemplation of all the evidence you brandished in the section called 'results'").

And so this master of experimental science concludes that "the inductive format of the scientific paper should be discarded . . . [and] scientists should not be ashamed to admit, as many of them apparently *are* ashamed to admit, that hypotheses appear in their minds along uncharted by-ways of thought." That last phrasing—"uncharted by-ways of thought"—can be taken, of course, to include the hypotheses generated by unanticipated and anomalous observations that one is making serendipitously en route to solution of the problem in hand.

In contrast to Medawar's mass audience was the audience of scientists and associates listening to the Nobel Prize address in Stockholm by the deeply and variously original physicist Richard P. Feynman just a few years later. Taking note that "there has been in these days some interest in this kind of thing," he began that address in his inimitable personal, terse, distinctively idiomatic, and utterly clear prose: "We have a habit in writing articles published in scientific journals to make the work as finished as possible, to cover up all the tracks, to not worry about the blind alleys or to describe how you had the wrong idea first, and so on, so there isn't any place to publish, in a dignified manner, what you actually did in order to get to do the work, although there has been in these days some interest in this kind of thing."[79]

---

[79] Feynman's Nobel address, "The Development of the Space-Time View of Quantum Electrodynamics," is reprinted in the readily accessible *Science* 153, no. 3737 (12 August 1966):699–708. As we know, "interest in this sort of thing"—i.e., SSA—was hardly confined to "these days." Still, it does seem to be of special interest to his fellow laureates. At the very time that Feynman was beginning his Nobel talk in this fashion, yet another laureate, the biochemical geneticist, George W. Beadle, was also beginning a paper in much the same way: "I have often thought how much more interesting science would be if those who created it told how it really happened, rather than reported it logically and impersonally, as they so often do in scientific papers." George W. Beadle, "Biochemical Genetics: Some Recollections," in *Phage and the Origins of Molecular Biology*, ed. John Cairns, Gunther S. Stent, and James D. Watson (Cold Spring Harbor, N.Y.: Cold Spring Harbor Laboratory of Quantitative Biology, 1966), pp. 23–32. A few years later, the laureate in physiology Alan Hodgkin was reminding his fellow scientists "that the record of published work conveys an impression of directness and planning which does not at all coincide with the actual se-

He then proceeds to distinguish between the two kinds of scientific publications—the SSA and the rare account that tells about the actual "sequence of ideas"—this, of course, still couched in his distinctive personal style:

> Since winning the prize is a personal thing, I thought I could be excused in this particular situation if I were to talk personally about my relationship to quantum electrodynamics, rather than to discuss the subject in a refined and finished fashion. . . . So, what I would like to tell you about today are *the sequence of events, really the sequence of ideas, which occurred,* and by which I finally came out the other end with an unsolved problem for which I ultimately received a prize. I realize that *a truly scientific paper* would be of greater value, but such a paper *I could publish in regular journals.* So, I shall use this Nobel Lecture as an opportunity to do something of less value, but *which I cannot do elsewhere.*[80]

One has only to contrast the styles and tones of the Nobel addresses published in the first century of the prize to recognize that this one was unique in both style and tone, and that it was unmistakably written by Richard Feynman. It required no signature.

His fellow laureates, the immunologist Medawar, the biochemical geneticist Beadle, the physiologist Hodgkin, and the physicist Feynman, had variously touched upon the significance of the expurgated SSA, but it is Roald Hoffmann, a laureate in chemistry (as well as a poet and self-taught sociologist of science), who finally comes to our specific notion of obliterated scientific serendipities (OSS): that this largely explains the infrequency—indeed, the rarity—of reported serendipitous moments in scientific inquiry. This he does, not in a talk broadcast to a mass audience or in a Nobel address directed to a composite audience of scientists and their associates, but in a scientific journal for his peers in chemistry, *Angewandte Chemie.* Having this sort of article appear in a scientific journal of this kind may seem ipso facto to refute the claim of constraints on the form and content of scientific articles, but that impression is more apparent than real, for the paper, tellingly titled "Under the Surface of the Chemical Article," was presented at a presumably rule-breaking celebration of the journal's hundredth anniversary.

Roald Hoffmann virtually recapitulates for us the entire argument we have been evolving over the years as we see from brief extracts drawn from that article on the latent content of the chemical article. He notes, in what has become familiar style, that "what is written in a scientific

---

quence of events," as he recalls in his book of reminiscences, *Chance and Design: Reminiscences of Science in Peace and War* (New York: Cambridge University Press, 1992).

[80] Feynman, "Development of the Space-Time View of Quantum Electrodynamics," p. 699. I have of course supplied the italics to make the obvious even more obvious.

article is not a true and faithful reproduction (if such a thing were possible) of what transpired. It is not a laboratory notebook, and one knows that that notebook in turn is only a partially reliable guide to what took place. It is a more or less (one wishes more) carefully constructed, man- or woman-made *text*. Most of the obstacles that were in the way of the synthesis or the building of the spectrometer have been excised from the text."[81] This observation on the SSA—we have ample, perhaps excessive, reason to know—has come to be reiterated, often independently, over the years.[82] But unlike those many others, Hoffmann joins in my focus on OSS, on how the conventional format of the article works to obliterate accounts of actual serendipity: "In order to present a sanitized paradigmatic account of a chemical study, one suppresses many of the truly creative acts. Among these are the 'fortuitous circumstance'—all of the elements of serendipity, of creative intuition at work."[83]

I had based the conclusion that such conventionalized omissions occur in the SSA (with unknown frequency) by contrasting it with the ample accounts of serendipitous moments recorded in the same scientists' memoirs, diaries, autobiographies, and other kinds of correlative writings and retrospections. By way of examining a truly significant case in point, we have only to contrast the austere form and substance of the article reporting the discovery that has been described as launching the "biological revolution of the twentieth century" with the later, full-fledged accounts of that discovery by its originators.

Restoration of Serendipitous Moments in the Discovery of the Double Helix

That revolutionary discovery was the double-helix model of the molecular structure of deoxyribonucleic acid (DNA) proposed in 1953 by the twenty-five-year-old statunitensean biologist, James D. Watson, and the thirty-seven-year-old English physicist turned biologist, Francis Crick,

[81] Roald Hoffmann, "Under the Surface of the Chemical Article," *Angewandte Chemie* 27 (December 1988):1593–1602, at p. 1597.

[82] In effect, such observations on the SSA correspond to what the philosopher of science Hans Reichenbach memorably described as the "context of discovery" (the way in which "the thinker" arrives at a result) and the "context of justification" (the way in which that finding is presented "before a public"), in *Experience and Prediction: An Analysis of the Foundations and the Structure of Knowledge* (Chicago: University of Chicago Press, 1938), pp. 6–7, 381–383. Tellingly, Reichenbach begins this influential book with the claim that "every theory of knowledge must start from knowledge as a sociological fact." It will be remembered that this is the work in which Kuhn serendipitously came upon the consequential reference to Fleck's *Entstehung und Entwicklung einer wissenschaftlichen Tatsache*.

[83] Hoffmann, "Under the Surface of the Chemical Article," p. 1598.

while they were working at the famed Cavendish Laboratory of Cambridge University. The severely condensed nine-hundred-word article in *Nature*[84] reporting that historic discovery can truly be taken as an archetype of the expurgated SSA and, more specifically in point, as an archetype also of OSS. For, of course, this paper has not a word to say about the "intuitive leaps, false starts, mistakes, loose ends, and *happy accidents* that actually cluttered up the inquiry."[85] It is devoted exclusively to "suggest[ing] a structure for the salt of deoxyribose nucleic acid (D.N.A.) [which] has novel features of considerable biological interest." After indicating why the triple-stranded helix proposed by Linus Pauling—"the world's greatest chemist," as Watson would later describe him—and R. B. Corey was "unsatisfactory," Watson and Crick "put forward a radically different structure for the salt of deoxyribose nucleic acid," and went on to present a schematic illustration of the double helix. Following upon further specification of the model, they introduce their grandly understated, seemingly casual, and now historic one-sentence paragraph that highlights the revolutionary implications of their "radically different structure": "It has not escaped our notice that the specific pairing we have postulated immediately suggests a possible copying mechanism for the genetic material." In short, Watson and Crick were more than hinting that it is the double-helical model of complementary hydrogen bonds that explains the process by which DNA replicates itself, that is, how it transmits genetic information from one generation to the next.

In terms long institutionalized even for the expurgated SSA, the closing paragraph acknowledges aid from three others. It begins with what in this generally understated paper is a relatively emphatic statement: "We are much indebted to Dr. Jerry Donohue for constant advice and criticism, especially on interatomic distances." And then it goes on, in rather more subdued tones, "We have also been stimulated by a knowledge of the general nature of the unpublished experimental results and ideas of Dr. M. H. F. Wilkins, Dr. R. E. Franklin and their co-workers at King's College, London." Others in the "invisible college"[86] of those at work on

---

[84] J. D. Watson and F. H. C. Crick, "Molecular Structure of Nucleic Acids: A Structure for Deoxyribose Nucleic Acid," *Nature* (25 April 1953):737–738.

[85] This inventory of expurgation, it will probably not be remembered, is quoted from one of my evolving versions of the SSA, which results in much of the public record of science being misleading with regard to the actual course of the reported investigation. Anticipating a bit, I call attention prematurely here to Crick's own reminiscent observation that "the path" to "the discovery of the double helix" involved "misleading data, false ideas, [and] problems of interpersonal relationships," although, understandably enough, these find no mention at all in the austere pathbreaking paper that appeared in *Nature*.

[86] In a felicitous stroke of terminological recoinage, the historian of science and pioneer of scientometrics Derek J. De Solla Price adopted and reconceptualized Robert Boyle's

the chemical structure of DNA would have instantly identified the bio-physicist Dr. Wilkins, who had begun his basic work a half-dozen years before Watson and Crick had ventured on the scene—in the event, as we know, they shared the Nobel Prize with him. His colleague, the crystal-lographer Dr. Rosalind Franklin (whose X-ray photographs of DNA were later described by the peerless J. D. Bernal as "among the most beautiful X-ray photographs of any substance ever taken") died four years before the prize was awarded. These closely knowing colleagues-at-a-distance would have instantly recognized the considerable debt owed to the prior and ongoing work of Wilkins and Franklin, but probably fewer of them would have known as much about the young American crystallographer Dr. Jerry Donohue and wondered why he had been singled out for such primary and emphatic notice. Thereby hangs a tale of the total oblitera-tion of a highly consequential serendipitous moment in the first pub-lished account of the discovery of the double helix, a tale told separately by the joint discoverers when, happily for our focus on OSS, the expur-gated milestone article was greatly expanded in their two reminiscent books. For each book provides a virtual cornucopia of serendipitous mo-ments in the complex course of that hugely consequential discovery.

The first of these was Watson's contentious and best-selling book *The Double Helix*,[87] a wonderfully candid self-portrait of the scientist as a young man in a hurry. Chattily written with pungent and ironic wit and at times with an almost clinical detachment, it tells of a variety of conse-quential but "completely unanticipated event[s]" and an "unexpected success" in getting "the X-ray pattern needed to prove that TMV (to-bacco mosaic virus) was helical," this coming about from "using a power-ful anode X-ray tube that had just been assembled in the Cavendish." Above all for our immediate purpose, it explains the special acknowledge-ment accorded Jerry Donahue in that otherwise terse milestone article. For as Watson not merely admits but repeatedly insists, Crick and he were at the outset simply ignorant about much they needed to know in

---

seventeenth-century term *invisible college* to designate informal collectives of closely inter-acting scientists, generally limited to a size "that can be handled by interpersonal relation-ships." Invisible colleges, he suggests, are consequential social and cognitive formations that "advance the research fronts of science." See Derek J. De Solla Price, *Little Science, Big Science . . . and Beyond* (New York: Columbia University Press, [1963] 1986).

[87] James D. Watson, *The Double Helix: A Personal Account of the Discovery of the Struc-ture of DNA* (New York: Atheneum, 1968). I draw freely here upon my review of the book, which appeared in the *New York Times Book Review*, 25 February 1968, pp. 1, 41–43. Both the book and a dozen or so essay reviews (including mine) are included in the more accessible anthology assembled by the molecular biologist Gunther S. Stent, *The Double Helix: Text, Commentary, Reviews, Original Papers* (New York: W. W. Norton, 1980). I now see that the crucial serendipitous role of Jerry Donohue has also been noted in some detail in Roberts, *Serendipity*, pp. 226–231.

order to search out the structure of DNA. The impressive inventory of this announced ignorance includes the techniques of X-ray diffraction, Pauling's work on the alpha-helix, Bragg's Law ("the most basic of all crystallographic ideas"), and, most specifically with regard to Donahue's decisive serendipitous role, the chemistry of hydrogen bonds. Yet, despite occasional qualms, these newcomers had the adventurous fortitude to acquire much of the knowledge they needed and the institutionalized good luck to have at their side the experts who could round out that knowledge sufficiently for them to do the job of imaginative scientific carpentry that led to their momentous model.

Here, then, we find Watson telling at length[88] of "the unforeseen dividend" of having the visiting Jerry Donohue, who "next to Linus [Pauling] knew more about hydrogen bonds than anyone else in the world, . . . occupy a desk in our office." Which is to say, more specifically, that the young Donohue, with his deep and authoritative knowledge, was able to show Watson that the crucial "tautometric forms I had [in rational ignorance] copied out of Davidson's book were, in Jerry's opinion, incorrectly assigned. My immediate retort that several other texts also pictured guanine and thymine in the enol form cut no ice with Jerry. Happily he let out that for years organic chemists had been arbitrarily favoring particular tautometric forms over their alternatives on only the flimsiest of grounds. In fact, organic-chemistry textbooks were littered with pictures of highly improbable tautometric forms."[89] This was a decisive turn that put Watson and Crick on the right track. Watson goes on to explain: "I couldn't kid myself that he did not grasp our problem. During the six months [N.B.] that he occupied a desk in our office, I never heard him shooting off his mouth on subjects about which he knew nothing."[90]

As we see, in telling us about that crucial serendipitous moment, Watson, evidently not a psychological reductionist, alerts us in virtually sociological fashion to its involving *both* a significant sociocognitive property of the microenvironment provided by the Cavendish Laboratory *and* their own personal acumen. As the sociologist would put it, he centers on the interaction between "structure" and "agency" as he makes it clear that, in his opinion, Crick and he could not have accomplished what they did and would probably have lost out in their race for priority to "the champion" Linus Pauling had it not been for the serendipitous environment provided by the Cavendish. He writes, "if [Jerry] had not been with us in Cambridge, I might still be pumping for a [thoroughly mis-

[88] Watson, *Double Helix*, pp. 189–209.
[89] Ibid., p. 190.
[90] Ibid., p. 192.

taken] like-with-like structure. Maurice [Wilkins] *in a lab* [at King's College in the University of London] *devoid of structural chemists* did not have anyone about to tell him that all the textbook pictures were wrong. But for Jerry, Pauling would have been likely to make the right choice and stick by its consequences."[91]

In his later retrospection, Crick is, if anything, all the more explicit about the role of serendipitous moments in the grand discovery.[92] To begin with, he takes pains to emphasize that their discovery, like most scientific discoveries, was far from having followed a unilinear course of inquiry: "I think what needs to be emphasized about the discovery of the double helix is that the path to it was, scientifically speaking, fairly commonplace. What was important was not the way it was discovered but the object discovered—the structure of DNA itself. You can see this by comparing it with almost any other scientific discovery. *Misleading data, false ideas, problems of personal interrelationships occur in much if not all scientific work.*"[93]

In stark contrast to the austere account in the SSA that appeared in *Nature*, Crick proceeds to underscore the crucial serendipitous moment in the evolving discovery: "The key discovery was Jim's determination of the exact nature of the two base pairs (A with T, G with C). He did this not by logic but by serendipity. . . . In a sense Jim's discovery was luck, but then most discoveries have an element of luck in them. The more important point is that Jim was looking for something significant and *immediately recognized the significance of the correct pairs when he hit upon them by chance*—"chance favors the prepared mind."[94]

And, like Watson, Crick elucidates the distinctive acknowledgment accorded Jerry Donahue in the *Nature* article by summarizing his truly pivotal role in getting the pair of them onto the right track: "This was not our only mistake. Misled by the term tautomeric forms, I assumed that certain hydrogen atoms on the periphery of the bases could be in one of several positions. Eventually Jerry Donahue, an American crystallographer who shared an office with us, told us that some of the textbook

---

[91] Ibid., p. 209. Watson notes in his final sentence here, as sporadically throughout much of the book, that multiple independent discoveries are an occupational hazard in science. As a result of peer recognition being given primarily for original work, competition in science is as old as modern science itself. Almost everyone placed in the pantheon of science, from the days of Galileo and Newton, has been caught up in the consequent race for priority. Cf. R. K. Merton, "Priorities in Scientific Discovery" (1957), reprinted in Merton, *The Sociology of Science.*

[92] Francis Crick, *What Mad Pursuit: A Personal View of Scientific Discovery* (New York: Basic Books, 1988).

[93] Ibid., p. 67. Italics added.

[94] Ibid., pp. 65–66. Here the italics are Crick's.

formulas were erroneous and that each base occurred almost exclusively in one particular form. *From that point it was easy going.*"⁹⁵

Crick conveys the significance of this conclusive serendipitous moment by going on to reiterate that final (italicized) observation: "Finally we see Jerry Donohue telling us that we had the wrong formulas [tautomeric forms] for the bases so that Jim was able to hit on the correct pairs. *After that the model was almost inevitable.*"⁹⁶ This concluding remark may bring to mind Thomas Kuhn's concluding remark about the decisive serendipitous moment in his work when he came to his concept of "paradigm" once he happened to notice that the cognitive behavior of social scientists at the Center for Advanced Study in the Behavioral Sciences differed basically from that of the "natural scientists" among whom he had been trained: "Once that piece of my puzzle fell into place, a draft of this essay [*The Structure of Scientific Revolutions*] emerged rapidly."⁹⁷

Thus we are led to understand why Jerry Donohue was singled out for brief but quite special acknowledgement in the SSA. And we also come to see that this relatively emphatic acknowledgement in the SSA nevertheless involved an OSS, for there is not a word there about that most consequential serendipitous moment that led at once to the conclusion of that monumental discovery. Furthermore, we learn from Watson's and Crick's extended accounts that the serendipitous component in the discovery was not the result merely of the "prepared minds" at work on the problem (as an extreme psychological reductionism would have us believe) nor was it the result merely of the sociocognitive microenvironment in which they worked (as an extreme sociological reductionism would have us believe). It was a complex psychosocial process, a joint product of both "agent" and "social structure." And so, once again, we find that just as the individual scientists engaged in the quest for the structure of DNA may have differed in their trained capacity for original discovery (serendipitous as well as hypothetico-deductive), so evidently did their sociocognitive microenvironments differ in the extent to which they made for the "accidental" component in the ongoing discovery.

But though the SSA does typically present a notably incomplete and consequently a misleading image of the course actually taken by the inquiry, it does not follow, as we have seen the consummate scientist Peter Medawar drastically proposing, that "the inductive format of the [standard] scientific paper should be discarded." After all, the SSA is not designed as a clinical or biographical account of the kind that holds prime

⁹⁵ Ibid., p. 65. Again, my italics.
⁹⁶ Ibid., p. 85. For another view of the discovery, see Anne Sayre, *Rosalind Franklin and DNA* (New York: W. W. Norton, 1975).
⁹⁷ See Kuhn, *Structure of Scientific Revolutions*, pp. vii–viii.

interest for historians, sociologists, psychologists, and philosophers of science. It is designed to inform fellow scientists about a claim to a new (small or great) contribution to a field of knowledge and to present the grounds for thinking that the claim is sound. All this is presumably understood by fellow scientists reading such articles (though, in that case, it becomes rather puzzling that attention should have been explicitly drawn to this aspect of the SSA over the years and, for unknown reasons, especially in the last two or three decades). Perhaps the problem can be solved by having scientific journals bear a "Warning to the Reader" that simply quotes Faraday's gentle statement in his historic *Experimental Researches in Electricity*: "These results I purpose describing, not as they were observed, but in such a manner as to give the most concise view of the whole."[98]

A Concluding Interlude

Two vastly different events have now intervened to signal that I had better bring this leisurely Afterword to a swift close. The first, merely personal, event was a rapid succession of four surgeries for what my surgeons assure me are quite discrete (i.e., not metastasizing) cancers, these being followed by cascades of drastic side effects culminating in a condition that my internist, Dr. Henry S. Lodge, describes as a "state of continuing exhaustion." And all the world knows of the second, earthshaking, event that occurred in the early morning hours of September 11, 2001: the barbaric terrorist attack on the United States of America focused on the airliner-missile immolation of thousands of human beings at work in the Twin Towers of the World Trade Center in New York City and in the Pentagon on the outskirts of Washington, D.C. The long-drawn-out global consequences of that event are of course beyond our here-and-now comprehension. These conjoint events lead me to conclude this Afterword with two brief sets of observations on our long-continued subject.

Serendipity: Bifurcation of the Arcane Word into Vogue Word and Psychosociological Epistemic Concept

From Arcane Word to Vogue Word

Faithful readers of our time-capsule text of 1958 will recall Elinor Barber's and my diligent search for the word serendipity in the germane English and statunitense literatures since the word's unique appearance

---

[98] Michael Faraday, *Experimental Researches in Electricity* (London: Bernard Quaritch, 1839), vol. 1, p. 2.

in Walpole's letter to Mann in 1754. So far as the documentary record shows, neither Walpole nor Mann ever again referred to the word. It did not have any public currency until the publication of Walpole's letters to Mann in 1833 and, later, the "complete correspondence of Walpole" in 1857 as "the best letter-writer in the English language." Even so, it was not until 1875 that Edward Solly, a chemist and antiquarian who had often experienced unexpected happy discoveries himself, first took note of Walpole's apt and engaging coinage and called attention to it in *Notes and Queries*, the Victorian journal designed for antiquarians, bibliophiles, and the erudite.

That laborious search of ours yielded a grand, but scanty, total of some 135 public usages of the word in the eighty-three years since Solly first brought it to light. This average of some 1.6 identified usages per year increased to some three usages per year in the 1950s. More on point, the increase was occasioned chiefly by appearance of the word in widely read publications such as the *New York Times*, the *Christian Science Monitor*, the *Times Literary Supplement*, and the *New Yorker*. However, we recognized that, diligent as our search may have been, it was bound to be only suggestive, since of course we could not possibly cover any representative portion of anglophone publications in that long period. Those most recent allusions to serendipity we uncovered back then were just straws in the wind but they did suggest that the word was beginning to diffuse from the relatively esoteric domains of antiquarians, bibliophiles, and a belated handful of scientists into the vernacular. Serendipity was apparently no longer wholly arcane.[99] As even a primitive sociological semantics would lead us to expect, the word took on quite diverse and distinctive meanings as it moved into different and distinctive sociocultural populations.

At least, that seemed to be the case as we set our manuscript aside for what proved to be almost a half-century of ghostly existence in that peculiar limbo of history, the world of unpublished documents. To our minds back then in 1958, the hesitant transition of serendipity from arcanum to popular science and the vernacular was most clearly demonstrated by the common practice of defining the word and occasionally giving a synoptic account of its history whenever it was employed. Thus, although the *New York Times* had virtually naturalized the word at least a decade before, it reverts to the earlier practice of indicating to readers that they were probably encountering the word for the first time when proposing a new

---

[99] I had evidently sensed the beginnings of that transition as early as 1949 when, as noted in this Afterword, I observed in the course of reprinting the paper that introduced *serendipity* into the social sciences, "Since the foregoing was first written in 1946, the word *serendipity* for all its etymological oddity, has diffused far beyond the limits of the academic community." Merton, *Social Theory and Social Structure*, pp. 376–377n.

"Serene Order of Serendipity" on its post-editorial page of June 7, 1958: "*Serendipity* is a useful word coined by Horace Walpole, English politico and novelist of the eighteenth century, to denote the faculty [N.B.] of making lucky and unexpected 'finds' by accident." Other kinds of asides that avoid both definition and history still manage to signal that it is a very new word that is both esoteric and elevating. Thus, the reviewer of a book by P. G. Wodehouse, the incomparable English master of "high comic art," writing (in 1958) in the long-since defunct (New York) *World Telegram and Sun*, a far less elevated newspaper than the *Times*, announced: "Then, by a stroke of serendipity (I knew I'd get a chance to use that word some day) . . ."

Still, despite our sense that serendipity had begun to move beyond its esoteric boundaries, Elinor Barber and I had not the least idea back then that it might eventually move from obscurity into the limelight and become a vogue word, let alone that it would acquire a riotous popularity. But I now note that serendipity appears in the *titles* of fifty-seven books published between 1958 and 2000 (as reported in *Books in Print*), although not a single such title appeared in the same source previously. Moreover, this current list includes only a handful of scholarly works,[100] the rest being metaphorical titles for books dealing with diverse popular subjects, ranging from genealogy and collectibles to cookbooks and tales for children to spiritual and self-help guides. This provides at least a hint of the considerable diffusion of our word into the vernacular after the late 1950s, when we put aside what has since turned into a time-capsule text.

However, comparative evidence provides a somewhat greater intimation of that massive diffusion as we examine the new kinds of data generated by current technologies of information search mechanisms now available on the Internet and in public and university libraries. Although these search mechanisms still do not generate anything like a representative sampling of comparative word usage over the decades, they do provide lenses for estimating the extent and character of the diffusion of our word that were of course not available in the 1950s. Nevertheless, despite these not inconsiderable imperfections, the increase in frequency of retrieved recent usages exhibited by their data is of a sufficiently ascending order of magnitude greater than what we were able to glimpse a half-

---

[100] One of these that bears the minimalist title *Serendipities* is by the polymath Umberto Eco, who, incidentally, wrote the Foreword to my *On the Shoulders of Giants: The Post-Italianate Edition* (Chicago: University of Chicago Press, 1993), pp. ix–xviii. However, although his book consists in large part of lectures given at the Columbia University Italian Academy for Advanced Studies in America on whose board of trustees I served, Umberto and I never happened to talk about my long-standing interest in the word and concept. This is not at all odd. After all, Tom Kuhn and I never discussed our respective notions of "paradigm," either.

century ago as to do more than merely hint at the word's virtual explosion into popularity. Thus, when we turn to the search capabilities of such new on-line information search engines as Lexis-Nexis and Google, we find that the frequency of appearances of serendipity in print vastly exceed those 135 instances we were able to assemble by the traditional modes of scholarly research we perforce employed back in the 1950s.

Lexis-Nexis, which provides a full-text database of more than eighteen thousand sources chiefly composed of newspapers and magazines, indicates this order of magnitude increase in usage since the close of our time-capsule text:

1960–1969	2
1970–1979	60
1980–1989	1,838
1990–1999	13,266

These figures of course exaggerate the degree of rapid increase in popular usage of serendipity since the coverage of Lexis-Nexis has grown every decade. Still, despite discounting the precision of those figures as I do, they nevertheless seem to provide strong intimations of the rapid diffusion of serendipity into popular discourse in recent decades. For the still (unknown) residual figures outstrip by orders of magnitude the paltry aggregate of 135 usages we were able to identify in the first eighty-three years since the word first came to public notice. By way of further evidence of like kind, the number of documents on the Worldwide Web in 2001 containing the word serendipity in the two-billion-document database provided by Google swells by a substantial order of magnitude to 636,000.

These newly available types of data do provide strong intimations of the increasingly rapid diffusion of serendipity in the vernacular. That impression is reinforced by other kinds of evidence that the word has reached ever-larger mass audiences. Thus, as I write these words (early in the year 2002), movie houses throughout the United States are showing an "escape romantic comedy" in which boy meets girl in "the mad holiday rush in New York City" and, though engaged "in a mutual, all-consuming passion," they go their separate ways only to have the Fates see to it that they find one another again ten years later. Its title? *Serendipity*, of course.[101]

Other rather more selective bits of evidence: The Boat Owners Asso-

---

[101] This came as no surprise to me, for, as is altogether fitting if not highly probable, Dr. James Shulman of the Andrew W. Mellon Foundation—where my wife, Dr. Harriet Zuckerman, is the senior vice-president—and the author of the telling Introduction to this long-gestating work, should have come serendipitously upon the filming of *Serendipity* on the streets of Manhattan.

ciation of the United States reports in the year 2000 that Serendipity is ranked as the tenth most popular name for boats. In that same year, the London Festival of Literature and the Bloomsbury Press held a self-selective poll asking for choices of "favorite words" in the English language. Serendipity heads the list of ten (against presumably powerful competition, since *Jesus* and *money* are tied for tenth place).[102]

Once having first infiltrated and then vastly extended its usage in public discourse and, it appears, in everyday speech, serendipity undergoes a sea change in its meaning. We may be witnessing here a uniformity in sociological semantics. As the word diffuses into various sociocultural strata, it is put to different uses and its meanings in action multiply. Thus, as early as December 1958 (just after our time-capsule manuscript had been put aside), we see a case in point that suggests a kind of self-elevating reason for its diffusion outside the ranks of antiquarians, book collectors, and scientists along with the correlative unwitting but considerable change of its presumed meaning. Here is Leslie Charteris, the author of the vastly popular detective stories starring the detective Simon Templar (otherwise known to movie audiences of the time as "The Saint"), declaring himself "by nature a sort of serendipitist myself" and going on to inform the aficionados of his *Saint Detective Magazine* that "any readers who would like to enrich their friends (meaning, of course, only those illiterates and low-life friends who are not also readers of this magazine) we are offering, gratis with this issue, the word *serendipity*, which means the collecting of weird and unrelated and seemingly pointless items of information."

In the decades of ever more rapid diffusion that follow, this process of attributing or implying new and successively differentiated meanings to serendipity gains force. Its initial unique and compendious meaning of a particular kind of complex phenomenon—the "discovery of things unsought" or the experience of "looking for one thing and finding another"—becomes ever more eroded as it becomes ever more popular. Ultimately, the word becomes so variously employed in various sociocultural strata as to become virtually vacuous. For many, it appears, the very sound of serendipity rather more than its metaphorical etymology takes hold so that at the extreme it is taken to mean little more than a Disneylike expression of pleasure, good feeling, joy, or happiness. For those who have consulted dictionaries for the word, its typical appearance between *serenade* and *serene* may bring a sense of tranquility and un-

---

[102] Bob Geldof, *The Book of Favourite Words* (London: Bloomsbury Publishing, 2001), p. 8. And, of course, in giving his definition of the prime favorite word, Geldof continues to adopt the psychologistic definition, thus: "Serendipity, *n.* a natural gift for making useful discoveries quite by accident."

ruffled repose. In any case, no longer a niche-word filling a semantic gap, the vogue word becomes a vague word.[103]

Confining ourselves to a very few instances of our word's appearance, we observe the multiplication of new tacit or explicit meanings in process. Here, early on in that process, appearing soon after we had put aside the 1958 text, is another casual but probably far more consequential transformation of its meaning by Herb Caen, the influential and nationally read daily columnist of the *San Francisco Chronicle*, described as the "voice and conscience" of the city when he was awarded the Pulitzer Prize for journalism in 1996: "SERENDIPITY and *lagniappe*—how can you beat 'em? Without resorting to the dictionary (it's across the room and I have a *rheum*) I think serendipity means an unexpected pleasure, and *lagniappe*. A little extra thrown in." And by way of more recent record, these few further semantic changes:

> 1992: the word *SERENDIPITY* is emblazoned on the cover of a catalogue of women's underwear, without further explanation in the presumably enticing pages that follow. The implication: there are unexpected treasures awaiting the reader.
>
> 1999: In a *New York Times* review of an autobiography by the illustrious British actor Sir Alec Guinness, attention is drawn to his "serendipitous writing style (sly, witty, elegant)."
>
> 2001: On the Internet we read: "SERENDIPITY: When love feels like magic . . . you call it *destiny*. When destiny has a sense of humor . . . you call it . . . Serendipity."
>
> 2002: And again on the Internet, we are welcomed "to *Serendipity Airedales*, home of the top winning Best in Show Airedale in the history of the breed."

These few instances of course only begin to suggest the diversity of meanings among the word's countless inept usages. It should be noted, however, that serendipity has been subject to a paradoxical semantic process in the course of its accelerated diffusion. At the very time that it has been losing its earlier specific and unique meaning in the course of be-

---

[103] Can it be that three grand masters of the mystery novel—the first statunitense, the other two British—are trying to stem the tide toward vacuity by reestablishing the Walpolean meaning of the word for their dedicated fans in possibly new social circles? At any rate, here is Lawrence Block's detective remarking in his 1989 mystery *Out on the Cutting Edge*: "I lucked out and found the solution almost by accident. Or whatever you called it. Serendipity." And here is the "serendipitous" plebian detective Joe Sixsmith in Reginald Hill's 1995 *Born Guilty*, responding to the sound as well as the substance of our word: "This negative capability sounded OK to Joe, like that other 'ity' some old lady had laid on him. . . . *Serendipity*. Finding things out by accident." And, finally, here is a dialogue by the subtle Ian Rankin in his 1997 mystery *Black and Blue*: "You know the word serendipity?" "I pepper my speech with it." "Dictionary meaning: the ability to make happy chance finds. Useful word." "Absolutely."

coming a vogue word in the vernacular, it is being transformed from a descriptive term into a psychosociological epistemic concept by psychologists, sociologists, and philosophers of science.

Serendipity as Psychosociological Epistemic Concept

As a mere novice in the art of browsing the Internet for apposite data, even I can bear witness that it is a veritable fount of serendipity. As an accomplished veteran in the use of the Internet, James Fallows, has observed: "If you start looking up information on Web sites, you almost never end up where you expected. There's a link to something you never heard of before. . . . The feeling is similar to that of going through library stacks—if there were no dust and you could instantly zoom from floor to floor."[104] Or as the Web expert David Weinberger notes: "It is the Web's hyperlinked nature" that makes for unanticipated and surprising exposure to new data and ideas.[105] Which may loosely put us in mind of the rather more elevated and concentrated centers of "institutionalized serendipity" that make for sociocognitive interaction among diverse talents, such as the Center for Advanced Study in the Behavioral Sciences (where Kuhn had his serendipitous moment) and the Cavendish Laboratory (where Watson and Crick had theirs). However, there is a fundamental difference between the Web at large and the world of published science, all apart from the transparently great difference in the quality of minds making use of the resources provided by these organizations and by the Web. Working scientists can with reasonable (though still not total) confidence assume that what they find in their journals is *reliable* knowledge, for this tacit assumption is rooted in the normative structure and practice of science as a social institution: a set of institutional norms that include the norm of what has been described as socially "organized skepticism." By that is meant, not that scientists are all inherently skeptical—though many of them may well be—but that the *community of scientists* requires that claims to new knowledge be adequately supported by what is then and there understood as credible evidence.[106] That type of socially organized skepticism finds its most institutionalized expression in

[104] James Fallows in *New York Review of Books* 49, no. 4 (2002), p. 7.
[105] David Weinberger, *Small Pieces, Loosely Joined: A Unified Theory of the Web* (Cambridge, Mass.: Perseus, 2002).
[106] R. K. Merton, "A Note on Science and Democracy," *Journal of Legal and Political Sociology* 1 (1942):115–126, and R. K. Merton, *The Sociology of Science: Theoretical and Empirical Investigations*, pp. 267–278. As is noted there (p. 269), "Although the ethos of science has not been codified, it can be inferred from the moral consensus of scientists as expressed in use and wont, in countless writings on the scientific spirit and *in moral indignation directed toward contraventions of the ethos*." Italics supplied.

the "peer review" system, in which scientists expert in the particular content of a manuscript submitted for publication act as referees of the reliability of its contents. This procedure is supplemented by having specialists in the field subject articles and books to further critical examination after their publication. The system of organized skepticism does not, of course, work to perfection but it does make for substantial reliability of what finds its way into scientific print. But there is no such institutionalized system governing what appears on the Web (except for those quite limited portions that provide similarly refereed publications).

One such proprietary resource, not available on the Web at large, is JSTOR, an on-line searchable archive that includes the entire runs (except for a few most recent years) of some 170 (in 2002) scholarly and scientific journals. So far as I know, there is nothing else quite like it.[107] Through access to JSTOR, we can readily and reliably identify 44 usages of the word serendipity for the decade of the 1950s alone and an accelerated 759 usages for the thirty-seven years between 1960 and 1996 (or 21 allusions per year). Thus, this one aggregated but delimited source provides some seventeen times the rate of appearance of the word serendipity found in our search in the time-capsule text through all the sources we were able to locate well beyond the journals JSTOR covers. This provides just another example of how the new technologies for information search have moved beyond even diligent searches for certain types of information in the past.

The JSTOR data for the 170 journals indicate this substantial decade-by-decade increase in the appearance of the word serendipity:

1930–1939	4
1940–1949	11
1950–1959	44
1960–1969	135
1970–1979	181
1980–1989	207
1990–1996 [7 years]	236

Although these figures suggest a steady increase, they do not begin to resemble the exponential increase in popular usages of serendipity registered in the Lexis-Nexis data set. The JSTOR data do, however, largely confirm the impression Elinor Barber and I reached in the time-capsule text about the diffusion of serendipity into the domain of science. On the American scene, it was primarily Walter B. Cannon, the eminent physi-

---

[107] But, of course, these 170 journals do not provide anything like a representative sample. I know of no well-grounded estimates of the total number of scientific and scholarly journals in the decades from the 1930s to the 1990s.

ologist at Harvard University medical school, who evidently took many occasions to introduce the term serendipity to his fellow scientists. Thus, as early as February 1932, Cannon delivered a major lecture at the Jefferson Medical College in Philadelphia. His lecture, one learns via JSTOR, "was entitled 'Serendipity.'" Again, during his tenure as president of the influential American Association for the Advancement of Science, Cannon delivered another public lecture, this one by the slightly extended title "Serendipity, or An Accidental Sagacity." In 1945, his series of serendipity lectures finds its most intensive recapitulation in a chapter of his widely read autobiographical reflections, *The Way of an Investigator: A Scientist's Experiences in Medical Research.*

Beyond any reasonable doubt, then, it was Cannon who first brought the term serendipity to the attention of his fellow statunitensean scientists. This he did most knowingly in the then traditional form of recounting the feats of "great men of science and their (serendipitous) discoveries," along with a reminder of the Pasteur dictum, "Dans les champs de l'observation, le hasard ne favorise que les esprits préparés."

Data from the JSTOR archive are also consistent with our impression recorded in the time-capsule text that the term and concept serendipity both found their way into sociology in the 1940s. The *term* first appeared in an article on "Sociological Theory" in 1945, as we have learned, where it was briefly noted and, so far as its origin is concerned, rather cryptically introduced with only an anonymous allusion to Walpole's language: "Fruitful empirical research not only tests theoretically derived hypotheses; it also originates new hypotheses. This might be termed the 'serendipity' component of research, i.e., the discovery, by chance and sagacity, of valid results that were not sought for."[108]

---

[108] Merton, "Sociological Theory," p. 469. I include the month of its appearance (May 1945, as JSTOR does) because I notice only now that my copy of Cannon's *The Way of an Investigator* is inscribed thus: "4 July 1945—To Bob, The best on your birthday. Bernie." "Bernie" is Bernard Barber, husband of Elinor Barber and, equally on point, a pioneer in the sociology of science who, as I have indicated, wrote, together with Renée C. Fox, an important article titled "The Case of the Floppy-Eared Rabbits: An Instance of Serendipity Gained and Serendipity Lost." All this now sets off a further reminiscence. I had been in touch with Walter Cannon in 1939, when I was an instructor and tutor at Harvard, but we did not discuss "serendipity" for the best of reasons: I had not yet encountered the word. Cannon was about to bring out a new edition of his 1932 book *The Wisdom of the Body*, with an epilogue that proposed an analogy between biological and "social organisms." A senior colleague, Gordon Allport, then the world doyen of social psychology, persuaded me to alert Cannon to the fact that many such analogies that had been proposed by "sociological organicists" over the years were thoroughly discredited. I did add that his own approach, which, unlike the earlier versions, centered on the *process* of homeostasis rather than on alleged structures, did have possible heuristic value. Cannon graciously acknowledged my would-be contribution but made no change in the epilogue for his new (1939) edition.

The JSTOR data also enable us to detect a phase in the nascent use of this bare allusion to serendipity, replete as it is with an appositive definition designed to clarify first-time encounters with this still very new word. For before long, Alvin W. Gouldner, destined to become a sociologist of some eminence but then still a graduate student of mine working on his doctoral dissertation while teaching at the University of Buffalo, quoted my brief allusion to the word and, as a student, made it abundantly clear to his colleagues that it was introduced by his mentor, "Dr. Merton."[109] Truly short-distance diffusion, which was duly noted in the time-capsule text. A few years later, however, a paper focused on serendipity (to which I have referred rather frequently in this Afterword) went on to open the "black box" of the term serendipity and to explore it as a concept. It is there that I introduced

> the serendipity pattern [which] refers to the fairly common experience [in science] of observing an *unanticipated, anomalous, and strategic datum* that becomes the occasion for developing a new theory or for extending an existing theory. Each of these elements of the pattern can be readily described. The datum is, first of all, *unanticipated*. A research directed toward the test of one hypothesis yields a fortuitous by-product, an unexpected observation which bears upon theories not in question when the research was begun.[110]
>
> Secondly, the observation is *anomalous*,[111] surprising, either because it appears inconsistent with prevailing theory or with other established facts. In either case, the seeming inconsistency provokes curiosity; it stimulates the investigator to 'make sense of the [new] datum,' to fit it into a broader frame of knowledge. . . .
>
> And thirdly, in noting that the unexpected fact must be *strategic*, i.e., that it must permit of implications which bear upon generalized theory, we are, of course, referring rather to what the observer brings to the datum rather than the datum itself. For it obviously requires a theoretically sensitized observer to detect the universal in the particular. After all, men had for centuries

---

See Stephen Cross and William Albury, "Walter Cannon, L. J. Henderson, and the Organic Analogy," *Osiris* 3 (1987):165–192, at pp. 183–184.

[109] Alvin W. Gouldner, "Industrial Sociology: Status and Prospects," *American Sociological Review* 13 (1948):396–400, at p. 397.

[110] I had obviously not yet recognized, as I did only a decade later, that the Walpolean pattern of an unexpected observation leading to a significant discovery not at all sought for was only one type of serendipitous experience. The other type, noted only later, is *the serendipitous moment in an ongoing research* leading unexpectedly to the solution of a well-defined problem, as with the cases we have examined of Kuhn's new conception of paradigmatic and non-paradigmatic disciplines and the Watson-Crick discovery of DNA.

[111] Just as I never got around to discussing the concept of "paradigm" with Tom Kuhn when he sent me his (1962) manuscript, so too the concept of "anomaly," which he put to new consequential use in his theory of paradigm shifts. Again, for reasons not accessible to me then or now, we never talked about this further pairing of concepts.

noticed such "trivial" occurrences as slips of the tongue, typographical errors, and lapses of memory, but it required the theoretic sensitivity of a Freud to see these as strategic data through which he could extend his theory of repression and symptomatic acts.[112]

But though this analysis began to open up the black box of serendipity as a concept, neither then in the late 1940s nor a decade later in our time-capsule text did I go on to distinguish between the psychological and sociological aspects of the complex phenomenon of serendipity. That theoretical distinction, as we have seen in some detail, is essential if serendipity is not to suffer from a misleading psychologistic reductionism and become merely an individual "gift or aptitude or capacity or talent." That distinction did not come to mind until the latter 1970s, when I found myself reflecting upon special sociocultural environments that evidently fostered scientific discoveries (including *fully serendipitous discoveries* and *serendipitous moments* in ongoing research that finally led to the solution of the initial problem). Organizations such as the Harvard Society of Fellows and the Center for Advanced Study in the Behavioral Sciences and a variety of scientific laboratories and university departments could then be identified as centers of "institutionalized serendipity" or as "serendipitous microenvironments" where diverse scientific talents were brought together to engage in intensive sociocognitive interaction.[113]

As we have seen, as the word serendipity became ever more popular, it became merely an attractive label designating ever more diverse pleasant phenomena and sensations. And for quite some time, further analyses of the sources and workings of serendipity in advancing scientific knowledge were virtually absent. It was enough to observe that serendipitous discoveries happened and to tell stories that described those happenings, especially narratives about the more fascinating examples of pure serendipity: for example, Wilhelm Roentgen's "accidental discovery" of X-rays (which was awarded the first Nobel Prize in physics), or Alexander Fleming's "accidental discovery" of penicillin (through that "contaminated" petri dish), or the Hewish-Bell "incidental discovery" of pulsars. In more recent years, however, a few sociologists, psychologists, and philosophers of science have gone on to develop an interest in the sources, mechanisms, and further implications of serendipity in scientific discovery. I can do no

---

[112] Merton, "Bearing of Empirical Research upon the Development of Sociological Theory."

[113] Merton, *Sociology of Science: An Episodic Memoir*, pp. 81–85, 102–105. Incidentally, in light of Paul Lazarsfeld's and my role in its origin, the Center for Advanced Study in the Behavioral Sciences could be considered an instance of "applied sociology of science." This though I had incurred Paul's bitter anger by rejecting his idea of a hierarchically organized center and concurred with other members of the planning group in its being instead a company of unstratified fellows of differing generations.

more than take brief note of the most recent sociological contributions by John Ziman, the theoretical physicist turned sociologist of science, and the contributions by the philosopher of science Aharon Kantorovich.[114]

It comforts me to see that Ziman and I are in accord with regard to the implications of serendipity for a further understanding of scientific discovery in general. He elects to organize his examination of "epistemic practices of science . . . around the Mertonian norms" but warns that this "does not imply that the sociological dimension is paramount. It just happens that these form the most convenient scheme for this purpose." He refers here to the norms of science as a social institution that I had singled out some sixty years ago: communism (later dubbed "communalism"[115]), universalism, disinterestedness, and organized skepticism, or, as that quartet of norms has since been acronymized, CUDOS).[116] Ziman notes further that "the Mertonian norms are not really blind to epistemic or cognitive values. . . . It is impossible, for example, to talk about 'criticism' [i.e., 'organized skepticism'] without reference to the knowledge being criticized or the ways in which individuals give it and take it in practice. [As Ziman might have added: and without reference to what are being taken as the working criteria for provisional acceptance of the claims to new knowledge.]"[117] Most on point for our immediate pur-

---

[114] Kantorovich and Ne'eman, "Serendipity as a Source of Evolutionary Progress in Science"; Kantorovich, *Scientific Discovery,* especially chapters 5–7. Extraneous pressures lead me to cite only the seminal work of the polymathic psychologist Donald T. Campbell, "Unjustified Variation and Selective Retention in Scientific Discovery," in *Studies in the Philosophy of Biology,* ed. F. J. Ayala and Theodosius Dobzhansky (London: Macmillan, 1974), pp. 139–161, and especially Donald T. Campbell, "Evolutionary Epistemology," in *The Philosophy of Karl Popper,* ed. F. A. Schilpp (La Salle: Open Court, 1974), pp. 413–463. Although the latter essay, prepared during Campbell's year as a fellow at the Center for Advanced Study in the Behavioral Sciences, uses the word *serendipity* only once, much of the dynamics of discovery that he develops there are propaedeutic for an understanding of the *phenomenon* of serendipity. Thus, he notes: "The research worker encountering a new phenomenon may change his problem to one which is thereby solved. Serendipity as described by Cannon and Merton, and the recurrent theme of 'chance' discovery, emphasize this double opportunism" (pp. 435–436). See also the extension of the Campbell legacy by Dean Keith Simonton, *Scientific Genius: A Psychology of Science* (New York: Cambridge University Press, 1988), passim, esp. pp. 35–38.

[115] That change in terminology was proposed by Bernard Barber in the 1950s during the Joseph McCarthy period of political witch-hunting in which countless members of government and academe were accused of subversion or disloyalty without evidence.

[116] R. K. Merton, "Science and the Democratic Social Structure" (1942), reprinted in *The Sociology of Science: Theoretical and Empirical Investigations,* pp. 267–278. I do not elucidate these norms of science further because they have been abundantly discussed in sociological and philosophical literatures during the last sixty years.

[117] John Ziman, *Real Science: What It Is and What It Means* (New York: Cambridge University Press, 2000), pp. 84–85. A fine summary of the norms that also treats them as a significant cultural context of serendipity is provided in this work (which I had intended to

poses, he goes on to observe that "true serendipity—the accidental discovery of something not sought for—does play a major role in the production of scientific knowledge" and expeditiously asks: "What is its epistemological significance?"[118]

In response to this telling question, Ziman identifies serendipity as a new line of inquiry when an unexpected phenomenon is noticed in the course of an ongoing research. More, the very fact that the observation of a phenomenon is "unexpected" strongly implies a departure from what was expected in terms of prior "established" knowledge. In effect, then, Ziman notes that serendipitous discoveries begin with observed "anomalies" in the course of inquiry, as I too have argued; and as Kuhn went on to observe, serendipitous discoveries were a consequential element in making for "paradigm shifts." This counters the fairly widespread belief among scientists that, as we have seen in the time-capsule text, serendipitous discoveries are somehow less "worthy" than those arrived at in accord with "the scientific method."[119] This echoes the occasionally expressed belief that "multiples"—that is, discoveries made independently by two or more scientists—were inevitable and therefore are less deserving than "singletons" (whose public appearance often forestalls others' claims to having arrived at the same results independently).[120] Moreover, as Ziman perceptively notes about the epistemological implications of multiples, "many scientists have personally experienced [multiples] and interpret these [as evidence] for the 'reality' of scientific knowledge. In effect, it illustrates the duality of scientific paradigms, as overarching so-

---

draw upon at greater length) in chapters 3 and 5. A longtime colleague-at-a-distance and academic friend, John briefly inscribes a gift copy of his latest book in the sociology of science thus: "Bob, look what you started."

[118] Ibid., p. 217. Three philosophically oriented sociologists, all of them, as it happens, young Italian scholars, have also recently noted epistemological implications of my concept of serendipity: Maniscalco, "Il concetto di 'serendipity' nell'opera di Robert K. Merton"; Angela Maria Zocchi Del Trecco, *Tra storia e narrazione: L'intenzione interpretativa in Robert K. Merton* (Milan: Franco Angeli, 1998), pp. 194–203; Riccardo Campa, *Epistemological Dimensions of Robert Merton's Sociology: The Debate in the Philosophy of Science of the Twentieth Century* (Torun, Poland: Nicholas Copernicus University Press, 2001), p. 197.

[119] Thus, a Nobel laureate in physics declared, "I think the Nobel Prize brought undue rewards. I got it for a purely accidental discovery. Anybody could have done that. This is often true in experimental physics. I think that you can happen to be in a position where an important discovery is right there." In an interview with Harriet Zuckerman, *Scientific Elite*, p. 211.

[120] On independent multiples, see Merton, *The Sociology of Science: Theoretical and Empirical Investigations*, pp. 343–370; for evidence on the frequent forestalling of multiples, see Warren Hagstrom, *The Scientific Community* (New York: Basic Books, 1965).

cial institutions and as underlying representations of the uniformity of nature."[121]

Ziman also notes that scientists often attribute serendipitous discoveries, like other discoveries, simply to their sense of "curiosity" as a personal characteristic. But as he points out and as we have seen in our sociological perspective on science as a social institution (in contrast to a psychologistic reductionism): "Scientific research is much more than the enlightened exercise of personal curiosity. [This was also recognized by the sociologically oriented psychologist Donald Campbell, who is rightly quoted as observing that 'science uses curiosity, it needs curiosity, but curiosity did not make science.'] Elaborate intellectual and institutional frameworks are required to harness this individual trait to the collective production of reliable knowledge. . . . [And, again, curiosity] is not so much a personal trait or an attitude of mind as a virtue associated with a social role."[122]

Finally and most tellingly, Ziman converges on the *sociological* element that we have seen in our analyses of "serendipitous microenvironments" is most distinctive of *serendipitous* discovery:[123] the occurrence of the perceived unanticipated anomaly as providing an *opportunity* for the discovery. As he puts it most emphatically, "The key point is that serendipity does not, of itself, produce discoveries: it produces *opportunities* for making discoveries. Accidental events have no scientific meaning in themselves: they only acquire significance when they catch the attention of someone capable putting them into a scientific context. [So much for Pasteur's *esprits préparés*.] Even then, the perception of an anomaly is fruitless unless it can be made the subject of deliberate research. In other words, we are really talking about discoveries made by the *exploitation of serendipitous opportunities* by persons already primed to appreciate their significance."[124] Which, in turn, is tacitly to raise the question of possible differences among sociocognitive microenvironments of scientists in the frequency with which such serendipitous opportunities do occur; in short, another sociological element in the seemingly simple but actually complex phenomenon of serendipitous discoveries. For, as I have suggested, those microenvironments do differ in their opportunity structures for such serendipitous discoveries.[125] A concluding reminder that if we are

[121] Ziman, *Real Science*, p. 216.
[122] Ibid., pp. 23, 217.
[123] See Merton, *Sociology of Science: An Episodic Memoir*, pp. 81–109.
[124] Ziman, *Real Science*, p. 217.
[125] See R. K. Merton, "Opportunity Structure: The Emergence, Diffusion, and Differentiation of a Sociological Concept, 1930s–1950s." In *Advances in Criminological Theory: The Legacy of Anomie Theory*, ed. Freda Adler and W. S. Laufer (New Brunswick, N.J.: Transaction Publishers), vol. 6. pp. 3–78.

to understand the phenomenon of serendipity in science and technology, it cannot be reduced to only its psychological *or* its sociological elements.

Having just been advised of yet another surgical bout impending—the fifth in the past few months—and beset by a variety of other ills, I must truly bring this incomplete and unedited draft of a planned penultimate section to a premature and abrupt close. Readers with further interest in the psychosociological epistemic character of serendipity can turn profitably to the sources cited in the preceding footnotes.

The Shulman Conjectures

Faithful readers of this volume will recall that the knowing and amiable author of its Introduction has in effect articulated and answered the questions that must have come to their discerning minds: "Why didn't the authors publish *The Travels and Adventures of Serendipity* in 1958?" "Why does it appear now and why has it not been changed with the times, updated to enfold, in equal detail, the events of the subsequent forty-plus years?"

Elinor Barber and I thoroughly subscribed to those Shulman answers. As we see from intimations in the 1958 version, it was a *preparazione OTSOGIA* and, as such, it may have been displaced—perhaps overwhelmed—in the queue of work demanding to be done.

It appears now in its original, virtually untouched, form as a time capsule (for how else could we reproduce in literal truth how the travels and adventures of serendipity appeared to us back then)? As such, the book chronicles the wandering of the word-and-concept in its journey from the small stage of arcana to the large stage of commonplace usage. As a nicely preserved time capsule, it captures the innocence of those adventures on the byways and highways of language. And so, encouraged by Laura Xella and Giovanna Movia, my ever-patient editor-friends at Il Mulino, Elinor Barber and I decided to send the book out into the world now as it was then, in the hope that it will be of some interest to some of those who had heard whispers of its existence. It is not what I would write now. But it is where the word was then and can never be again.

Select References[1]

Austin, James H. *Chase, Chance and Serendipity: The Lucky Art of Novelty.* New York: Columbia University Press, 1978.

Barber, Bernard, and Renée Fox. "The Case of the Floppy-Eared Rabbits: An Instance of Serendipity Gained and Serendipity Lost." *American Journal of Sociology* 64 (1958):128–136.

Campa, Riccardo. *Epistemological Dimensions of Robert Merton's Sociology.* Torun, Poland: Nicholas Copernicus University Press, 2001.

Campanario, J. M. "Using Citation Classics to Study the Incidence of Serendipity in Scientific Discovery." *Sociometrics* 37 (1996):3–24.

Comroe, Julius H. *Retrospectroscope: Insights in Medical Discovery.* Menlo Park, Calif.: Von Gehr Press, 1977.

Dean, Colin. "Are Serendipitous Discoveries a Part of Normal Science?" *Sociological Review* 25 (1977):73–86.

Dri, Pietro. *Serendippo—Come nasce una scoperta: La fortuna nella scienza.* Rome: Editori Riuniti, 1994.

Eco, Umberto. *Serendipities: Language and Lunacy.* Italian Academy Lectures. New York: Columbia University Press, 1998.

Eliel, Ernest Ludwig. *Science and Serendipity: The Importance of Basic Research.* Washington, D.C: American Chemical Society, 1992.

Friedlander, Michael W. *At the Fringes of Science.* Boulder, Colo.: Westview Press, 1995.

Goodman, Leo. "Notes on the Etymology of SERENDIPITY and Some Related Philological Observations." *Modern Language Notes* 76 (1961).

Halacy, Daniel Stephen, Jr. *Science and Serendipity: Great Discoveries by Accident.* Philadelphia: Macrae Smith, 1967.

Hamilton, J. Wallace. *Serendipity.* Westwood, N.J.: Fleming H. Revell, 1965.

---

[1] Earlier references to serendipity are contained in the notes of our "time-capsule" manuscript of 1958, with other references in the Afterword.

Hodges, Elizabeth Jamison. *The Three Princes of Serendip.* New York: Atheneum, 1964.

Hoffmann, Roald. "Serendipity, the Grace of Discovery." *Drug Innovation and Approval* 1 (1998):28–31.

———. "Under the Surface of the Chemical Article." *Angewandte Chemie* 27 (1988):1593–1602.

Kantorovich, Aharon. *Scientific Discovery: Logic and Tinkering,* chapters 5–7. Albany: State University of New York Press, 1993.

Kantorovich, Aharon, and Yuval Ne'eman. "Serendipity as a Source of Evolutionary Progress in Science." *Studies in History of Philosophy of Science* 20 (1989):505–529.

Koestler, Arthur. *The Act of Creation.* London: Hutchinson, 1964.

Kohn, Alexander. *Fortune or Failure: Missed Opportunities and Chance Discoveries.* Cambridge: Basil Blackwell, 1989.

Lazlo, Pierre. "A Camel a Lady and a Dimetallacyclobutane." *Noveau Journal de Chimie* 7 (1983):675–677.

Maniscalco, Maria Luisa. "Il concetto di 'serendipity' nell'opera di Robert K. Merton." In *L'Opera di R. K. Merton e la sociologia contemporanea,* ed. Carlo Mongardini and Simonetta Tabboni, pp. 283–293. Genoa: Edizione Culturale Internazionale Genova, 1989.

Merton, Robert K. *On the Shoulders of Giants.* New York: Free Press, 1965.

Merton, Robert K., and Simona Fallacco. *La 'serendipity' nella ricera sociale e politica: Cercare una cosa e trovarne un'altra.* Rima: Luiss Edizioni, 2002.

Olson, Everett Claire. *The Other Side of the Medal: A Paleobiologist Reflects on the Art and Serendipity of Science.* Blacksburg, Va.: McDonald and Woodward Publishing, 1990.

Reichardt, Jasia. *Cybernetic Serendipity: The Computer and the Arts.* New York: Praeger, 1969.

Remer, Theodore G., ed. *Serendipity and the Three Princes: From the Peregrinaggio of 1557.* Norman: University of Oklahoma Press, 1965.

Rescher, Nicholas. *Luck: The Brilliant Randomness of Everyday Life.* New York: Farrar Straus Giroux, 1995.

Richardson, Robert S. *The Stars and Serendipity.* New York: Pantheon Books, 1971.

Roberts, Royston M. *Serendipity: Accidental Discoveries in Science.* New York: John Wiley and Sons, 1989.

Root-Bernstein, Robert Scott. *Discovering: Inventing and Solving Problems at the Frontiers of Scientific Knowledge.* Cambridge, Mass.: Harvard University Press, 1989.

Schwabe, Calvin W. *Knot Tying, Bridge Building, Chance Taking: The Art of Discovery.* Davis: University of California, 1984.

Small, Jocelyn Penny. "Plautus and the Three Princes of Serendip." *Renaissance Quarterly* 29 (1976):183–194.

Sproul, Robert C. *Not a Chance: The Myth of Chance in Modern Science and Cosmology.* Grand Rapids, Mich.: Baker Books, 1994.

van Andel, Pek. "Anatomy of the Unsought Finding—Serendipity: Origin, History, Domains, Traditions, Appearances, Patterns, and Programmability." *British Journal of the Philosophy of Science* 45, no. 2 (1994):631–648.

Ziman, John. *Real Science: What It Is and What It Means.* New York: Cambridge University Press, 2000.

Name Index

General Index